Spectroscopy from Space

NATO Science Series

A Series presenting the results of scientific meetings supported under the NATO Science Programme.

The Series is published by IOS Press, Amsterdam, and Kluwer Academic Publishers in conjunction with the NATO Scientific Affairs Division

Sub-Series

I. Life and Behavioural Sciences	IOS Press
II. Mathematics, Physics and Chemistry	Kluwer Academic Publishers
III. Computer and Systems Science	IOS Press
IV. Earth and Environmental Sciences	Kluwer Academic Publishers

The NATO Science Series continues the series of books published formerly as the NATO ASI Series.

The NATO Science Programme offers support for collaboration in civil science between scientists of countries of the Euro-Atlantic Partnership Council. The types of scientific meeting generally supported are "Advanced Study Institutes" and "Advanced Research Workshops", and the NATO Science Series collects together the results of these meetings. The meetings are co-organized bij scientists from NATO countries and scientists from NATO's Partner countries – countries of the CIS and Central and Eastern Europe.

Advanced Study Institutes are high-level tutorial courses offering in-depth study of latest advances in a field.
Advanced Research Workshops are expert meetings aimed at critical assessment of a field, and identification of directions for future action.

As a consequence of the restructuring of the NATO Science Programme in 1999, the NATO Science Series was re-organized to the four sub-series noted above. Please consult the following web sites for information on previous volumes published in the Series.

http://www.nato.int/science
http://www.wkap.nl
http://www.iospress.nl
http://www.wtv-books.de/nato-pco.htm

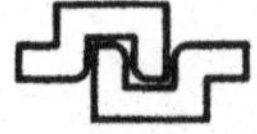

Series II: Mathematics, Physics and Chemistry – Vol. 20

Spectroscopy from Space

edited by

Jean Demaison
CNRS, Université de Lille I,
Lille, France

Kamil Sarka
Comenius University,
Bratislava, Slovakia

and

Edward A. Cohen
Jet Propulsion Laboratory,
California Institute of Technology,
Pasadena, U.S.A.

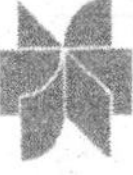

Kluwer Academic Publishers
Dordrecht / Boston / London

Published in cooperation with NATO Scientific Affairs Division

Proceedings of the NATO Advanced Research Workshop on
Spectroscopy from Space
Bratislava, Slovakia
October 31–November 4, 2000

A C.I.P. Catalogue record for this book is available from the Library of Congress.

ISBN 0-7923-6992-0 (HB)
ISBN 0-7923-6993-9 (PB)

Published by Kluwer Academic Publishers,
P.O. Box 17, 3300 AA Dordrecht, The Netherlands.

Sold and distributed in North, Central and South America
by Kluwer Academic Publishers,
101 Philip Drive, Norwell, MA 02061, U.S.A.

In all other countries, sold and distributed
by Kluwer Academic Publishers,
P.O. Box 322, 3300 AH Dordrecht, The Netherlands.

Printed on acid-free paper

Printed in the Netherlands.

Table of Contents

Preface

Many satellites operating from the microwave to the infrared ranges have been recently launched or are in preparation. Their main goals are the observation of the earth atmosphere or of interstellar clouds. The analysis of their data requires extensive laboratory work because the number of species for which the spectrum is known accurately enough (line positions, line intensities, and line profiles) is still quite small, especially for the whole spectral range. Furthermore, the needs of the observing community for laboratory data are continually evolving, particularly as technological developments open new possibilities (e. g. observations in the submillimeterwave range). Since the research on these subjects involves many different specialized areas, the scientists doing the work have seldom had the opportunity to meet, to discuss their problems and therefore, to have a global overview of the field. This is particularly detrimental because molecular spectroscopy is an extremely active field of research and a wealth of data are published. Thus, it is difficult for the end user of these data to keep track of the latest information and, more important, to assess the reliability of the published results.

Therefore an idea emerged to organize a meeting where the laboratory and space spectroscopists would have a chance to discuss their mutual problems and also to get a second half of the whole picture. The NATO Science Programme provided just the right opportunity for organization of such a meeting. We have contacted in this connection several scientists and met with very positive response.

As a result, a NATO Advanced Research Workshop intitled: "spectroscopy from space" was organized in Bratislava from October 31 to November 5, 2000. The main objective of the workshop was to put together radio astronomers, atmospheric scientists, and spectroscopists. People responsible for the field observations explained which results they are expecting from their measurements and how laboratory people could help them in analyzing the satellite data. Laboratory spectroscopists explained what they can do right now and which kind of experimental as well as theoretical developments they could undertake to meet the needs of the remote sensing community. A very efficient flow of information was established thanks to the fact that the right level of background knowledge already existed and plenty of new material was presented and discussed. Another important goal of the workshop was to make aware physicists of central and eastern Europe of these new needs and to retrain them to work in this new field.

53 participants from 18 countries attended the conference. Dr. Jean Demaison (CNRS, University of Lille) served as Director and Professor Kamil Sarka (Comenius University, Bratislava) did an outstanding job as Director of Local Arrangements. Professors P. Encrenaz (Paris), H.-P. Röser (Berlin) and Drs. E. A. Cohen (JPL, Pasadena) and J. K. G. Watson (Ottawa) joined the team as members of the Scientific Committee. 16 invited talks and 30 poster papers were presented. In addition, three round tables were organized. The first one, conducted by P. Bernath (Canada) and B. Carli (Italy) allowed us to discuss the scientific and political problems of our communities and the different ways to improve the links between these communities as well as the efficiency of the collaborations. The second one, conducted by J. E. Boggs (USA) and J.-M. Flaud (France) provided a forum for the generation of concrete

proposals which were discussed and clarified during the third round table, conducted by E. Cohen (USA) and P. Encrenaz (France). Among the actions taken as a result of the forums are the following:

i) The astrophysicists and atmospheric scientists were asked to prepare lists of important problems that should be studied by the spectroscopists. The co-directors of the ARW would be responsible for the circulation of these lists.

ii) It was pointed out in several presentations that the tabulated spectroscopic data are not reliable enough and that there is no uniform method of recommending "best" data sets. It was suggested that a scientific committee be created that will be co-chaired by the two co-directors of the ARW. The goal of this committee should be to analyze the problems and to propose concrete solutions. Another goal of this committee should be the maintenance of efficient links between our different communities in following both the evolution of spectroscopic needs and laboratory capabilities. It was also proposed that this committee submit a proposal to IUPAC in order to get funding for a continuing effort. Its work could be helped by establishment of a web-oriented, open (but controlled) platform whose goals should be **a)** to exploit available databases and to help their development with critical evaluation of the data; **b)** to get feedback from end-users of the data; **c)** to provide tools for making the information relevant to a wider research community.

These proceedings contain the texts of most of the invited talks as well as a few posters. The chapters in this volume are divided among the various different areas.

We would like to express our gratitude to the NATO for the generous grant that made it possible to organize the workshop on a world-wide basis and to the Ministry of Foreign Affairs of the Slovak Republic for the significant help in organizing the meeting in Bratislava.

INTERSTELLAR CHEMISTRY

Eric HERBST
Department of Physics, The Ohio State University,
174 W. 18th Ave, Columbus, OH 43210-1106 USA

1. INTRODUCTION: SPECTRAL OBSERVATIONS

Interstellar space contains matter that is concentrated into cool regions known as interstellar clouds, which can be many light years in extent [1]. These clouds consist of material in the form of gas and tiny dust particles. The matter is mainly gaseous, and the dominant species in the gas is hydrogen, in either atomic or molecular form. Clouds come in a variety of gas densities ranging from 10^1–10^4 cm^{-3} with temperatures mainly in the range 10–100 K. The denser clouds are often associated with star formation; local portions of matter, known as protostars, collapse and heat up until nuclear fusion reactions begin to burn. From spectroscopic studies of the interstellar medium, astronomers have determined the composition of the gas and of the dust particles. In the more diffuse clouds, the gaseous matter can be studied by optical absorption spectroscopy using background stars as sources of light. The gas appears to be mainly atomic, although concentrations of diatomic molecules, chiefly H_2, are detectable. The gas is overwhelmingly neutral; the fractional ionization of $\sim 10^{-4}$ is due mainly to carbon atoms ionized by UV radiation. Many unidentified spectral lines remain, chiefly of a larger width than the assigned features. These lines, known as the diffuse interstellar bands, represent one of the major unsolved mysteries of spectroscopy and may be caused by large but particularly stable molecules [2].

J. Demaison et al. (eds.), Spectroscopy from Space, 1–21.

Visible light does not penetrate the denser clouds, due to scattering and absorption by the dust particles. Dense clouds can be studied at longer wavelengths since extinction by the dust lessens as the wavelength increases [1]. For many years, radio astronomers have taken the lead in studying dense interstellar clouds, using wavelengths better described as millimeter-wave, or even submillimeter-wave [3]. The atmosphere of the earth is rather transparent at frequencies through the millimeter-wave, and there are still windows of relative transparency in the submillimeter-wave, so that ground-based observatories can be utilized. These have been supplemented at shorter wavelengths by aircraft and even satellites [4]. In the last several years, infra-red astronomy has become almost equally important in the study of dense interstellar clouds although ground-based observations are not facile. Particularly important has been the European-based satellite named ISO [5].

Spectroscopic observations show that the gaseous matter in dense interstellar clouds is no longer atomic but is overwhelmingly molecular in nature. Upwards of 120 different molecular species ranging in size from 2-13 atoms have been unambiguously detected, mainly via high-resolution, rotational emission spectra that can be compared with laboratory absorption spectra [6]. Table 1 contains a current list of these species. Many isotopomers have also been detected, especially involving the isotopes deuterium, carbon-13, and oxygen-18. The molecules are predominantly organic and are quite unsaturated in nature. About half of the molecules are well-known species in the laboratory, but about half are quite unusual by terrestrial standards. The unusual species include positive molecular ions (e.g. HCO^+, H_3^+, H_3O^+), radicals (e.g. C_nH through n=8), three-membered rings (e.g. C_3H, C_3H_2), and isomers (e.g. HNC). The fractional ionization is considerably lower than in diffuse clouds, but is also highly uncertain. Besides the spectral emission lines, there is a weak continuum arising from black body emission from the dust particles.

In addition to what molecules are present, astronomers can learn much about the physical conditions in clouds from the spectral transitions [7]. If the transitions are optically thin, relative concentrations (known as fractional abundances) can be determined. It is found that, except for H_2, molecules are trace constituents of the gas; the second most abundant molecule, CO, has a fractional abundance 10^{-4} that of H_2, while larger molecules have even smaller fractional abundances. Still, the large amounts of matter in interstellar clouds mean that the absolute amounts of these molecules are far greater than those found on earth! The fractional abundances of molecules in typical interstellar clouds are not the same throughout the clouds; near star-forming regions, the abundances

Table 1.1 Gas Phase Interstellar/Circumstellar Molecules - High Resolution (10/00)

H_2	CH	CH^+	NH	OH
HF	C_2	CN	CO	CSi
CP	CS	NO	NS	SO
HCl	NaCl	KCl	AlCl	AlF
PN	SiN	SiO	SiS	CO^+
SO^+	H_3^+	CH_2	NH_2	H_2O
H_2S	C_2H	HCN	HNC	HCO
HCO^+	HOC^+	HN_2^+	HNO	HCS^+
C_3	C_2O	C_2S	SiC_2	SO_2
CO_2	OCS	MgNC	MgCN	N_2O
NaCN	CH_3	NH_3	H_2CO	H_2CS
HCCH	$HCNH^+$	H_2CN	C_3H (lin)	c-C_3H
HCCN	HNCO	$HOCO^+$	HNCS	C_2CN
C_3O	H_3O^+	C_3S	CH_4	SiH_4
CH_2NH	H_2C_3(lin)	c-C_3H_2	CH_2CN	NH_2CN
CH_2CO	HCOOH	C_4H	HC_2CN	HCCNC
HNCCC	C_4Si	H_2COH^+	C_5	CH_3OH
CH_3SH	C_2H_4	CH_3CN	CH_3NC	HC_2CHO
NH_2CHO	HC_3NH^+	H_2C_4 (lin)	C_5H	C_5N
CH_3NH_2	CH_3CCH	CH_3CHO	C_2H_3CN	c-CH_2OCH_2
C_6H	HC_4CN	C_7H	$HCOOCH_3$	CH_3COOH
CH_2OHCHO	CH_3C_2CN	H_2C_6 (lin)	C_8H	CH_3OCH_3
C_2H_5OH	C_2H_5CN	CH_3C_4H	HC_6CN	CH_3COCH_3
CH_3C_4CN?	NH_2CH_2COOH?	HC_8CN	$HC_{10}CN$	isotopomers

change in a variety of ways depending on the mass of the star that is forming and the amount of energy that is being emitted. Since one-line identifications are not the rule, rotational temperatures can be determined from comparison of the intensities of different lines from the same molecule. Internal rotor species such as methanol are especially useful in tracing temperatures, since they have particularly dense spectra, with many transitions of widely different excitation. In particular, rotational transitions in excited torsional states are useful in probing regions above 100 K. The very detection of emission lines is made possible by collisional excitation, so that the existence of lines for any given molecule can be used to determine a minimum, or critical, density for excitation given knowledge of cross sections for rotationally inelastic collisions with the dominant species - H_2. Weakly polar species such as CO are excited at very low densities (10^2 - 10^3 cm^{-3}) since the Einstein A coefficient is rather small, especially for low-lying rotational transitions, whereas polar species such as HCN are excited at higher densities, which are found only in selected areas of the clouds known as "dense cores." Some dense cores are quiescent while others are collapsing to form stars [7]. Maps of interstellar clouds made by plotting contours of equal intensity points for a given rotational transition of a given molecule are thus quite dependent on the molecule and on the transition [3]. High spatial resolution in these maps, necessary to study star formation regions, can be obtained via radio astronomy despite the low angular resolution inherent in long-wavelength observations, by using numbers of radio telescopes in interferometric mode. Finally, positions of the line maxima and the line shapes are due to the Doppler effect, and so tell us about the velocities of the clouds and the large scale motions inside them. In the main, clouds appear to be turbulent rather than thermal in nature in their velocity structures.

Since low density conditions can often lead to non-thermal situations, population inversions are occasionally obtained. Interstellar clouds are even the sites of masers, which are confined to regions of considerable excitation in the neighborhood of newly formed and forming stars. Some of the molecules seen in maser emission are OH, SiO, H_2O, and methanol [8]. The excitation of the maser emission can be achieved by radiation, collisions, or both.

It is difficult to excite vibrational spectra of molecules under typically cold interstellar conditions, and consequently, infrared spectra are often seen in absorption. Two types of continuum sources are used as lamps - "field stars", which are stars behind the interstellar cloud that emit sufficient radiation, and protostellar sources within the clouds themselves. The protostellar sources can be either stars in the process of formation or newly formed stars; in either case the infrared radiation comes from heated dust surrounding the objects. Infrared absorption spectra are particularly important for detecting non-polar molecules - only one of which, H_2, is sufficiently abundant to be detected by quadrupolar rotational transitions. In recent years, the most important infrared identification has been that of the most basic polyatomic species - H_3^+ [9].

In regions of sufficient excitation, infrared transitions can be detected in emission. For example, surrounding young high-mass stars are very hot and bright volumes known as H II regions, in which the gas consists mainly of protons and in which molecules are not present. Surrounding these regions are rather warm portions of clouds in which the gas is mainly neutral. These so-called "photon-dominated regions" show a variety of broad emission features throughout the infrared. Known previously as the UIR (unidentified infrared) features, these spectral lines have been assigned [10] to a collection of polycyclic aromatic hydrocarbons, closely akin to automobile soot on earth. Although the assignments, based on functional group analysis, do not point at any one particular species, they indicate that the total fractional abundance of these PAH's far exceeds all small gas-phase heavy species except for CO. Other interpretations of the spectra include the idea that they are not caused by individual PAH molecules but by small dust particles of amorphous carbon.

In general, the nature of the dust particles is more difficult to determine than the nature of the interstellar gas, but from the scattering, polarization, and re-emission of radiation, the collection of interplanetary dust particles, the depletions of certain elements from the gas phase, and broad spectroscopic features in the infrared, astronomers have deduced that the particles are roughly 0.1 μ in size or smaller and consist of metallic silicates and/or carbonaceous matter [11]. In dense clouds, additional broad infrared absorption features indicate mantles of icy materials containing mainly water, carbon monoxide, carbon dioxide, and methanol. The features have been reproduced by laboratory experiments, which yield not only the carriers but their likely environments (e.g. polar or non-polar) [5]. Mantles of up to 100 monolayers are inferred.

2. CHEMICAL PROCESSES

The gaseous and particulate material in interstellar clouds comes from previous generations of stars that, late in life, have parted with much of their matter [12]. High mass stars part explosively with matter, via supernovae, whereas lower mass stars part with it more gently by blowing it out in the form of extended atmospheres, or envelopes. Most stars at this stage are said to be "oxygen-rich," by which is meant that there is more oxygen than carbon. Under both "oxygen-rich" and "carbon-rich" conditions, dust particles form as material is ejected and begins to at least partially condense. The dust particles in O-rich sources are silicaceous in nature except perhaps in supernovae whereas those in C-rich sources are carbonaceous. It is thought that PAH's are formed in the extended atmospheres of C-rich stars at $\approx$ 1000 K [13]. Smaller molecules are formed mainly in cooler layers of the atmospheres of older stars, especially in C-rich sources. No matter the source, as the molecules stream out and the density diminishes, the interstellar radiation field, which consists of the light from all other stars, is no longer attenuated and the molecules are photodestroyed. For example, the lifetime of CO in the unshielded interstellar medium is thought to be little more than 100 years [14]. Consequently, the matter emanating from stars is mainly in the form of atoms and bare dust particles, so that when clouds are formed by the influence of gravity, the gas is atomic in nature. In addition, there may be some large molecules such as PAH's that are particularly stable against photodegradation.

What is the elemental composition of the matter emanating from stars? The composition of stellar atmospheres can be measured by atomic absorption spectroscopy. It is found that stars have differing amounts of each element, but the cosmic (number) average, as represented by nearby stars, is as follows for selected elements: H(1), He(0.05), O(5×10^{-4}), C(2×10^{-4}), N(7×10^{-5}), S(1×10^{-5}), Si (2×10^{-5}), Fe (3×10^{-5}) [15]. In the gas-phase of diffuse clouds, all heavy elements are depleted to a greater or lesser extent to form the dust particles. The important elements C, N, and O are depleted by at most a factor of a few, while certain heavy elements such as Si and Fe are depleted by orders of magnitude. The elemental abundances in dense clouds cannot be directly determined since the gas-phase matter is molecular not atomic. The extent of depletion of some of the metallic elements is particularly important because these elements, which are easily ionized, contribute the most to the small ionization fraction unless they are too heavily depleted. Despite these uncertainties, it is clear that the chemistry will occur in a strongly hydrogen-rich environment, and that the heavy elements C, N, O will play important roles.

The synthesis of the many molecules seen in interstellar clouds takes place under the seemingly inhospitable conditions of these regions, which possess very low densities and temperatures. These conditions constrain gas-phase reactions to be exothermic two-body processes without activation energy. The dominant processes appear to be exothermic ion-molecule reactions since these reactions typically possess no activation energy barriers, and are known to occur rapidly at low temperature, unlike most reactions involving neutral species [16]. Ions are formed in dense interstellar clouds chiefly by bombardment of the clouds by cosmic rays, which are high-energy protons and heavier nuclei travelling at relativistic speeds [1]. The rate coefficient $k_{\mathrm{i-m}}$ for ion-molecule reactions involving a non-polar neutral reactant is often found to agree with a simple expression first derived by Langevin [16]:

$$k_{\mathrm{i-m}} = 2\pi e\sqrt{\alpha/\mu} \sim 10^{-9}\ \mathrm{cm^3\,s^{-1}} \tag{1.1}$$

where e (esu) is the electronic charge, α (cm^3) is the polarizability, and μ (gm) is the reduced mass of the reactants. This expression, which essentially yields the rate coefficient for strong collisions, is independent of temperature. Rate coefficients $k_{\mathrm{n-n}}$ for neutral-neutral reactions, on the other hand, are typically of the Arrhenius form

$$k_{\mathrm{n-n}} = A(T)\exp(-E_{\mathrm{a}}/k_{\mathrm{B}}T) \tag{1.2}$$

where the activation energy, E_{a}, is on the order of eV, so that low temperature reactions are very slow, occurring on only a small fraction of collisions. Exceptions to the Arrhenius rule are known to occur for reactions involving radicals and atoms, and even for some systems involving one atom or radical and a so-called stable molecule [17]. The pre-exponential factor $A(T)$ can have a weak temperature dependence, which is typically but not always given by transition state theory [18].

Although gas-phase reactions can produce many if not most of the observed gas-phase interstellar molecules, there are some glaring exceptions, the most important of which is H_2 – the dominant interstellar molecule. Under cool and rarefied interstellar conditions, the only mechanism for H_2 formation in the gas phase is via radiative association. In this process, the two H atoms collide and manage to stick together by giving off sufficient energy in the form of a photon. Radiative association is probably an important type of reaction for larger species, but it is very inefficient for two H atoms and cannot explain the total conversion of atomic into molecular hydrogen in dense clouds. Instead, the formation of H_2 must occur on the surfaces of the dust particles, most probably by a diffusive, or Langmuir-Hinshelwood, mechanism in which two H atoms collide with and stick to a grain, then undergo a two-dimensional random-walk motion before finding one another and giving off sufficient energy to the grain to remain together [19]. This diffusive reaction has been studied in the laboratory on two representative surfaces - olivine (polycrystalline silicate) and amorphous carbon and found to occur [20], albeit not as rapidly as astronomers had estimated. In addition, the experiments have shown that on about half of the formation reactions, the newly produced H_2 immediately returns to the gas phase. The diffusive mechanism requires weakly-bound adsorbates that can move about a grain surface; an alternative mechanism for strongly-bound adsorbates is the Eley-Rideal process, in which a gas-phase species lands atop a stationary adsorbate and reacts with it. It appears from theoretical studies that the formation of H_2 on graphite particles occurs via this latter mechanism since graphite binds H atoms strongly [21].

The production of heavier species on grain surfaces and their possible removal back into the gas are not well understood and are often ignored. The most likely processes to occur, besides H_2 formation, are association reactions between slow-moving heavy atoms or radicals and H atoms, which are thought to move much more quickly on materials such as silicates, amorphous carbon, or even ice that may form the surfaces of low temperature grains. Via such reactions, O atoms landing on a grain can be hydrogenated into OH and subsequently H_2O, while N and C atoms can be hydrogenated into NH_3 and CH_4 respectively. Some investigators even think that methanol can be formed by the successive hydrogenation of species starting from CO [22]. Since heavy species produced on cold grains will probably not be ejected back into the gas, and since thermal evaporation of such species at low temperatures is very slow, it is unclear how or if these species will cycle back into the gas. Assuming that most remain on the grains, these then constitute the ice mantles detected via infrared absorption.

3. GAS-PHASE PATHWAYS TO POLYATOMIC MOLECULES

Once H_2 is formed on grains and ejected into the gas, a rapid and complex gas-phase chemistry ensues that is dominated by ion-molecule reactions [23,24]. The primary process is the ionization of hydrogen, which cannot occur directly via UV photons (not that many are present inside clouds) but does occur via collisions with cosmic rays, which are sufficiently energetic to penetrate dense clouds. As well as directly, ionization occurs indirectly through secondary electrons produced in the primary process. A first-order rate coefficient for the ionization of H_2 can be deduced by measuring the flux of cosmic rays above the earth, assuming this flux to be representative of an interstellar cloud, and calculating the cross sections for direct and indirect ionization. This analysis [1] leads to a rate coefficient ζ (s^{-1}) of $\approx 1 \times 10^{-17}$. The cosmic ray flux is not large enough to ionize a significant fraction of the dense cloud gas, but does lead to an ion fraction (10^{-6}–10^{-8}) sufficient for ion-molecule reactions to play a crucial role in the subsequent chemistry.

The H_2^+ produced by cosmic rays reacts with the first hydrogen molecule it approaches to form the polyatomic ion H_3^+ via the well-studied ion-molecule reaction

$$H_2^+ + H_2 \longrightarrow H_3^+ + H. \tag{1.3}$$

The abundance of this ion can be directly related to the cosmic ray ionization rate ζ; observations of H_3^+ in dense clouds yield an ionization rate similar to that discussed above [9]. The H_3^+ ion can be destroyed by a variety of exothermic reactions. In diffuse clouds, the major destruction mechanism is probably dissociative recombination with electrons:

$$H_3^+ + e \longrightarrow H + H_2, 3H. \tag{1.4}$$

This reaction has been studied most recently in storage rings as well as in a variety of bench-top pieces of apparatus including a flowing afterglow [25]. Most dissociative recombination reactions have rate coefficients in the range $1\times10^{-7} - 1\times10^{-6}$ cm^3 s^{-1} at 300 K and possess a weak inverse dependence on temperature, which typically goes as $T^{-1/2}$. Although the storage ring results for H_3^+ + e appear to be typical in this regard, the flowing afterglow results show a rate coefficient perhaps an order of magnitude lower. The discrepancy between high and low values is long-lasting and appears to show no signs of being resolved. Theoretical treatments have up to now proven unhelpful.

In dense clouds, H_3^+ is destroyed more rapidly by reactions with heavy species such as the atoms C and O. These reactions are of the type

$$H_3^+ + X \longrightarrow XH^+ + H_2, \tag{1.5}$$

where X is a heavy species. Let us consider the specific reaction with O:

$$H_3^+ + O \longrightarrow OH^+ + H_2. \tag{1.6}$$

The products of this "proton transfer" reaction themselves react efficiently via an H-atom transfer reaction:

$$OH^+ + H_2 \longrightarrow H_2O^+ + H. \tag{1.7}$$

A subsequent H-atom transfer reaction with ubiquitous molecular hydrogen then leads to the hydronium ion:

$$H_2O^+ + H_2 \longrightarrow H_3O^+ + H. \tag{1.8}$$

This ion, which has been detected via its rotation-inversion spectrum, does not react with molecular hydrogen, and so is relatively abundant. It is depleted by a dissociative recombination reaction with electrons:

$$H_3O^+ + e \longrightarrow H_2O + H; OH + 2H; OH + H_2; O + H_2 + H. \tag{1.9}$$

The product branching fractions for this process have been measured in two storage rings and in a flowing afterglow apparatus [26]. The agreement is less than spectacular; the flowing afterflow results indicates little or no water product, while the two latest storage ring results indicate 19% and 24% water, respectively. Most independent observers favor the storage ring results, showing that gas-phase water (as well as the hydroxyl radical OH) can be produced via a chain of ion-molecule and dissociative recombination reactions. These species are depleted by gas-phase reactions: water by ion-molecule reactions and hydroxyl by both ion-molecule and neutral-neutral reactions.

The reaction between H_3^+ and neutral atomic carbon proceeds via a proton-transfer mechanism:

$$H_3^+ + C \longrightarrow CH^+ + H_2. \quad (1.10)$$

"Carbon fixation" can also occur from the atomic ion C^+. This ion does not undergo a normal exothermic reaction with H_2, but does undergo a radiative association:

$$C^+ + H_2 \longrightarrow CH_2^+ + h\nu, \quad (1.11)$$

which occurs on roughly one out of every 10^{6-7} collisions according to experiment and theory [27]. Once CH^+ and CH_2^+ are produced, H-atom transfer reactions with H_2 lead to the synthesis of CH_3^+:

$$CH^+ + H_2 \longrightarrow CH_2^+ + H, \quad (1.12)$$

$$CH_2^+ + H_2 \longrightarrow CH_3^+ + H. \quad (1.13)$$

As is the case for C^+, the methyl ion does not undergo a normal exothermic reaction with H_2, but does undergo a radiative association reaction:

$$CH_3^+ + H_2 \longrightarrow CH_5^+ + h\nu, \quad (1.14)$$

which has been studied in the laboratory [27] and proceeds on one out of every 10^4 collisions at low temperatures. The dissociative recombination reaction between CH_5^+ and electrons leads mainly to the methyl radical; methane is formed more efficiently by the reaction

$$CH_5^+ + CO \longrightarrow CH_4 + HCO^+. \quad (1.15)$$

Carbon monoxide, a very stable molecule, is formed by a wide variety of exothermic ion-molecule (through HCO^+) and neutral-neutral reactions.

The formation of small hydrocarbons such as the methyl ion, methyl, and methane sets the stage for the production of more complex species via reaction sequences that have only partially been studied in the laboratory. Larger hydrocarbons are produced via two types of processes – (i) carbon insertion reactions, and (ii) condensation reactions. The first class of reactions is illustrated by

$$C^+ + CH_4 \longrightarrow C_2H_3^+ + H; C_2H_2^+ + H. \quad (1.16)$$

The acetylenic ions formed in this reaction can undergo dissociative recombination with electrons to form species such as C_2H and C_2H_2. The second class of reactions is illustrated by

$$C_2H_2^+ + C_2H_2 \longrightarrow C_4H_3^+ + H. \quad (1.17)$$

Interestingly, reactions between hydrocarbon ions with more than one carbon atom and molecular hydrogen tend for the most part not to occur, so that the degree of saturation remains low. Other classes of molecules can be produced by analogous reactions. Nitrogen atom insertion reactions lead to organo-nitrogen ions; e.g.,

$$N + C_3H_3^+ \longrightarrow HC_3NH^+ + H, \tag{1.18}$$

which, via dissociation recombination, produce nitrogen-bearing neutrals; viz.,

$$HC_3NH^+ + e \longrightarrow HC_3N + H. \tag{1.19}$$

Except for a small number of reactions, the products of dissociative recombination reactions have not been studied and must, sometimes crudely, be guessed at. Radiative association reactions may also be of importance although much laboratory work remains to be done on such systems. One example is part of the only known gas-phase synthesis of methanol:

$$CH_3^+ + H_2O \longrightarrow CH_3OH_2^+ + h\nu. \tag{1.20}$$

The protonated methanol ion can then possibly form methanol via dissociative recombination.

Ion-molecule reactions are not the only processes leading to molecular complexity. Some newly studied neutral-neutral reactions are of great importance [17]. The reaction

$$C + C_2H_2 \longrightarrow C_3H + H \tag{1.21}$$

not only produces the radical C_3H efficiently at low temperatures, it apparently produces both the linear and cyclic isomers. The reaction

$$CN + C_2H_2 \longrightarrow HC_3N + H \tag{1.22}$$

is the dominant process leading to the formation of cyanoacetylene, a well-known interstellar molecule.

The growth of organic molecules in dense clouds is countered mainly by ion-atom reactions involving atomic oxygen, which is abundant in models in which (as is customary) the gas is oxygen-rich. Neutral-neutral reactions between organic radicals and O also tend to reduce complexity. In diffuse clouds, external UV photons play a most important role in limiting the growth of complexity so that only diatomic species have to the present been detected. In dense clouds, there are fewer photons, although some exist as a result of excitation of hydrogen molecules by secondary electrons produced, in turn, by cosmic ray ionization [28]. In general, photodestruction in dense clouds is slower than destruction by reaction and accretion onto the dust grains. In purely gas-phase models, the net result of formation and destruction reactions is the production of steady-state abundances. The time scale for reaching steady state in dense clouds is so long, however, that accretion onto the dust particles occurs first so that, in the absence of non-thermal desorption processes, there is hardly any gas phase around. Whether to include the rather uncertain desorption processes, ignore the grains completely, or consider results at times much shorter than that necessary to reach steady state remains a matter of dispute. Whichever strategy is followed, models with gas-phase reactions do appear capable of reproducing many if not most of the observed molecular abundances.

4. CHEMICAL MODELS OF DENSE INTERSTELLAR CLOUDS

Current model networks of the gas-phase chemistry of interstellar clouds consist of thousands of reactions involving hundreds of molecules and a wide assortment of different elements [29]. In our research group, we use three sets of reaction networks in our model calculations. The networks are distinguished mainly by the extent of rapid neutral-neutral reactions that are included. Since only a small number of rapid reactions involving an atom or radical and a stable molecule have actually been measured at low temperatures, we are left with the problem of how to extrapolate these results to the much larger number of possible reactions that can occur in the intestellar medium. In our standard set of reactions (the new standard model), we tend to be very conservative and not to add reactions that have not been studied. In another network (the new neutral-neutral model), we take the opposite approach and add all reactions similar to those that have been studied. Finally, in a third network, we take a more centrist view. In general, for reasons not yet understood, our most successful network is the new standard model.

Whatever the reaction network, one constructs models by incorporating the reactions into coupled kinetic equations, which must then be integrated as a function of time and possible changes of physical conditions. In the simplest cases – quiescent dense cloud cores with no changes in physical conditions – the integration of hundreds of coupled differential equations to over 10^8 yr of astronomical time can be accomplished in minutes with a variety of computers.

Our models of quiescent cores, in which the hydrogen is assumed to be in its molecular form initially, are generally able to reproduce the concentrations of many of the observed molecules outside of star-forming regions accurately and to predict the concentrations of others not yet detected. Analogous models can be used to study the chemistry occurring in stellar envelopes, which surround old, low-mass stars. That the results of these models are not fictional is seen by a qualitative and successful comparison of their salient predictions with the observed types of molecules.

First, the models predict, not surprisingly but successfully, that molecular ions will be present in dense clouds and that these ions will in the main be protonated species. Such predictions actually predated the observations of these species.

Secondly, the models predict that metastable isomers and radicals will be produced at rates comparable to stable forms. The classic example of metastable formation is the dissociative recombination of the linear ion $HCNH^+$ to produce both HCN and its metastable isomer HNC:

$$HCNH^+ + e \longrightarrow HCN + H; HNC + H. \tag{1.23}$$

Although the branching fraction between these two sets of products has not yet been measured, groups at several storage rings are preparing for just such a measurement using position-sensitive detectors. Theory indicates a branching fraction near unity. Dissociative recombination reactions also lead to radicals as well as stable species; for example, the dissociative recombination of CH_5^+ leads mainly to methyl (CH_3) rather than methane. So detectable abundances of radicals are also predicted, in agreement with observations.

Thirdly, the models predict quantitatively the amount of isotopic fractionation which, in the case of deuterium, can be very large. In dense clouds, HD is the major carrier of deuterium and possesses a fractional abundance of $\approx 10^{-5}$. For trace molecular species, the abundance ratio of the singly deuterated isotopomer to the normal species can be much larger than this, an effect known as fractionation. Consider the well-studied reaction system

$$H_3^+ + HD \rightleftharpoons H_2D^+ + H_2. \tag{1.24}$$

The left-to-right reaction is exothermic by about 230 K (astronomers prefer to use temperature as a unit of energy) due to the difference in zero-point vibrational energies and the Pauli Exclusion Principle, which shows that the ground rotational state of H_3^+ is forbidden. For clouds at temperatures well below 200 K, the back reaction is rather slow and a large abundance of H_2D^+ can be built up [24]. Indeed, in a 10 K cloud, the ratio of H_2D^+ to H_3^+ can reach 0.10, a four order-of-magnitude effect! The deuterated ion can then react with other species to spread the deuteration around the network of reactions. For example, DCO^+

can be produced in high abundance via

$$H_2D^+ + CO \longrightarrow DCO^+ + H_2. \tag{1.25}$$

There are actually very few exchange-type reactions involving an ion and HD that do not possess activation energy barriers, so that it is relatively easy to map out the formation pathways of the deuterated ions and their subsequent ion-molecule reactions [30]. An extension of this analysis to doubly-deuterated isotopomers has recently been made and used to explain the observation of doubly deuterated ammonia [31].

Fourthly, models predict correctly that the chemistry will be rather unsaturated despite the large abundance of molecular hydrogen, because most reactions between complex molecular ions and H_2 are either endothermic or have barriers.

If one wishes to discuss the extent of quantitative agreement between theory and observation, one must consider individual sources. The best-studied dense cloud core is known as TMC-1, where the initials stand for Taurus Molecular Cloud. This rather quiescent cloud core is located in the constellation Taurus in the midst of a giant dense molecular cloud of complex topology. Over 50 molecules have been detected in this source alone, some of which are the most complex species seen unambiguously in space. Using as a criterion of success the reproduction of molecular abundances to within an order of magnitude, we can reproduce 80-90% of the molecular abundances with our most successful network (the new standard model) over time intervals that depend strongly on the carbon and oxygen elemental abundances chosen [29]. With the normal oxygen-rich abundances (C/O = 0.4), large organic molecules tend to exist with reasonable concentrations only at times ($\leq 10^6$ yr) well before steady state is reached (10^7 - 10^8 yr) because they eventually 'combust" to form CO. Of course, for times longer than the so-called "early times" of best agreement, the gas phase begins to deplete onto the dust grains. If one considers the possibility of subsequent non-thermal desorption of species back into the gas via cosmic ray heating of grains, the period of resonable agreement with observation can be lengthened. Still, it is clear that chemical models do not predict 100% of the molecular abundances correctly. Although there are obviously a variety of uncertainties concerning the chemistry (including the role of surface processes), it is probably the astronomical conditions and their temporal variations that are the main source of the discrepancies. In particular, sources such as TMC-1 are not as quiescent as might at first appear; evidence of the initial stages of star formation is apparent.

In addition to the problem of star formation, many unanswered questions remain. One current area of interest is to understand the role of surface chemistry in the production of molecules more complex than H_2. The latest models with both gas-phase and granular chemistry are able to reproduce the development of icy mantles in dense clouds and show interesting departures from purely gas-phase models in their predictions of gas-phase abundances [32]. There are, however, many uncertainties concerning the nature of the chemistry that occurs on grain surfaces. These involve both our understanding of surface chemistry and our knowledge of the grains. Among the problems and uncertainties are:

(i) *the relatively unknown size distribution and topological nature of grains.* The surface chemistry is strongly dependent on the size of the grains and their topological nature.

(ii) *the poorly constrained chemical nature of the surface on which reactions occur.* Surface reactions are clearly determined by the nature of the surface and the strength of the binding.

(iii) *the multiplicity of mechanisms of chemical reaction.* Clearly, whether the chemistry involves the Langmuir-Hinshelwood mechanism, the Eley-Rideal mechanism, some combination of the two, or yet a third mechanism, will affect the results.

(iv) *the various types of chemical reactions.* Are surface chemical reactions mainly associative in nature (e.g. $A + B \rightarrow AB$) or do normal exothermic channels compete?

(v) *mathematical problems associated with small particles.* On small surfaces with few reactive particles, standard rate equations are not necessarily correct for diffusion and should probably be replaced with stochastic or probabilitistic methods, yet this has only been done for small, artificial models [31]

(vi) *the highly uncertain rate of non-thermal desorption of heavy species.*

5. STAR FORMATION AND MOLECULES

Perhaps the most exciting area of current interest is the fate of interstellar molecules as stars and planetary systems form inside clouds [34]. A myriad of phenomena attend the process of star formation, depending on the mass of the material. Perhaps the simplest of these is the starless core, an object in the initial stages of star formation. In a starless core, there is evidence of collapse of material but there is no evidence that the core at the center of collapse is heating up at all. From the abundance dependence of different molecules as a function of radial distance from the center, one can gain information on the dynamics of the collapse.

A much later stage of star formation, especially associated with low-mass stars, occurs when the collapse eventually leads to a new-born stellar object, which is still quite unstable. At this stage, the so-called T

Tauri star is surrounded by a flat mass of swirling gas and dust, known as a "protoplanetary" disk, which revolves around the star in Keplerian motion, slowly accreting into the center as it revolves. The density of matter, which is much higher than in the parent molecular cloud, decreases with increasing distance from the star. Physical models indicate that protoplanetary disks are not entirely flat, but have a height; as one goes up or down from the so-called midplane of the disk, the density decreases. This decrease is less pronounced at longer distances from the star so that the disk becomes less flat at such distances. These disks are presumably similar to the Solar System before the condensation of the dust into planets, comets, and meteors. Therefore, an understanding of the chemistry occurring in disks should tell us much about the history of the early Solar System.

In recent years, radioastronomers have begun to study the abundances of molecules in protoplanetary disks; in general gas-phase molecules show lower average abundances than in the parent molecular clouds, because much of the gaseous material in the midplane is condensed onto the growing dust particles. More recently, interferometric observations have begun to probe the radial dependence of gas-phase molecules in the outer portions of disks. At the current time, modellers are ahead of observers, however, since current generation models are two-dimensional in character [35]; the abundances of molecules are calculated as a function of both height and radial distance. In the outer portion of a disk, the distribution of molecules as a function of height shows three distinct regions: in the midplane, most molecules are condensed onto dust particles, while at intermediate heights there are much larger gas-phase molecular abundances. Finally, as the height continues to increase, radiation from external stars as well as from the central stellar object photodissociate the molecules so that the gas becomes mainly atomic. This type of region resembles a "photon-dominated" area, and is especially difficult to model because the T Tauri star emits lots of X-rays in addition to normal stellar radiation. Once radioastronomical observations are able to resolve both the height and radial distributions of molecules, the models will be able to constrain important parameters such as the age of the star and disk, the X-ray output of the star, the density and temperature dependence of the disk, and the state of coagulation of the dust particles. Infra-red observations of the dust particles will add even more to our knowledge.

The stages of formation of a high-mass star are not as well defined as those of its low-mass cousin. In the early stages of collapse, these rather large objects emit enough heat to produce a continuum of infra-red radiation useful to studying cool foreground gaseous and condensed-phase material. Eventually, an object known as a "hot core" is formed; in such an object the concentrations of molecules in the gas-phase change markedly and larger abundances of some saturated organic molecules are detected. Molecules such as methyl alcohol are much more abundant in hot cores than in normal quiescent sources and more complex species such as ethyl alcohol, methyl formate, and dimethyl ether are only seen in hot cores. What causes the chemistry to be so altered? The dominant current explanation [36] is that the temperatures in hot cores (100-300 K) are warm enough to evaporate the grain mantles so that the products of granular chemistry during the previous low temperature era are gradually released into the gas phase, where they can undergo further chemical processing. Despite great uncertainties in our knowledge of dust chemistry, it is clear that this scenario leads to more saturated species than does ion-molecule chemistry. In particular, the surface hydrogenation of CO via successive reactions with atomic hydrogen leads to large abundances of methanol. More complex species may also be formed on grain surfaces, although it is more likely that they are formed in the gas with desorbed methanol as a precursor.

There is an alternative explanation for hot core chemistry – that it is driven by shock waves. Shocks are a frequent occurrence in the interstellar medium, where they are often associated with star formation [37]. Although the details of the shock structure are very dependent on parameters such as the local magnetic field, the basic picture of a one-fluid or J-shock is one in which passage of a shock leads abruptly to very high temperatures (up to perhaps 4000 K) followed by gradual cooling. Chemical reactions with activation energy barriers and even endothermic reactions can occur during the high temperature phase. Saturated species can be produced by H-atom exchange reactions between neutrals and molecular hydrogen. More modern treatments of shock processes using magnetohydrodynamics show that even a small magnetic field in the right direction can cause the characteristics of the shock to be quite different. In particular, a two-fluid or C-shock can be produced, in which temperature rises are much more moderate but in which large-scale motions of ions with respect to neutrals can occur. This relative velocity can then power endothermic ion-molecule but not neutral-neutral reactions.

Although shock models have been applied to hot cores, they are mainly associated with explanations for one molecule seen in diffuse clouds but totally inexplicable by standard treatments of gas-phase and even grain chemistry. The molecule - CH^+ - was first detected in absorption in diffuse clouds more than half a century ago. The only reaction that can produce it at its observed abundance is the process

$$C^+ + H_2 \longrightarrow CH^+ + H, \tag{1.26}$$

which is endothermic by 0.4 eV. Although the temperatures in J-shocks, or the ion-neutral drift velocities in C shocks, can be high enough to power the reaction, shock models must also avoid worsening the good agreement between theory and observation for other molecules in diffuse clouds. This has not proven to be an easy task, and currently other suggestions are being made to explain the CH^+ abundance, including high temperatures produced by intermittent turbulence [38].

Molecules are not only detected in diffuse and dense interstellar clouds and around young stars in dense clouds, but around normal stars, and, more pronouncedly, among older stars. In a special class of such older stars, known as AGB carbon-rich objects, hydrocarbons and cyanopolyynes are produced in rather high abundance as material streams out of the stellar atmospheres to form large circumstellar envelopes. Detailed models of the chemistry of these objects are similar to their interstellar brethren [39], although the carbon-rich conditions lead to a brief period of high molecular abundances. The analysis of molecules in normal stellar atmospheres relies on high-temperature thermodynamic equilibrium and not on kinetic treatments.

6. SUMMARY

Molecules and molecular spectroscopy have in the last two decades become an integral part of astronomical science. Molecules are detected in a wide class of objects, in our own galaxy, in galaxies close to our own, and in galaxies at the most distant parts of the universe. In all of these objects, molecules are important both as signposts that chemistry is not unique to the Solar System and as tools for understanding the physical conditions under which they thrive. The chemistry of molecules in quiescent interstellar clouds is only partially understood, and the chemistry of molecules in assorted regions associated with star formation still needs much more research to reach a significant level of understanding. Still, the field of interstellar chemistry is a rather young one, and much has been learned in the quarter-century since radioastronomical observations first began to detect polyatomic molecules in space. With the continuing interest of spectroscopists, kineticists, and astronomers of differing types, the role of molecules in astronomy will remain a major and fruitful subject of investigation.

Acknowledgments

Acknowledgment is made to the National Science Foundation (U.S.) for support of my research program in astrochemistry.

References

1. Spitzer, L. (1978) *Physical Processes in the Interstellar Medium*, John Wiley & Sons, New York.
2. Tielens, A.G.G.M. and Snow. T.P. (eds.) (1995) *The Diffuse Interstellar Bands*, Kluwer Academic Publishers, Dordrecht.
3. Wall, W.F., Carraminana, A., Carrasco, L. and Goldsmith, P.F. (eds.) (1999) *Millimeter-Wave Astronomy: Molecular Chemistry & Physics in Space*, Kluwer Academic Publishers, Dordrecht.
4. Melnick, G.J. et al. (2000) The Submillimeter Wave Astronomy Satellite: Science Objectives and Instrument Description, *Astrophysical Journal* **539**, L77-L85.
5. Ehrenfreund, P. and Schutte, W.A. (2000) Infrared Observations of Interstellar Ices, in Y.C. Minh and E.F. van Dishoeck (eds.), *Astrochemistry: From Molecular Clouds to Planetary Systems*, Sheridan Books, Chelsea, Michigan, pp. 135-146.
6. Herbst, E. (1999) Molecules in Space and Molecular Spectroscopy, in W. F. Wall et al. (eds.) *Millimeter-Wave Astronomy: Molecular Chemistry & Physics in Space*, Kluwer Academic Publishers, Dordrecht, pp. 329-340.
7. Winnewisser, G., Herbst, E. and Ungerechts, H. (1992) Spectroscopy Among the Stars, in K. N. Rao and A. Weber (eds.), *Spectroscopy of the Earth's Atmosphere and Interstellar Medium*, Academic Press, New York, pp. 423-517.
8. Elitzur, M. (1999) Masers, in W. F. Wall et al. (eds.) *Millimeter-Wave Astronomy: Molecular Chemistry & & Physics in Space*, Kluwer Academic Publishers, Dordrecht, pp. 127-142.
9. Geballe, T.R. and Oka, T. (1996) Detection of H_3^+ in Interstellar Space, *Nature* **384**, 334-335.
10. Allamandola, L.J., Tielens, A.G.G.M. and Barker, J. (1989) Interstellar Polycyclic Aromatic Hydrocarbons: The Infrared Emission Bands, The Excitation/Emission Mechanism, and the Astrophysical Implications, *Astrophysical Journal Supplement Series* **71**, 733-775.
11. Witt, A.N. (2000) Overview of Grain Models, in Y. C. Minh and E. F. van Dishoeck (eds.), *Astrochemistry: From Molecular Clouds to Planetary Systems*, Sheridan Books, Chelsea, Michigan, pp. 317-330.
12. Abell, G.O. (1982) *Exploration of the Universe*, Saunders College Publishing, Philadelphia.
13. Frenklach, M. and Feigelson, E.D. (1989) Formation of Polycyclic Aromatic Hydrocarbons in Circumstellar Envelopes, *Astrophysical Journal* **341**, 372-384.

14. Van Dishoeck, E.F. (1988) Photodissociation and Photoionisation Processes, in T. J. Millar and D. A. Williams (eds.), *Rate Coefficients in Astrochemistry*, Kluwer Academic Publishers, Dordrecht, pp. 49-72.
15. Snow, T.P. and Witt, A.N. (1996) Interstellar Depletions Updated: Where All the Atoms Went, *Astrophysical Journal*, **468**, L65-L68.
16. Rowe, B.R., Rebrion-Rowe, C. and Canosa, A. (2000) Low Temperature Experiments on Gas-Phase Chemical Processes, in in Y. C. Minh and E. F. van Dishoeck (eds.), *Astrochemistry: From Molecular Clouds to Planetary Systems*, Sheridan Books, Chelsea, Michigan, pp. 3237-250.
17. Sims, I.R. and Smith, I.W.M. (1995) Gas-Phase Reactions and Energy Transfer at Very Low Temperatures, *Annual Review of Physical Chemistry* **46**, 109-138.
18. Weston, Jr., R.E. and Schwarz, H.A. (1972), *Chemical Kinetics*, Prentice-Hall, Englewood Cliffs, New Jersey.
19. Hollenbach, D. and Salpeter, E.E. (1971) Surface Recombination of Hydrogen Molecules, *Astrophysical Journal* **163**, 155-164.
20. Katz, N., Furman, I., Biham, O., Pirronello, V. and Vidali, G. (1999) Molecular Hydrogen Formation on Astrophysically Relevant Surfaces, *Astrophysical Journal* **522**, 305-312.
21. Farebrother, A.J., Meijer, A.J.H.M., Clary, D.C. and Fisher, A.J. (2000) Formation of molecular hydrogen on a graphite surface via an Eley-Rideal mechanism, *Chemical Physics Letters* **319**, 303-308.
22. Charnley, S.B., Tielens, A.G.G.M. and Rodgers, S.D. (1997) Deuterated Methanol in the Orion Compact Ridge, *Astrophysical Journal* **482**, L203-L206.
23. Herbst, E. and Klemperer, W. (1973) The Formation and Depletion of Molecules in Dense Interstellar Clouds, *Astrophysical Journal* **185**, 505-533.
24. Smith, D. (1992) The Ion Chemistry of Interstellar Clouds, *Chemical Reviews* **92**, 1473-1485.
25. Larsson, M. (2000) Experimental studies of the dissociative recombination of H_3^+, *Philosophical Transactions of the Royal Society: Mathematical, Physical & Engineering Sciences* **358**, 2433-2444.
26. Vejby-Christensen, L., Andersen, L.H., Heber, O., Kella, D., Pedersen, H.B., Schmidt, H.T. and Zajfman, D. (1997) Complete Branching Ratios for the Dissociative Recombination of H_2O^+, H_3O^+, and CH_3^+, *Astrophysical Journal* **483**, 531-540.
27. Gerlich, D. and Horning, S. (1992) Experimental Investigations of Radiative Association Processes as Related to Interstellar Chemistry, *Chemical Reviews* **92**, 1509-1540.

28. Prasad, S.S. and Tarafdar, S.P. (1983) UV Radiation Field Inside Dense Clouds - Its Possible Existence and Chemical Implications, *Astrophysical Journal* **267**, 603-609.
29. Terzieva, R. and Herbst, E. (1998) The Sensitivity of Gas-Phase Chemical Models of Interstellar Clouds to C and O Elemental Abundances and to a New Formation Mechanism for Ammonia, *Astrophysical Journal* **501**, 207-220.
30. Millar, T.J., Bennett, A. and Herbst, E. (1989) Deuterium Fractionation in Dense Interstellar Clouds, *Astrophysical Journal* **340**, 906-920.
31. Tiné, S., Roueff, E., Falgarone, E., Gerin, M. and Pineau des Forêts, G. (2000) Deuterium fractionation in dense ammonia cores, *Astronomy and Astrophysics* **356**, 1039-1049.
32. Ruffle, D.P. and Herbst, E. (2000) New models of interstellar gas-grain chemistry - I. Surface Diffusion Rates, *Monthly Notices of the Royal Astronomical Society*, in press.
33. Caselli, P, Hasegawa, T.I. and Herbst, E. (1998) A Proposed Modification of the Rate Equations for Reactions on Grain Surfaces, *Astrophysical Journal* **495**, 309-316.
34. Minh, Y.C. and Van Dishoeck, E.F. (eds.) (2000) *Astrochemistry: From Molecular Clouds to Planetary Systems*, Sheridan Books, Chelsea, Michigan.
35. Aikawa, Y. and Herbst, E. (1999) Molecular evolution in protoplanetary disks, *Astronomy and Astrophysics*, **351**, 233-246.
36. Millar, T.J. and Hatchell, J. (1998) Chemical models of hot molecular cores, *Faraday Discussions* **109**, 15-30.
37. Shull, J.M. and Draine, B.T. (1987) The Physics of Interstellar Shock Waves, in D.J. Hollenbach and H.A. Thronson, Jr. (eds.), *Interstellar Processes*, Reidel, Dordrecht, pp. 283-320.
38. Pety, J. and Falgarone, E. (2000) The elusive structure of the diffuse molecular gas: shocks or vortices in compressible turbulence?, *Astronomy and Astrophysics* **356**, 279-286.
39. Millar, T.J., Bettens, R.P.A. and Herbst, E. (2000) Large molecules in the envelope surrounding IRC+10216, *Monthly Notices of the Royal Astronomical Society* **316**, 195-203.

THE MILLIMETER, SUBMILLIMETER AND FAR-INFRARED SPECTRUM OF INTERSTELLAR AND CIRCUMSTELLAR CLOUDS

J. Cernicharo
CSIC. Instituto de Estructura de la Materia. Departamento de Física Molecular. C/Serrano 121. 28006 Madrid. Spain

Abstract Infrared spectroscopy allow us to study the molecular content of interstellar and circumstellar clouds through the observation of the ro-vibrational transitions of molecular species. In particular, some species can only be studied at infrared wavelengths due to their lack of permanent dipole moment and, hence, of pure rotational lines. In addition, polar light molecular species have their pure rotational lines in the far-infrared and can be studied only using infrared telescopes. In this paper I present the results obtained with ISO related to the detection of CH^+, FH, CH_3, C_3, CO_2, C_4H_2, C_6H_2 and of benzene from observations of their rotational and/or their ro-vibrational lines in the near, mid, and far infrared.

Keywords: ISO – New Molecules – IRC+10216 – CRL618, CRL2688, NGC7027 – Infrared Spectroscopy – Radioastronomy – Molecular lines.

1. INTRODUCTION

Most molecules discovered in the space have been detected at radio wavelengths, mainly in the millimeter domain. However, important molecular species without permanent dipole moment have escaped detection because they can only be observed in the infrared domain. The vibrational stretching frequencies are mostly below 10 μm for C-H, C=O, C-C, C=C, C=-C, and C=-N bonds, while those of bending modes of poliatomic species such as HCN, HNC, HCO^+, NNH^+, and CCH are between 10 and 30 μm. Diatomic species with silicon and other refractory elements have their vibrational frequencies mostly above 10 μm. Heavier

J. Demaison et al. (eds.), Spectroscopy from Space, 23–41.

molecules have bending modes at larger wavelengths (HC_3N and C_4H_2 at 45 μm, C_6H_2 at 96 μm, C_3 at 158 μm, C_4, C_5, ...).

Despite the important astrophysical output that could be obtained from infrared observations, the wavelength range 2-200 μm has remained poorly studied due to the absorption of radiation produced by the terrestrial atmosphere. The Infrared Space Observatory (ISO, Kessler et al., 1996) was equipped with two spectrometers covering the 2.5-197 μm range. The Short Wavelength Spectrometer (SWS, de Grauuw et al., 1996) covered the 2.5-45 μm range and provided a spectral resolution up to 2000 in its grating mode and 30000 in its Fabry-Perot mode (FP). The Long Wavelength Spectrometer (LWS, Clegg et al., 1996) covered the 43-197 μm range with a spectral resolution of 300-600 in its grating mode and up to 10000 in its FP mode. In some cases the grating resolution of the SWS allows to resolve the ro-vibrational band structure for molecules with rotational constants larger than 0.2-0.3 cm^{-1}.

The two spectrometers on board ISO have permitted completion of spectral surveys of several proto-typical interstellar and circumstellar sources. These surveys give us a detailed spectroscopic information that can lead to the identification of new molecular species, some of them playing an important role in the chemistry of these objects. Molecules such as CO_2, CS_2, NCCN, CH_3, CH_4, carbon chains, complex hydrocarbons, H_3^+, ..., could be detected in the SWS and LWS spectral scans. Among the expected new molecules it is worth noting the N-atomic carbons C_3, C_4, C_5, C_6, C_7, which are efficiently formed in the external layers of circumstellar envelopes (hereafter CSE). The detection, identification and spectroscopic characterization of these species will bring up key information about the chemistry of evolved stars. These species are formed in a large variety of physical conditions (flames, pirolyse of C_2H_2, etc.) and are responsible, together with acetylenic chains, for the growth of cyclic carbons and the formation of PAHs.

2. AGB and Post-AGB OBJECTS

The evolution of transition objects between the asymptotic giant branch (AGB) stage and the planetary nebulae (PN) phase is known to be exceptional, probably the all-decisive moment, where very efficient mass loss processes determine the basic morphological, kinematical and chemical layout of PN in the process of formation. The global reshaping of the circumstellar envelopes is dictated in a crucial way by the copious mass ejection which occurs towards the end of the AGB phase. Simultaneously with the erosion of the AGB envelope by these high velocity winds, UV photons from the central star impinge on the neutral envelope ionizing the gas and driving an interesting photochemistry which may lead to an efficient build up of new, large molecules. It is perhaps at these early evolutionary stages of C-rich proto-planetary nebulae (PPN), when important amounts of C_2H_2 and CH_4 are still available in the gas phase, that a new series of small hydrocarbons are formed. These species could be the small "blocks" from which the large C-rich molecules responsibles for the emission in the Unidentified Infrared Bands (UIBs) could be created.

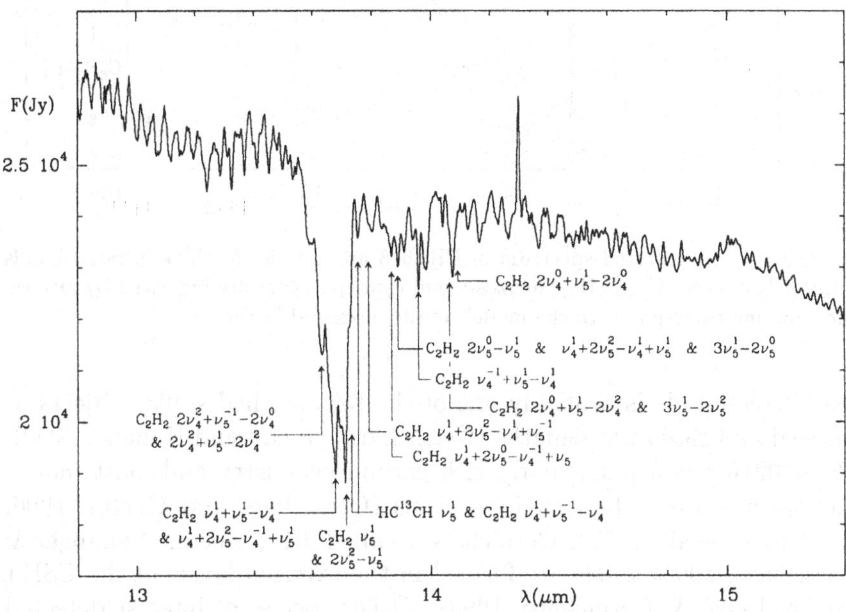

Figure 1.1. Observed spectrum around 15 μm in the direction of IRC+10216 (from Cernicharo et al., 1999). The position of the C_2H_2 and HCN bands are indicated. Vertical lines indicate the positions of the individual R and P lines of the fundamental bending mode of C_2H_2. Note that the HCN bands are in emission as indicated by the shape of the P-branch of C_2H_2 relative to the R one, and by the strong Q-branch of the $2\nu_2$ l=0 – ν_2 l=1 transition.

The medium and far infrared spectrum of IRC+10216 (Cernicharo et al., 1996a, Cernicharo et al., 1999), the post-AGB object CRL2688 (Cox et al., 1996; Justtanont et al., 2000), the proto-planetary nebula CRL618 (Cernicharo et al., 2001a,b; Herpin & Cernicharo, 2000; Justtanont et al., 2000), and the bright planetary nebula NGC7027 (Liu et al., 1996, Cernicharo et al., 1997a, Liu et al., 1997, Beintema et al., 1996, Justtanont et al., 2000) have been observed with ISO. In this work I will summarize the main results from these data.

IRC+10216 is the brightest C-rich, evolved star in the sky. It has an extended circumstellar envelope where some 50 molecular species have

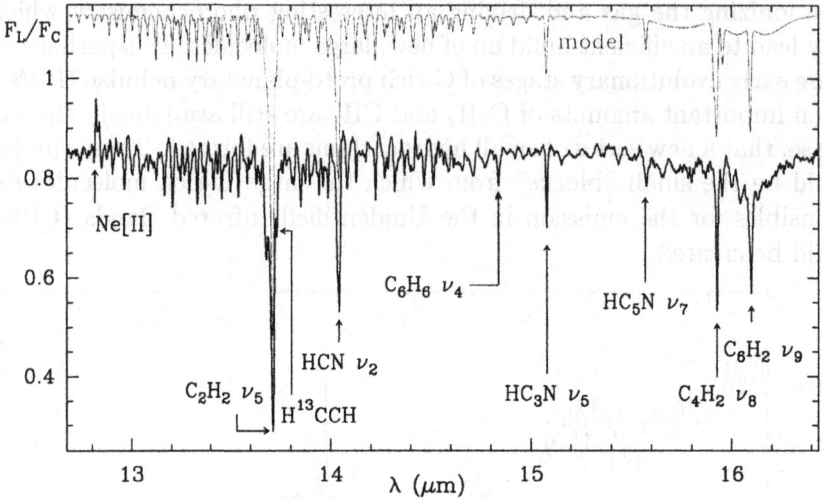

Figure 1.2. Mid-Infrared spectrum of CRL618 around 15 μm. The infrared bands of the polyynes (C_4H_2, C_6H_2), benzene, and cyanopolyynes are indicated by arrows. The thin line corresponds to the model results discussed in the text.

been detected. It is probably one of the best studied stellar objects at infrared and radio wavelengths (Cernicharo, Guélin and Kahane, 2000). IRC+10216 has a particularly rich carbon chemistry and most molecular species are carbon chain radicals (Cernicharo and Guélin, 1996, Cernicharo et al., 1991a, Cernicharo et al., 1991b, Guélin, Neininger & Cernicharo, 1998) which are formed in the external layers of the CSE (Guélin, Lucas & Cernicharo, 1993). Other species of interest detected in IRC+10216 include silicon carbide (Cernicharo et al., 1989a), the metal halides NaCl, AlCl, AlF, and KCl (Cernicharo and Guélin, 1987). Some metal-bearing species are also detected in the external shell of IRC+10216, for example MgNC (Guélin, Lucas & Cernicharo, 1993).

Figure 1 shows the 14 μm spectrum of IRC+10216. The absorption features are due to the ν_5 bending mode of acetylene and its associated hot and combination bands. This figure shows that the HCN bands appear in emission with the Q-branch of the $2\nu_2^0$-ν_2^1 transition being particularly strong. The HCN emission was modelled by Cernicharo et al., 1999 using a non-LTE code which includes the rotational transitions of $2\nu_2+\nu_3$ (l=0,2), $\nu_2+\nu_3$, ν_3, $\nu_3+\nu_2-\nu_2$, $\nu_3-\nu_2$, $2\nu_2$ (l=0,2), ν_2, $2\nu_2-\nu_2$, and $3\nu_2 - 2\nu_2$ bands of HCN. Only radiative excitation between the vibrational levels was considered. The results (see Figure 4 of Cernicharo et al., 1999) predict that the ν_3, $\nu_3+\nu_2$ and the $2\nu_2$ bands of HCN will be in absorption while the ν_2 and its overtone bands will be in emission. The model also predicts that the ν_3-ν_2 band will be in emission. The pumping mechanisms are different for the different regions of the envelope. In the innermost regions, before the dust formation zone, absorption of photons at 3 μm in the ν_3 mode is very efficient. Radiative decay to the $2\nu_2$ l=0 and ν_2 l=1 states would then be an important mechanism to populate both levels. Also in this region the ν_3-ν_2 band appears in emission. Outside the dust formation region the flux at 3 μm decreases but it increases considerably at 7.1 μm. The pumping from the ground state to the $2\nu_2$ l=0 vibrational level, followed by radiative decay to the ν_2 l=1 level, is the main mechanism in producing the observed emission at 14.3 μm. These calculations show that resonant scattering through the envelope is very efficient for HCN, and also probably for C_2H_2 and other polyatomic molecular species.

Besides the C_2H_2 and HCN vibrational bands no others features are seen in the spectrum of Figure 1. However, the situation is complitely different for CRL618. Figure 2 (from Cernicharo et al., 2001a)) shows the mid-infrared spectrum of this object as observed by ISO. In contrast with IRC+10216 where many hot bands of acetylene and HCN have been detected (see Figure 1), CRL618 shows only the fundamental bands of these molecules. This fact means that the kinetic temperature of the gas is much lower in this source than in IRC+10216.

CRL 618 is an extreme example of a proto-PN with a thick molecular envelope (Bujarrabal et al., 1988 surrounding a B0 star and an ultracompact HII region from which UV radiation, by several orders of magnitude larger than that in CRL2688, impinges on the envelope. The brightening of the HII region in the 1970s (Kowk & Feldman, 1981; Martín-Pintado et al., 1993) and the discovery of molecular gas with velocities up to 200 kms^{-1} (Cernicharo et al., 1989b) illustrate the rapid evolution of the central star and its influence on the surrounding AGB ejecta. Herpin & Cernicharo, 2000 have reported on the detection of H_2O and OH. These O-bearing species, together with H_2CO (Cernicharo et al., 1989b), are not present in the envelope of IRC+10216 (Cernicharo, Guélin & Kahane, 2000), the prototype of AGB C-rich circumstellar envelopes. In CRL618 these O-bearing molecular species are produced under very different physical conditions from those prevailing in its previous AGB phase.

As the innermost region of the envelope of CRL618 is occupied by an HII region the neutral gas we see in absorption at infrared wavelengths is at much larger distances from the central star than in IRC+10216. Cernicharo et al., 2001a have derived from the ISO data a temperature of 200 K for C_2H_2 and HCN and column densities of $2\,10^{17}$ and $1.5\,10^{17}$ cm^{-2} respectively. An upper limit of 250 K for the kinetic temperature of the absorbing gas was obtained from the limits to the intensity of the hot bands of HCN and C_2H_2 in Figure 2. The individual lines of the P and R-branches of HCN and C_2H_2 are clearly seen in the spectra of Figure 2 and agree in intensity and position with the results of the model of Cernicharo et al., 2001a. All the other features in this Figure correspond to the Q-branches of the bending modes of heavy species. Two of these features at 15.9 and 16.1 μm arise from the fundamental bending modes ν_8 of C_4H_2 and ν_{11} of C_6H_2. Di- and tri-acetylene have also been found in the direction of CRL2688 (see Figure 3). The identification of C_4H_2 and C_6H_2 is further confirmed by the detection of the C_4H_2 combination band $\nu_6+\nu_8$ at 8 μm and the $\nu_8+\nu_{11}$ combination band of C_6H_2 at 8.11 μm (see Figure 5 and Cernicharo et al., 2001b).

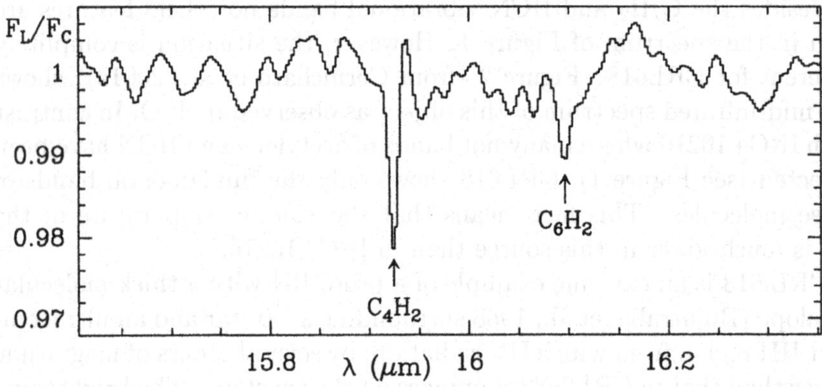

Figure 1.3. ISO/SWS spectrum of CRL2688 around the bending modes of di- and tri-acetylene (from Cernicharo et al. 2001a).

The kinetic temperature is a key parameter in modeling the absorption of the infrared active modes ν_8 of C_4H_2 and ν_{11} of C_6H_2 because the presence of hot bands arising from low energy bending modes of these species. The best fitting model obtained by Cernicharo et al., 2001a corresponds to N(C_4H_2)= 1.2 10^{17} cm^{-2} and N(C_6H_2) = 6 10^{16} cm^{-2}. Hence, the abundance of C_6H_2 is only a factor of 2 lower than that of C_4H_2 which in turn is also only a factor of 2 below that of C_2H_2. These large abundances are rather surprising and unpredicted by any chemical model so far. Besides the absorptions from these species three other features are prominent in the spectrum of Figure 2. Two of them can be easily identified as the ν_5 and ν_7 bending modes of HC_3N and HC_5N. Cernicharo et al., 2001a have obtained N(HC_3N)=5 10^{16} cm^{-2} and N(HC_5N)= 1.5 10^{16} cm^{-2}. The last feature in Figure 2, which is as strong as the ν_7 Q-branch of HC_5N, corresponds to benzene.

Figures 2 & 4 show that a narrow feature is centered within 0.0015 μm of the expected Q-branch of the ν_4 band of benzene. This narrow feature can not be associated with any of the lines of the P-branch of C_2H_2 and HCN bending modes (see Figure 2). The P(13) line of HCN at 14.847 μm could contribute with a maximum of 1% of absorption in the red wing of the ν_4 feature. In addition, the feature that Cernicharo et al., 2001a attribute to benzene is not only visible in each up and down scans individually, but also in the off-band extra data obtained by the SWS #4 detector. This provides an independent confirmation of the absorption. C_6H_6 has 20 vibrational modes, (Herzberg 1964) but only four of them are infrared active. The bending mode ν_4 is the strongest infrared band of benzene. Cernicharo et al., 2001a obtained the best fit to the data for a benzene column density of 6 10^{15} cm^{-2} and a kinetic temperature of 200 K.

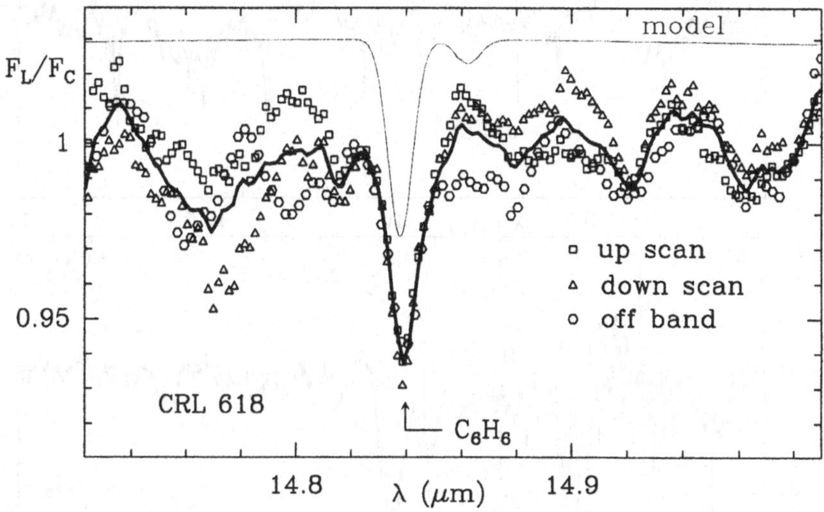

Figure 1.4. ISO/SWS spectrum of CRL618 around the ν_4 bending mode of benzene. The individual markers corresponds to up, down and offband data. The averaged spectrum from these data is plotted as a thick line. The modelled spectrum is shown by the thin line (from Cernicharo et al. 2001a).

The mid-infrared spectrum of CRL618 between 7.2 and 9.7 μm is shown in Figure 5. Figure 5a shows that the region between 9 and 11 μm is full of absorption bands. Many of these bands correspond to NH_3 and C_2H_4 that have a large number of Q-subbranches. The modeling of this spectral region by Cernicharo et al., 2001b indicate a column density for ammonia and ethylene of 3 10^{16} and 5 10^{16} cm^{-2} respectively. Although ammonia and ethylene are responsible for many of these bands many broad absorption features still remain unidentified. The same happens in the region between 7 and 9 μm. Here the combination bands $\nu_6+\nu_8$ of C_4H_2 and $\nu_8+\nu_{11}$ of C_6H_2 are clearly detected (see Figure 5b). Methane also presents strong absorption lines in this wavelength domain with $N(CH_4) = 7\ 10^{16}$ cm^{-2}. The model shown in Figure 5b also includes the combination bands $\nu_4+\nu_5$ of C_2H_2 and the $2\nu_2^0$ band of HCN (see Cernicharo et al., 2001b). These features appear between the absorption produced by CH_4 and that of the poly-acetylenes. The region above 8 μm (see Figure 5b) presents some weak bands that certainly arise from heavy species.

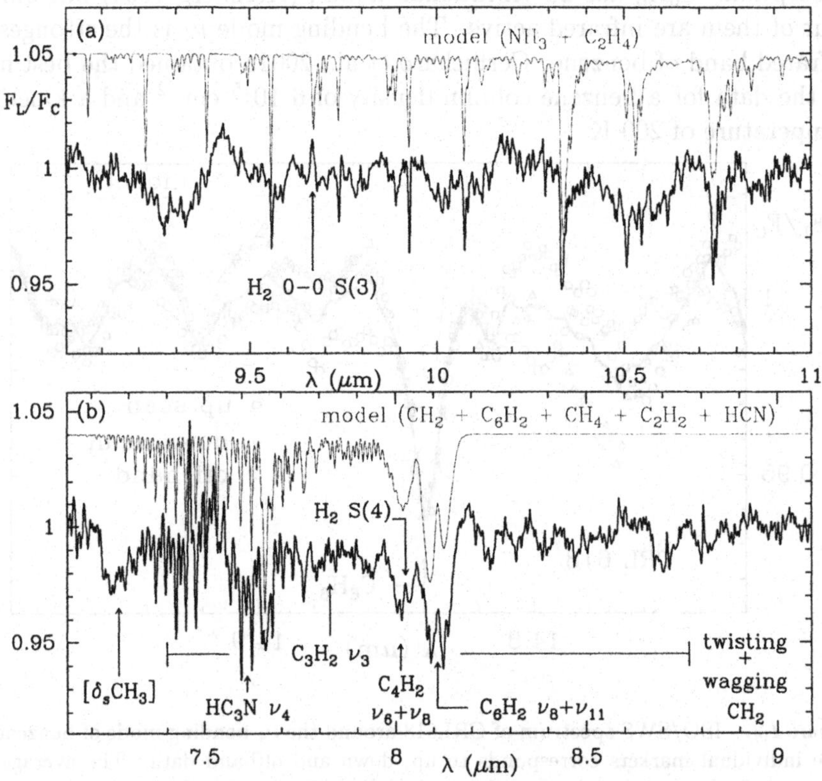

Figure 1.5. Mid-Infrared spectrum of CRL618 at selected wavelengths. The thin line in each panel corresponds to the model of Cernicharo et al. (2000b). The feature at 7.27 μm corresponds to the symmetric bending of CH_3. The broad range of absorption bands produced by wagging and twisting of CH_2 in small hydrocarbons is indicated by the horizontal line (from Cernicharo et al. 2001b).

Figure 6 shows the 5.3-7.2 μm spectra of CRL 618, NGC7027, and CRL 2688. The spectrum of CRL 618 shows a weak and broad absorption feature (10%) centered around 6.25 μm that appears in emission in the other two objects. They have generally attributed to the IR fluorescence of UV pumped PAH molecules (see Tielens et al. 1999 for a recent review). Many abundant molecular species present in CRL618, such as CCH, C_3H_2, H_2CO (Cernicharo et al. 1989), CH_3CN, NH_3 (Martín-Pintado et al. 1993; this paper) , ..., have vibrational transitions in this range and we estimate that they will produce absorption deeps of $\simeq$ 1%. Even H_2O (Herpin and Cernicharo 2000) could produce $\simeq$ 1-1.5% narrow line absorptions in its 6 μm bending mode. However, the observed absorption is too large and intense to be explained alone by these molecules.

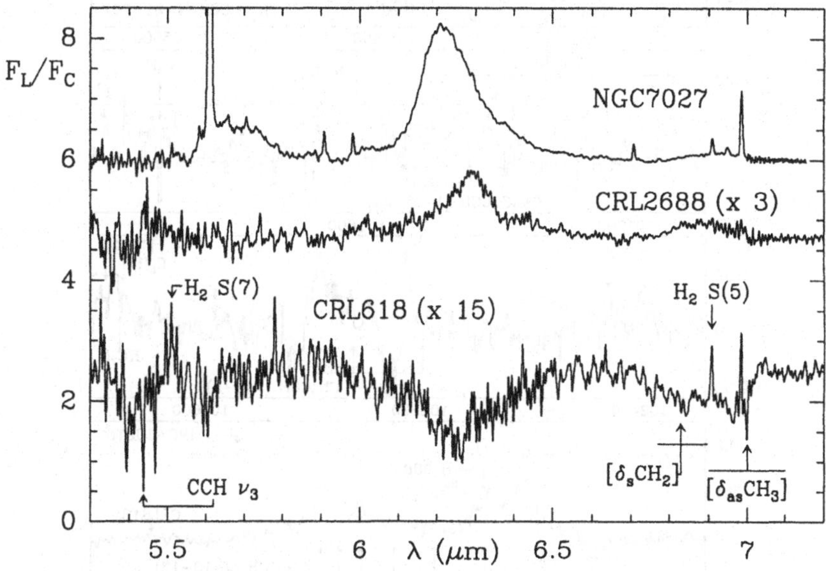

Figure 1.6. Mid-infrared spectra of NGC7027, CRL618 and CRL2688 around 6.2 μm. Pure rotational lines of H_2 in its ν=0 state are indicated. The typical wavelength range for the CH_3 asymmetrical bending and for scissoring methylene (CH_2) vibrations of alkanes are indicated (from Cernicharo et al. 2001b).

Among the three objects, only CRL618 shows a forest of absorption bands above 5.5 μm (see Figures 5 & 6) which suggests that several gas phase molecular species are producing absorption in the mid-infrared and contributing also to the broad 6.2μm absorption. Three of these bands can be easily identified as the symmetrical (δ_s) and asymmetrical (δ_{as}) bending of methyl groups and as the scissoring δ_{sc} of methylene groups in small hydrocarbons. The wagging and twisting of the methylene group in many hydrocarbons produce a large number of weak bands in the range 1100-1350 cm^{-1} (7.4-9.1 μm). It is thus tempting to look for molecules containing these groups. The best way to identify them, assuming they have a permanent dipole moment, is through the observation of their pure rotational lines. Cernicharo et al., 2001b have discovered two of these species (see Figure 7), CH_3CCH and CH_3CCCCH in their line survey at millimeter wavelengths of CRL618. They obtained $N(CH_3CCH)$=1.8 10^{16} cm^{-2}, and $N(CH_3C_4H)$= 8 10^{15} cm^{-2}.

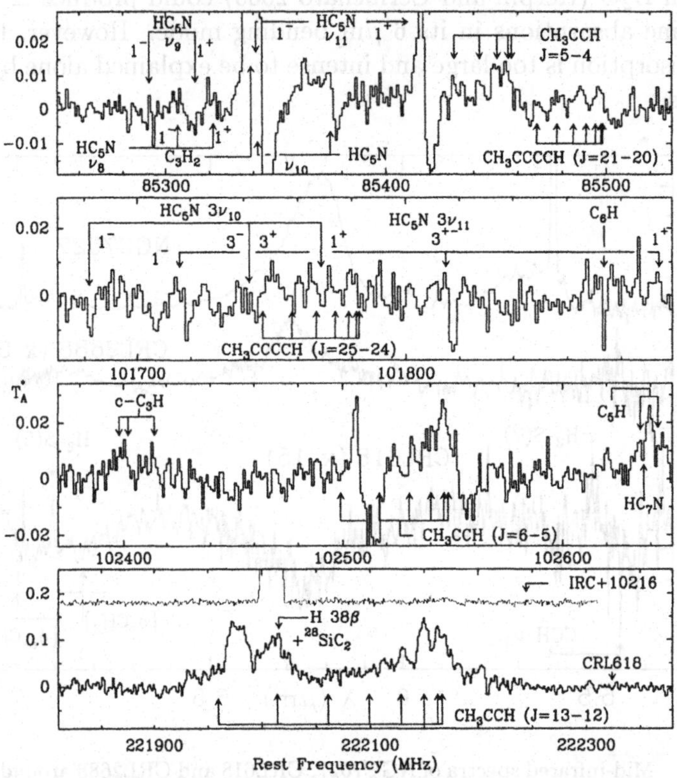

Figure 1.7. 30-m IRAM telescope spectra of CRL618 at selected frequencies. The bottom panel shows, for comparison purposes, the spectra of IRC+10216 and CRL618 at the same frequency, with an arbitrary vertical axis shift. The identified species are indicated by arrows. Note the P-Cygni profiles of the pure rotational lines of HC_5N in the ν_{11} and $3\nu_{11}$ vibrational states (From Cernicharo et al., 2001b)

The formation of long polyynes in the peculiar environment of CRL618 can be understood from the photolysis of acetylene. The reactions C_2H_2 + h$\nu \rightarrow C_2H$ + H, C_2H+ $C_{2n}H_2 \rightarrow C_{2n+2}H_2$+ H are the path for the growth of acetylenic chains. Polyynes longer than C_6H_2 could condensate or polymerize easily. The photolysis of methane CH_4 + h$\nu \rightarrow CH_2$ + H_2 could lead to the formation of methyl-polyynes through the reactions CH_2 + $C_{2n}H_2 \rightarrow H_2CC_{2n-1}CH_2$, followed by H + $H_2CC_{2n-1}CH_2 \rightarrow$

$CH_3C_{2n}H$ + H. Other mechanisms involving shocks, grain desorption, or grain-surface chemistry could be important as well.

Benzene, the basic aromatic unit, has never been found outside the solar system. Large policyclic aromatic hydrocarbons have been proposed as the carriers of the UIBs, but identification of specific PAH molecules has been elusive. The detection of benzene and of the small species C_4H_2, C_6H_2, CH_3CCH and CH_3C_4H indicate that the conditions in CRL 618 are favourable to activate a rich and diverse chemistry. Taking into account the high H_2 density in the inner layers of CRL618, and the high UV flux, it is reasonable to expect a fast evolution of the chemical composition of its AGB envelope during its transition towards the PN phase. Likely, this chemical processing will lead in short time scales to the formation of much larger C-rich molecular complexes. Then, at some moment of its evolution, the infrared spectrum of CRL618 will be dominated by the emission of the UIBs.

The infrared spectrum of PN is dominated by the presence of strong atomic lines (Liu et al., 1996; Liu et al., 1997; Beintema et al., 1996 and the bands of UIBs. In the far-infrared, these objects show strong CO emission lines (Liu et al., 1996; Herpin & Cernicharo, 2000; Justtanont et al., 2000) and a forest of water vapour lines (see, e.g., Barlow et al., 1996) and of CO_2 lines (Justtanont et al., 1996) in the case of O-rich objects. The far-infrared spectrum of PN could also contain lines from small molecules. Among the light molecular species, CH^+ was one of the first molecules detected in the space through its $^1\Pi$-$^1\Sigma$ electronic transition at optical wavelengths. This molecule is abundant in the diffuse interstellar medium. During many years the chemical reactions involved in its formation have been subject of important controversy. Cernicharo et al. (1997a) have reported the discovery of several pure rotational transitions of CH^+ in the LWS spectrum of NGC 7027. They derived a rotational temperature for the CH^+ lines of 160 K, a volume density of a few 10^7 cm^{-3}, and a CH^+/CO abundance ratio of 0.2-1 10^{-3}. Given the carbon-rich nature of NGC 7027 and the strong UV radiation field in its inner regions is not surprising to find CH^+ in this prototypical PN. However, this is the first time that CH^+ has been seen through its pure rotational spectrum. These transitions constitute a unique tool to trace the physical conditions of the PDRs of the interstellar and circumstellar medium.

3. THE INTERSTELLAR MEDIUM

The interpretation of molecular lines in the far infrared and submillimeter domains presents much more problems than at millimeter wave-

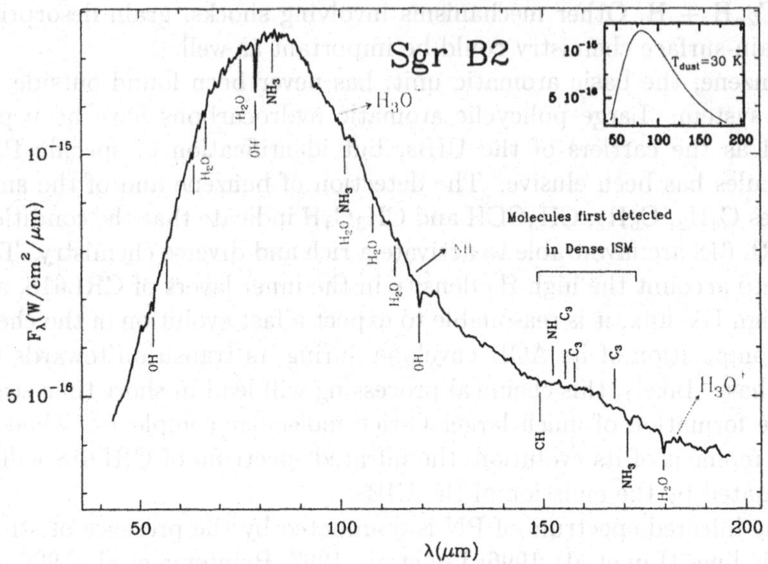

Figure 1.8. **The ISO LWS grating spectrum of Sgr B2 central position between 43 and 196 μm. Main features are labelled. Upper panel: gray-body fit to the continuum dust emission.**

lengths. The reason is that dust grain emission becomes very important in molecular clouds. The photons emitted by the dust grains can excited molecular transitions and the dust grains can absorb the photons emitted by the molecules. The interpretation of the data is not obvious and detailed modeling is necessary (see, e.g., Cernicharo et al., 1999; González-Alfonso et al., 1998). A good example of this problem is shown in Figure 8 (taken from Goicoechea and Cernicharo 2001) where the far infrared spectrum of the galactic center cloud Sgr B2 is shown. Obviously the grating observations shown in Figure 8 are only sensitive to the broadest molecular features because the spectral resolution of the grating LWS spectrometer is rather low, $\lambda/\Delta\lambda \simeq 300$. All detected molecular Far-IR features in Sgr B2 (see Figure 8) are seen in absorption against the continuum emission of the dust grains. This means that these transitions come from regions where excitation temperatures are lower than the dust temperature ($T_{exc} \leq T_d$; note that the dust is optically thick even at long wavelengths).

Due to the high column density of Sgr B2 this cloud is a very opaque source at optical but with high luminosities in far-IR, millimeter and radio wavelengths. Because of its physical and chemical complexity (clumped structure, cloud envelope, hot cores, ultracompact HII regions, etc) is a unique object for spectroscopic studies and for the searching of new molecules. We carried out a search for some molecular species, in particular carbon chains, in this cloud using the LWS Fabry Perot spectrometer ($\lambda/\Delta\lambda \simeq$ 7000-10000). Among the carbon chain radicals, triatomic carbon has played an important role in astrophysics since its detection in cometary tails in the last century and its observation at optical wavelengths in the atmospheres of cool stars (see, e.g., Zuckerman et al., 1976). It has been identified in the envelope of IRC+10216 through its ν_3 antisymmetric stretching mode in the mid-infrared by Hinkle, Keady & Bernath, 1988. Recent results with the Infrared Space Observatory (ISO) show that this molecule has a very high abundance in C-rich evolved stars and that its ro-vibrational transitions in the mid-infrared are one of the main sources of line-opacity in these objects. However, nothing is known about the appearance and abundance of triatomic carbon in the interstellar medium. It has been suggested that C_3 could be involved in the formation of the diffuse interstellar bands (Douglas, 1977; Clegg & Lambert, 1982), but optical observations have been unsuccessful so far (Snow, Seab & Joseph, 1988).

Recently, Cernicharo, Goicoechea & Caux, 2000c have reported on the detection of nine lines of the ν_2 bending mode of triatomic carbon, C_3, in the direction of SgrB2 (see Figure 9). The R(4) and R(2) lines of C_3 were also detected in the carbon-rich star IRC+10216. They derive an abundance for C_3 in the direction of Sgr B2 and IRC+10216 of $\simeq$3 10^{-8} and $\simeq 10^{-6}$ respectively. They also reported on the detection of the 2_3-1_2 line of NH and derived an abundance for this species of a few 10^{-9}.

Most lines in the grating spectrum of Sgr B2 (see Figure 8) are due to water vapor and OH. H_2O is one of the most important and abundant molecules in the ISM. It plays a critical role in the radiative cooling of dense molecular gas : star forming regions, envelopes of evolved stars (from Asymptotic Giant Branch stars to Planetary Nebula), etc. Hence, the determination of its abundance, distribution and formation is decisive for our understanding of chemical and physical proccess taking place in a great variety of astronomical environments. Figure 10 shows the detection of several thermal transitions of water vapour taken by ISO, confirming the previous detections reported with ISO and proving its ubiquitous presence in the cloud (Cernicharo et al., 1997b). Moreover, the 183.31 GHz 3_{13}–2_{20} maser transition of para-H_2O (Cernicharo et al., 1990, 1994, 1996b, 1999; González-Alfonso et al. 1995; González-Alfonso and Cernicharo 1999) has been also detected in this cloud (see Figure 10; Cernicharo, Goicoechea and Pardo, 2001c).

Other features in Figure 8 where also observed with the LWS/FP. Several lines of H_3O^+ have been detected (Goicoechea and Cernicharo 2001). This species is an oblate symmetric molecule, isoelectronic to

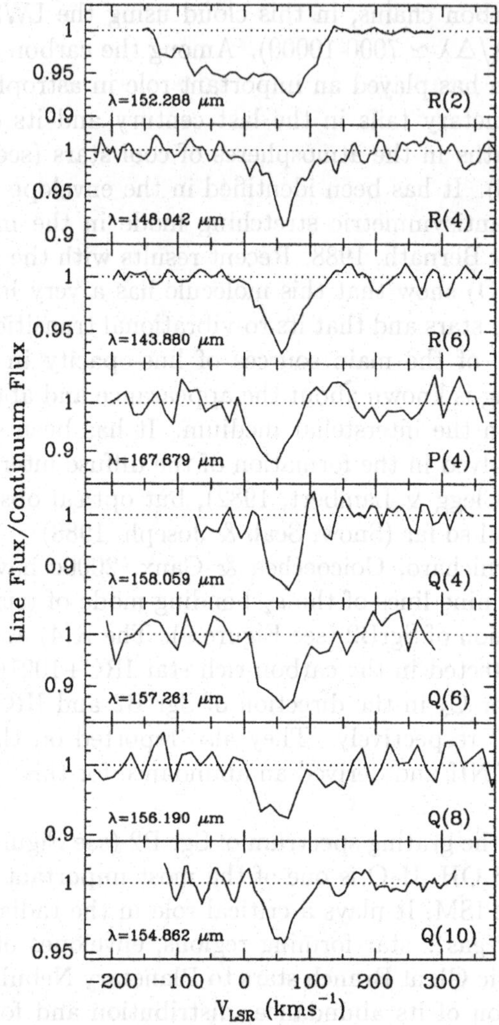

Figure 1.9. **LWS F–P observations of C_3 in Sgr B2. The intensity scale corresponds to the continuum normalized flux. The wavelength of each transition is indicated at the bottom left corner of each panel. The Q(10) spectrum could also contain the J=5/2-3/2 transitions of SH at 154.864 and 154.915 μm. Both lines of SH could have similar intensities and will appear separated by 100 km^{-1}. The SH lines remain undetected while the C_3 Q(10) line is detected with a signal to noise ratio of 7. (from Cernicharo et al. 2000c).**

NH_3 with the same physical structure, i.e., pyramidal geometry. Like ammonia it has a double minimum vibrational potential so that the oxygen can tunnel through the plane of the hydrogen atoms. In such a kind of molecules (also AsH_3 and PH_3), this leads to a inversion splitting of the rotational levels that can be observed from radio to infrared wavelengths depending on the potential barrier heigth. Surprisingly, the

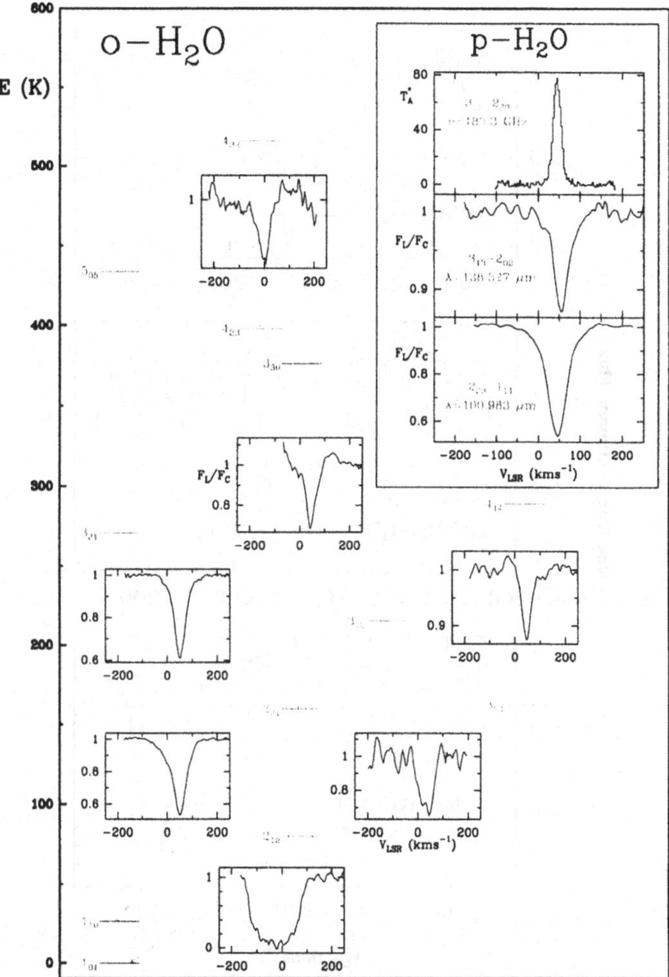

Figure 1.10. LWS F–P observations of water vapour in Sgr B2. All Far-IR H_2O lines are observed in absorption. The emission feature comes from the 3_{13}–2_{20} maser transition of para-H_2O observed with the 30-m IRAM radiotelescope at 183.3 GHz (from Cernicharo, Goicoechea and Pardo 2001c).

inversion splitting of H_3O^+ is very large, $\simeq$55 cm^{-1}, and its fundamental transitions of the ν_2 (umbrella) mode, lay at much higher frecuencies than those of the NH_3 splitting ($\simeq$1 cm^{-1}).

H_3O^+ is thought to be formed in the first stages of interstellar oxygen based chemistry and plays a very important role in gas-phase reactions prevailing in dense molecular clouds. Its dissociative recombination leads to the formation of important molecules OH and H_2O with an imprecise branching ratio for each channel of the dissociation. However, this ratio is a crucial parameter for the interstellar chemistry and for the mod-

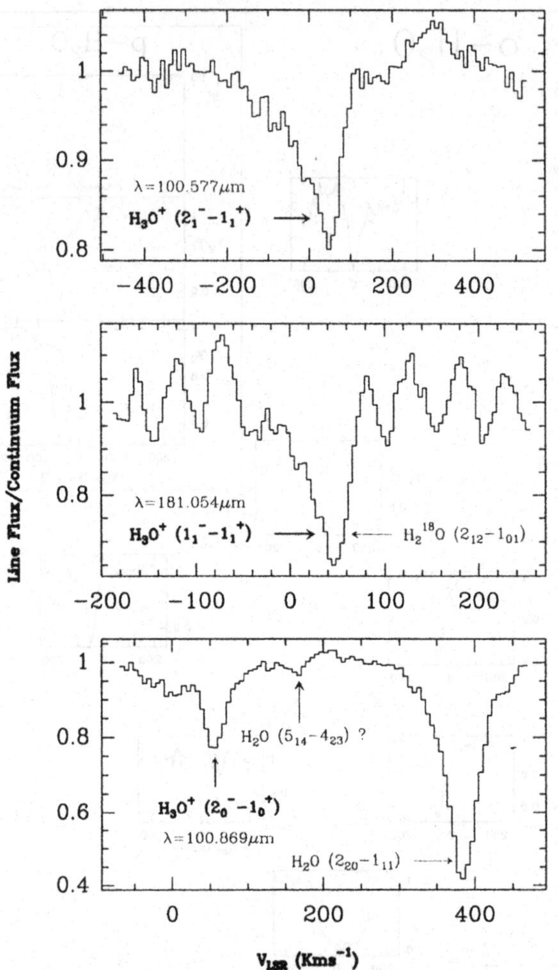

Figure 1.11. LWS F–P observations of H_3O^+ in Sgr B2. The upper box contains the broad R(1,1) absorption line. Middle, Q(1,1) transition is blended with 2_{12}-1_{01} transition of $H_2{}^{18}O$. Bottom, spectrum including the R(1,0) and two water vapour features.

elling of molecular clouds. Goicoechea and Cernicharo have detected three lines arising from the ν_2 ground-state inversion mode ($0^+ \rightarrow 0^-$) at 55.3 cm^{-1} (see Figure 11). These detections suggest a scenario in which water ice may be also formed by surface reactions on dust grains and then released back into the gas by young stars radiation or sputtering. This may explain the high abundance of water in Sgr B2. The derived abundance is $\simeq$(1-5)$\cdot 10^{-9}$ in agreement with submm observations (Phillips et al. 1992).

Together with CH^+, H_2O, and H_3O^+, other light species, such as CH_2, NH (Cernicharo, Goicoechea & Caux, 2000c), SH, SiH, FH, NH_3 could be detected in the far-infrared. Neufeld et al., 1996 have reported the detection of the J=2-1 line of FH in the direction of SgrB2 with an abundance of 10^{-9}. Another Fluorine-bearing molecule, AlF, has been previously reported by Cernicharo and Guélin, 1987 in the direction of the circumstellar envelope of IRC+10216, but it is the first time that a Fluorine bearing molecule has been detected in the interstellar medium.

The methyl radical has been recently detected in the mid-infrared with ISO by Feuchtgruber et al., 2000 in the direction of SgrA. The abundance they derive for this species is 1.3 10^{-8}.

The number of abundant light molecular species with significant dipole moment producing emission/absorption in the far–IR is rather reduced (CO, HCN, NH_3, H_2O, NH_2, NH, CH, OH, CH^+, CH_2, ...). One could expect a feature–free far–infrared spectrum in interstellar and circumstellar clouds compared with the richness of the millimeter and submillimeter spectrum of these objects. Polyatomic molecules such as C_3, C_4, C_5, ..., C_3H, C_4H, ..., C_4H_2, C_6H_2 however, have vibrational bending modes at very low energy. The ro-vibrational transitions between these levels, as it is the case for C_3, could contribute to the far-infrared spectrum of interstellar and circumstellar clouds. Obviously, the species having large amplitude movements in their bending modes, i.e., large oscillator strengths for the vibrational transitions, will produce large absorption and/or emission features. Moreover, the detection of the bending mode of triatomic carbon opens new possibilities for identifying molecular species without permanent dipole moment (symmetrical species) in the far–IR. Although these species could also be observed in the mid and near-infrared through their stretching vibrational transitions, not all molecular clouds are strong emitters at these wavelengths. The detection of C_3, C_5 (Bernath, Hinkle & Keady, 1989), and of C_4H_2, C_6H_2 and of benzene (Cernicharo, 1999; Cernicharo et al. 2001a,b) clearly indicated that symmetric polyatomic species are created in a very efficient way in interstellar and circumstellar clouds.

Future heterodyne space missions such as the Far Infrared Space Telescope (FIRST) will provide important insights in the chemical complexity of the ISM and CSM through the observation in the submillimeter and far-infrared domain of large molecular species. Instruments onboard *FIRST* will provide high spectral resolution observations of the Far–Infrared and Submillimeter spectrum of dense molecular clouds. Not only physical, chemical and kinematic environments will be scrutinized. Detection of new polyatomic molecules and the use of known molecules in a broader range of transitions will offer unique tools for the understanding of dense star forming regions such as Sagittarius B2 molecular cloud. However, determination of the dominant carriers of emission/absorption features in new spectral bands requires an exhaustive knowledge of molecular spectroscopy and the experience obtained with *ISO* and other space-based infrared facilities will be of extreme importance for the future. A very close collaboration with laboratory spectroscopists will be absolutely needed to improve the expected frequencies above 1 THz of the rotational transitions of most abundant molecular species. Such a collaboration could also provide a molecular database, as large and complete as possible, to help us in identifying the new molecular species that FIRST will certainly discover in interstellar and circumstellar clouds.

Acknowledgments

I acknowledge Spanish DGES for this research under grants PB96-0883 and ESP98-1351E. I thank Juan R. Pardo and Javier R. Goicoechea for useful comments and suggestions.

References

Barlow M.J., Nguyen-Q-Rieu, Truong-Bach, et al., A.&A., 315, L241

Beintema D.A., van den Ancker M.E., Molster F.J., et al., 1996, A.&A., 315, L369

Bernath P.F., Hinkle K.W., Keady J.J., 1989, Science, 244, 562

Bujarrabal V., Gómez-González J., Bachiller R., Martín-Pintado J., 1988, A.&A., 204, 242

Cernicharo J., Guélin M., 1987, A.&A., 183, L10

Cernicharo J., Gottlieb C.A., Guélin M., et al., 1989, ApJ, 341, L25

Cernicharo J., Guélin M., Peñalver J. et al., 1989, A.&A., 222, L20

Cernicharo J., Thum C., Hein H., et al., 1990, A.&A., 231, L15

Cernicharo J., Gottlieb C.A., Guélin M., et al., 1991a, ApJ, 368, L39

Cernicharo J., Gottlieb C.A., Guélin M., et al., 1991b, ApJ, 368, L43

Cernicharo J., González-Alfonso E., Alcolea J., et al., 1994, ApJ, 432, L59

Cernicharo J., Barlow M., González-Alfonso E., et al., 1996a, A.&A., 315, 201

Cernicharo J., Bachiller R., and González-Alfonso E., 1996b, A.&A., 305, L5

Cernicharo J., Guélin M., 1996, A.&A., 309, L27

Cernicharo J.. X-Liu, González-Alfonso E., et al. 1997a, ApJ, 483, L65

Cernicharo J.. Lim T., González-Alfonso E., et al. 1997, Astronomy and Astrophysics, 323 L25

Cernicharo J., 1999, IAU Symposium 197, "Astrochemistry: from Molecular Clouds to Planetary Systems", Ed. Y.C. Minh and E.F. van Dishoeck
Cernicharo J., Yamamura I., González-Alfonso E., et al., 1999, ApJ, 526, L21
Cernicharo J., Guélin M., Kahane C., A.&A. Suppl. Series, 142, 181
Cernicharo J., Goicoechea J.R., Caux E., 2000, ApJ, 534, L199
Cernicharo J., Heras A., Tielens A.G.G.M., et al., 2001a, ApJ, 546, L123
Cernicharo J., Heras A., Pardo J.R., et al., 2001b, ApJ, 546, L127
Cernicharo J., Goicoechea J., Pardo J.R., et al., 2001c, ApJ, submitted
Clegg M.F., et al., 1996, A.&A., 315, L38
Clegg R.E.S., Lambert D.L.:1982, MNRAS, 201, 723
Cox P., González-Alfonso E., Cernicharo J., et al., 1996, A.&A., 315, L265
Douglas A.E.:1977, Nature 269, 130
de Graauw Th., et al., 1996, A.&A., 315, L49-L54
Feuchtgruber H., Helmich F.P., van Dishoeck E.F., Wright C.M., 2000, ApJ, 535, L111
Goicoechea J., and Cernicharo J., 2001, ApJ, submitted
González-Alfonso E., Cernicharo J., Bachiller R., Fuente A., 1995, A.&A., 293, L9
González-Alfonso E., Cernicharo J., van Dishoeck, E.F., et al., 1998, ApJ, 502, L169
González-Alfonso E. and Cernicharo J., 1999, ApJ, 525, 845
Guélin M., Lucas R., Cernicharo J.:1993, A.&A., 280, L19
Guélin M., Neininger N., Cernicharo J., 1998, A.&A., 335, L1
Herpin F., Cernicharo J., 2000, ApJ, 534, 199
Hinkle K.W., Keady J.J., Bernath P.F., 1988, Science, 241, 1319
Kessler M.F., et al., 1996, A.&A., 315, L27
Kwok S., Feldman P.A., 1981, ApJ, 267, L67
Justtanont K., deJong T., Helmich F.P., et al., 1996, A.&A., 315, L217
Justtanont K., Barlow M.J., Tielens A.G.G.M., et al., 1996, A.&A., 360, 1117
Liu, X.-W., Barlow M.J., Nguyen-Q-Rieu, et al., A.&A., 315, L257
Liu, X.-W., Barlow M.J., Dalgarno A., et al., MNRAS, 290, L71
Martín-Pintado J., Gaume R., Bachiller R., Johnston K., 1993, ApJ, 419, 725
Neufeld D.A., Chen W., Melnick G.J. et al., 1996, A.&A., 315, 237
Phillips, T.G., van Dishoeck, E.F., et al. 1992, ApJ, 399, 533
Snow T.P., Seab C.G., Joseph C.L., 1988, ApJ, 335, 185
Zuckerman B., Gilray D.P., Turner B.E., et al, 1976, ApJ, 205, L15

SUBMILLIMETER AND FAR-INFRARED OBSERVING PLATFORMS FOR ASTRONOMY

H.-W. HÜBERS and H. P. RÖSER
German Aerospace Center (DLR)
Institute of Space Sensor Technology and Planetary Exploration
Rutherfordstr. 2
12489 Berlin
Germany
heinz-wilhelm.huebers@dlr.de

Abstract

During the next decade several airborne and space-borne platforms for far-infrared and submillimeter astronomy and spectroscopy will become operational. These observatories will provide sensitivity and spatial resolution several times better than previous missions. New observations in this relatively unexplored part of the electromagnetic spectrum will be possible. They will greatly improve our understanding of the universe. In this article the spectroscopic capabilities of future airborne and space-borne platforms for astronomy will be described and compared.

1. Introduction

The submillimeter (300 μm – 1 mm) and far-infrared (30 μm – 300 μm) part of the electromagnetic spectrum bears an amazing scientific potential for spectroscopy and its applications in astronomy. Most fundamental absorption and emission lines of astrophysically and astrochemically important molecules and atoms occur in this wavelength region. Up to now more than one hundred molecules and atoms have been detected in space. There are fairly simple molecules such as CO but also molecules consisting of ten or more atoms have been detected [1]. Beside their existence and abundance valuable information about the physical conditions such as temperature, density, and dynamics of the observed astronomical object can be obtained from high resolution spectroscopy.

There are several reasons to use high flying platforms or satellites for FIR/ submm astronomy. The most important is the poor transmission of Earth's atmosphere. From ground observations of the universe in this spectral region are only can be observed at a few frequencies where the atmosphere of the Earth is sufficiently transparent. Water vapor, CO_2, ozone and other molecules absorb most of the FIR and submm radiation

J. Demaison et al. (eds.), Spectroscopy from Space, 43–58.

from outside the Earth. Fig. 1 shows the atmospheric absorption. The wavelength region between 30 μm and 300 μm is completely absorbed. Above 300 μm only a few transmission windows are available. In order to avoid the absorption by the Earth's atmosphere one has to use high flying platforms like aircrafts, balloons or satellites. In addition, the emission of Earth's atmosphere is much lower for high flying observatories and does not exist for satellite platforms.

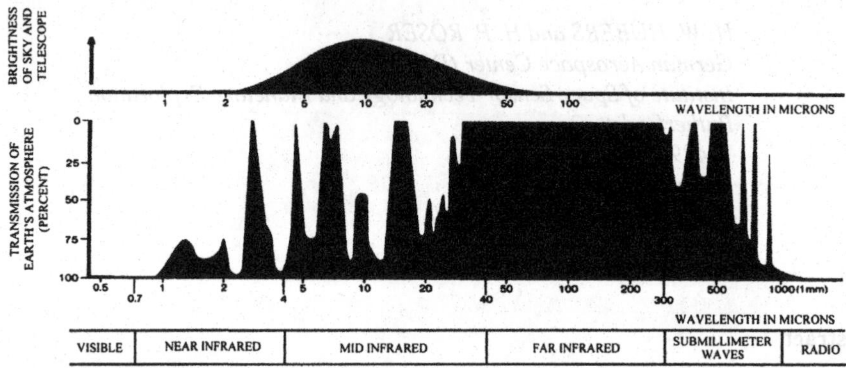

Figure 1. Background continuum radiation emitted by a warm telescope and the atmosphere (upper panel). Absorption of infrared, far-infrared and submm radiation by the atmosphere of Earth (lower panel).

Up to now three major satellite missions devoted to astronomy in the FIR and submm region of the electromagnetic spectrum have been completed. Fig. 2 gives an overview of FIR/submm missions in the period from 1980 to 2025. Only missions which were operational for more than half a year have been included. The **I**nfra**R**ed **A**stronomical **S**atellite (IRAS) has mapped the entire sky in four bands centered around 12 μm, 25 μm, 60 μm and 100 μm with an angular resolution of a few arcmin.

The **CO**smic **B**ackground **E**xplorer (COBE) was launched in 1989 and in operation until 1992. The instrument package of COBE consisted of microwave radiometers, Michelson interferometers and infrared photometers. It covered wavelengths from the near infrared to the cm range. The main goal was to measure the black body spectrum of the sky and the anisotropy of the cosmic background. The spatial resolution was limited to seven degrees. From November 1995 to May 1998 the **I**nfrared **S**pace **O**bservatory (ISO) was in operation. ISO had a 0.6 m diameter telescope and was equipped with an IR camera, a photometer and two spectrometers. Based on the surveys provided by IRAS, detailed maps and spectra of selected astronomical objects at wavelengths between 2.5 μm and 200 μm were obtained. ISO was the first observatory with the ability to observe spectra over the entire IR and much of the FIR spectral region. Because of that it was a milestone in the observation of the IR/FIR universe.

Currently there is only one space-based submm observatory in operation. NASA's **S**ubmillimeter **W**ave **A**stronomical **S**atellite (SWAS) is a small satellite which was launched in December 1998. A similar mission, the Swedish-French ODIN satellite will be launched soon (2002). Both missions carry telescopes with a diameter of about 1 m and are equipped with heterodyne receivers which are optimized for the observation of molecular oxygen and water in galactic sources. ODIN will also be used for the investigation of Earth's atmosphere [2].

Within the next ten years four major satellite missions are in preparation. The **S**ubmillimeter and **I**nfra**R**ed **T**elescope **F**acility (SIRTF) will be launched in 2002. It is NASA's equivalent of ISO. SIRTF will be equipped with larger array detectors than ISO and it will carry a 0.85 m diameter telescope which is cooled down to 5.5 K. ASTRO-F (IRIS) is the Japanese **I**nfra**r**ed **I**maging **S**urveyor. With a 0.7 m cryogenically cooled telescope it is somewhat similar to SIRTF. There are two major ESA missions which will be launched by the end of this decade. The Planck mission is devoted to the measurement of the black body background in the universe. It is named in honor of Max Planck who first discovered the equation which describes the emission spectrum of a black body. With its 1.5 m telescope and an angular resolution of about 9 arcminutes it will measure anisotropies of the cosmic background radiation with very high precision. The second ESA mission is the **F**ar-**I**nfra**R**ed and **S**ubmillimetre **T**elescope, FIRST. It is one of the four cornerstones in ESA's science program. Equipped with a 3.5 m telescope and with several photometers and spectrometers FIRST will perform observations between approximately 60 μm and 670 μm.

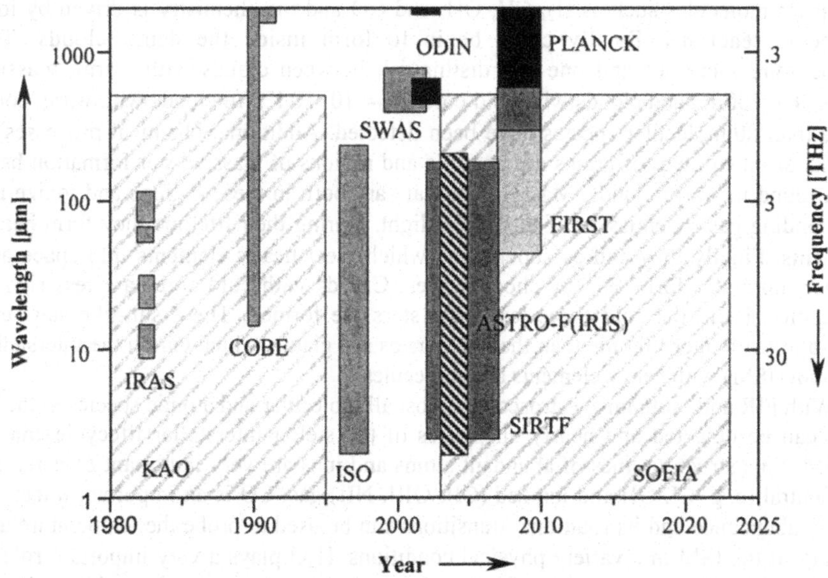

Figure 2. Summary of airborne and space-borne FIR/submm observatories for the period 1980-2025.

In parallel to satellite observatories two airborne observatories have been operated or will be operated soon. The **K**uiper **A**irborne **O**bservatory (KAO) was operational from 1974 to 1996. It was a follow up of the pioneering Lear-Jet observatory. The KAO had a 0.9 m diameter telescope and different kinds of spectrometers which were steadily improved during the lifetime of the KAO have been used. Because of these highly successful airborne observatories, NASA and the German Aerospace Center (DLR) decided to build a new airborne telescope, the **S**tratospheric **O**bservatory **F**or **I**nfrared **A**stronomy, SOFIA. First light of this observatory will be in 2002. With an operation time of minimum 20 years it will be a complementary observatory to all future FIR/submm satellite observatories.

This article is organized as follows. In the following section the significance of spectroscopy for astronomy in the FIR and submm spectral region will be highlighted. Following that the most important airborne and space-borne FIR/submm observatories will be described. Finally, the different observatories will be compared. Throughout this article emphasis is put on the discussion of capabilities and objectives in the field of line spectroscopy and on missions with a long phase of operation. Shorter missions or balloon-borne observatories are not considered here.

2. Astronomical Questions Addressed by FIR and Submm Spectroscopy

Questions associated with the stellar-interstellar lifecycle are the main scientific drivers for spectroscopy at FIR and submm wavelengths. Diffuse interstellar clouds gather and grow in size until they finally evolve into dense molecular clouds. These clouds are rich in simple molecules such as H_2, CH, OH, and CO and the chemistry is driven by ion-molecule reactions. Finally cores begin to form inside the dense clouds. The temperature increases and one can distinguish between clouds with warm, massive cores ($T \approx 200$ K) and clouds with cold cores ($T \approx 10$ -50 K). In clouds with warm cores more than 80 molecular species have been detected. Additional chemical processes in shocks or on surfaces of grains are possible and regions of massive star formation have been found in warm clouds. Massive hot stars are born in these regions and ionize the surrounding gas by emission of ultraviolet light. During their lifetime they form heavy elements. Finally they end as supernovae which eject heavy elements into space and trigger more star formation by shock waves. Clouds with cold cores are less rich in molecules. From the cold cores low mass stars are formed. These sun-like stars end after a much longer lifetime than the hot stars as red giants, which enrich the interstellar medium (ISM) with heavy elements and molecules.

With FIR and submm spectroscopy almost all molecular and atomic species in the ISM can be detected and nearly all phases of the stellar-interstellar lifecycle can be probed. Fine structure lines of abundant atoms and ions are very important coolants for the neutral (e. g. CI, OI) and ionized (CII, OIII, NII) gas. CO is an important tracer for the H_2 abundance and its rotational transitions can be used to probe the temperature and density of the ISM in a variety physical conditions. H_2O plays a very important role in the physics and chemistry of interstellar clouds. It has a large number of lines in the FIR/submm region and the populations of these levels are very sensitive to the physical conditions in the cloud. In addition, it can be excited by FIR emission from dust and it contributes significantly to the cooling of the dense, warm gas. The observation of oxygen containing species, especially H_2O, O_2 and OI, is important to determine the major reservoirs of gas-phase oxygen, still an unsolved problem since up to now observations indicate very low O_2 abundances. Because none of these species (except for O_2 under rare circumstances) can be observed from the ground airborne and space-borne instruments are required. Complete spectral line surveys in the FIR/submm region are needed. Compared with more limited spectral coverage they have the advantage to yield a complete inventory of the interstellar species. Because many lines for each species and its isotopes are covered, more accurate abundance determinations and more accurate constraints on the physical parameters of the region are possible. By comparing the lineshapes of species with different excitation and chemistry dynamical processes can be studied in more detail. Finally, spectral line surveys provide the opportunity to detect new species.

Beside spectroscopy of the ISM in our galaxy the observation of other galaxies is of prime interest. Examples are active galactic nuclei (AGN) and highly red-shifted galaxies. X-rays from the AGN excite atoms which reemit the energy in their fine structure lines (e. g. CII, OI) and CO rotational lines at 50 – 200 μm can be used for the detection of molecular gas around AGN's. In the case of highly red-shifted galaxies IR emission lines are shifted into the FIR/submm region. Their observation yields information about the early stages of galaxy formation. Observations of HD in galactic and extragalactic sources are important for cosmological questions, because it can be used to constrain the original amount of deuterium created in the big bang. Finally, high resolution spectroscopy of planetary atmospheres yield valuable information of the vertical distribution of molecules and the temperature profile.

Detailed discussions of these and other topics in FIR/ submm astronomy can be found in Ref. [3].

3. SOFIA Observatory

The **S**tratospheric **O**bservatory **F**or **I**nfrared **A**stronomy is an airborne observatory for IR, FIR, and submm astronomy. It is a joint project of NASA and DLR, the German Aerospace Center. The specifications and spectroscopic instruments are discussed in this paragraph.

3.1 AIRCRAFT

The aircraft to be used for SOFIA is a Boeing 747 SP, a slightly shorter version than the normal one. Fig. 3 shows the aircraft during a test flight in 1997. The black painted part at the rear side of the aircraft marks the location of the telescope and resembles the size of the opening for the telescope. Fig. 3 shows a cut-away view through the aircraft. A bulkhead separates the warm and pressurized cabin from the telescope cavity which is at outside atmospheric conditions during operation. The scientific instruments are mounted to the Nasmyth tube inside the cabin and therefore accessible during flight. The scientific instrument, the biggest part of the Nasmyth tube, some control electronics and additional weights counterweight the telescope. The total weight is supported by the wall of the bulkhead. The telescope cavity is sealed from the outside by a barrel door which opens as soon as the cruising altitude is reached. Prior to each observing flight the telescope will be precooled. The devices for this procedure are located in the aft section. Warming up of the telescope after each observing flight is done by floating the cavity with dry nitrogen gas.

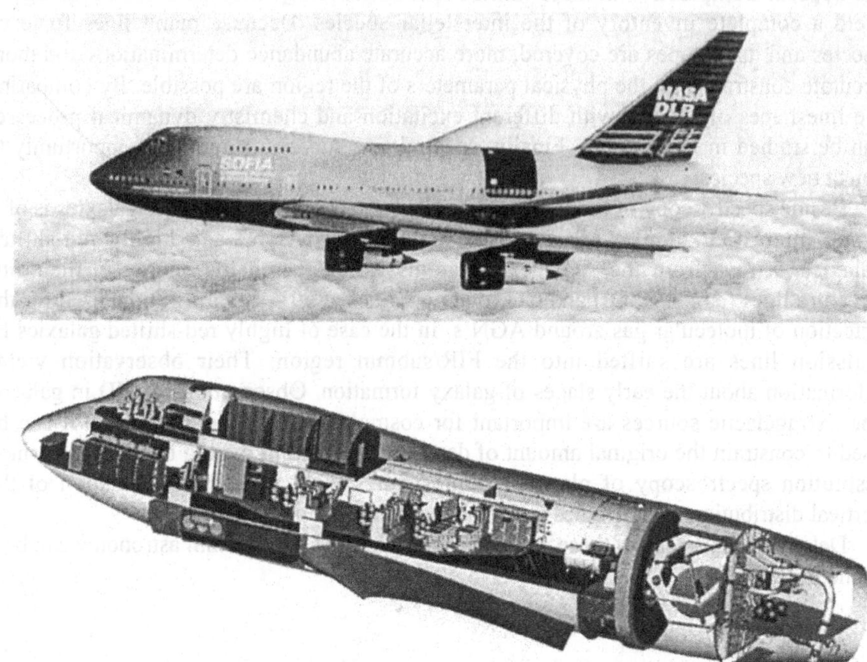

Figure 3. The SOFIA aircraft during a test flight in 1997. The black painted area marks the location of the telescope (upper panel). Cut-away view through the aircraft (lower panel).

3.2 TELESCOPE

The SOFIA telescope is a Cassegrain system with Nasmyth focus (see Fig. 4). The primary mirror (M1) has a parabolic shape (f-number: 1.2) and a physical diameter of 2.7 m. The diameter of the system clear aperture is 2.5 m. The mirror is made from Zerodur (Schott). This material has an extremely low thermal expansion and provides sufficient stiffness for use in an aircraft environment. A honeycomb structure is milled into the rear side of the mirror. This makes it the largest lightweighted mirror in the world. The surface quality of the mirror is sufficient to observe wavelengths as short as 0.3 μm. The hyperbolic secondary mirror (M2) has a diameter of 35 cm. It is made from silicon carbide. The secondary mirror is attached to a chopping mechanism which provides chop amplitudes up to ±5 arcmin at chop frequencies up to 20 Hz. The tertiary mirror (M3-1) is a flat dichroic which reflects the IR and FIR radiation towards the scientific instrument. The visible radiation is transmitted by the dichroic mirror and reflected by a second tertiary mirror (M3-2) to the focal plane imager.

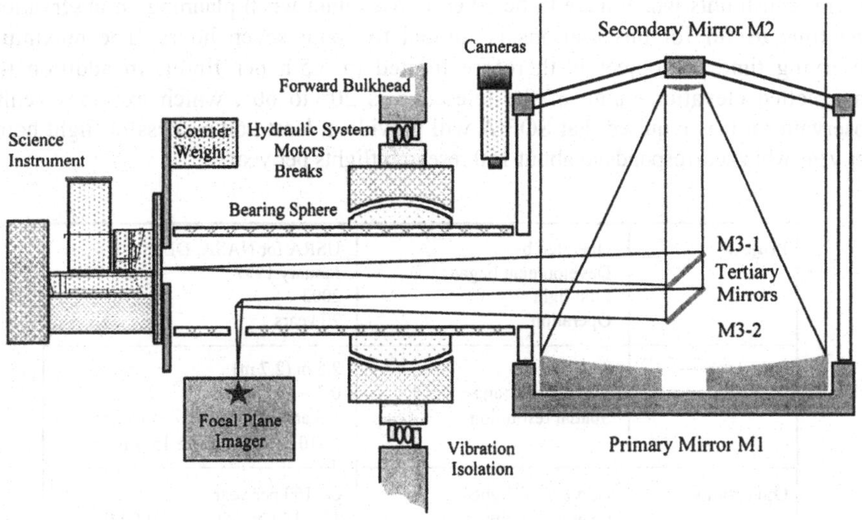

Figure 4. Scheme of the telescope assembly and the optical path.

The telescope structure is designed like a dumbbell which is supported in its center of gravity by a bearing sphere. The bearing sphere is part of the forward bulkhead. A vibration isolation assembly, a passive spring-damper system, minimizes the influence of aircraft vibrations on the telescope pointing. In order to maximize the useful observing flight duration it is mandatory to make every component as light as possible. Therefore, the telescope structure is made from carbon fiber reinforced plastic (CFRP). Additional advantages of this material are its high stiffness and its high breaking strength.

Above about 15 μm the telescope will be diffraction limited (see Fig. 7). This limit was chosen because at shorter wavelengths large ground based telescopes are available which offer diffraction limited image quality except between 5 μm and 8 μm where the atmosphere blocks ground based observations. Below 15 μm seeing limits the image quality. The seeing has two major contributions. Disturbances of the wavefront by air motions inside the telescope cavity and density fluctuations in the shear layer above the cavity are limiting. The required pointing stability is 0.2 arcsec. However, it should be mentioned that this requirement will be met only after certain improvements in the operational phase of SOFIA. In summary, the image quality and the image stability requirements permit nearly diffraction limited imaging as well as superresolution down to about 15 μm.

3.3 OPERATION

SOFIA's home base will be at NASA Ames Research Center in Moffett Field south of San Francisco. Since from this location only the northern hemisphere is observable it is planned to deploy SOFIA for certain periods of time in Auckland, New Zealand. From here observation of the southern hemisphere will be performed. Beside this there are several constraints which have to be taken into account when planning an observation. The time useful for observations is limited to about seven hours. The maximum observing time per object is therefore limited to 3.5 h per flight. In addition the unvignetted elevation range of the telescope is 20° to 60°, which excludes zenith observations. It is required that SOFIA will provide at least 960 successful flight hours per year which corresponds to about 150 research flights per year.

Project	Operated by Development began First light Operation	USRA for NASA, DLR January 1997 2002 20 years
Telescope	Aperture Wavelength range Spatial resolution	2.5 m (2.7 m) 0.3 – 1600 μm 1-3 arcsec for $0.3 < \lambda < 15$ μm (λ/10 arcsec) for $\lambda > 15$ μm
Operations	Number of flights Operating altitudes Nominal observing time Crew	ca. 160 per year 12 – 14 km, 39000 – 45000 ft 6 – 9 hours ca. 12: pilots, operators, technicians, and scientists
Instrumentation	Instrument changes Research flights in-flight access to instrumentation	15 – 25 per year ~150 per year continuous

Table 1. Summary of the SOFIA characteristics.

4. First light instruments for SOFIA

Ten scientific instruments have been approved and will be available when the operational phase of SOFIA will begin (Table 2). The instruments cover a wide spectral range from about 600 μm down to less than 1 μm with a spectral resolution up to 10^7 (see Fig. 5). There are four cameras and six spectrometers. However, this distinction is to a certain extend arbitrary since some cameras have additional low resolution spectroscopic capabilities and the spectrometers often use array detectors to improve spatial multiplexing. In the following we will give a short description of the spectrometers and the main scientific objectives. More detailed descriptions can be found in Refs. [4,5,6].

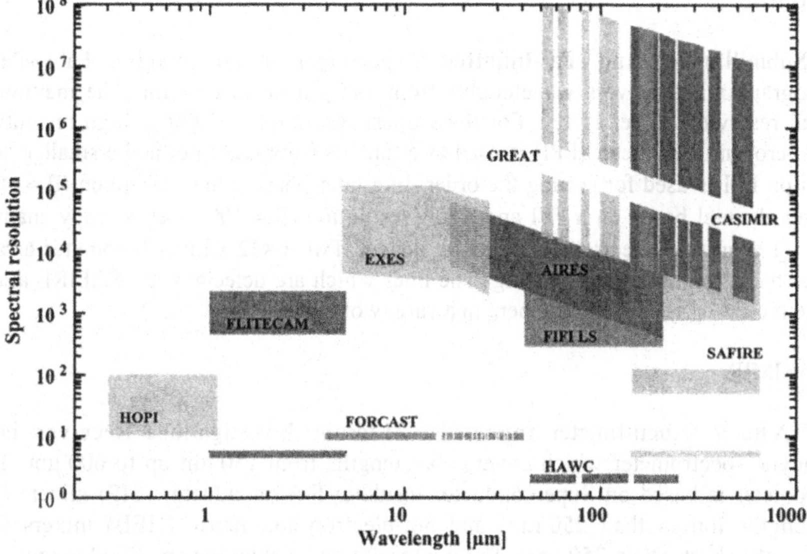

Figure 5. Spectral resolution and wavelength coverage of the SOFIA first light instruments.

4.1 AIRES

The **A**irborne **I**nfra**R**ed **E**chelle **S**pectrometer is a high resolution (R ≈ 2000 – 35000) spectrometer. The heart of the spectrometer is an about $250 \times 1100\,mm^2$ large, cryogenically cooled echelle with a line spacing slightly less than 1 mm and a blaze angle of 76°. A rotating K-mirror allows simultaneous observations of the spectrum along a user-selectable strip of the sky. At first light its operating wavelength for spectroscopy is 30 – 120 μm. A 16×25 element Ge:Sb array will be used as detector. At a later stage a stressed Ge:Ga array will be implemented in order to extend the wavelength range to 210 μm. AIRES will be ideal for spectral imaging of the gas-phase in the interstellar medium.

4.2 EXES

The **E**chelon-**C**ross-**E**chelle **S**pectrograph is a high resolution grating spectrograph for wavelenghts from 5.5 μm to 28.5 μm. It has three spectroscopic modes yielding a resolution of 10^5, 2×10^4 and 4000. High resolution is achieved with an echelon, a coarsely ruled and steeply blazed diffraction grating. An echelle is used in low order for cross-dispersion. The highest resolution is achieved by using both gratings while for medium and low resolution only the echelle is used. The detector is a 256×256 Si:As array optimized for low background absorption. EXES is primarily designed for high resolution studies of interstellar molecules such as H_2, H_2O and CH_4. In the high resolution mode it is possible to resolve the lines which improves analysis and interpretation of the data significantly.

4.3 SAFIRE

The **S**ubmillimeter **A**nd **F**ar-**I**nfra**R**ed **E**xperiment is an imaging Fabry-Perot spectrograph which covers wavelenghts from 145 μm up to 655 μm. The maximum spectral resolving power is 10^4. For the highest resolution (>1000) a high resolution Fabry-Perot Interferometer (FPI) is used as a tunable bandpass filter and a smaller, low resolution FPI is used for sorting the order. In a later phase a low resolution (R ≈ 100) imaging channel based on a FPI and a low resolution (R ≈ 100) spectroscopy channel based on a grating spectrometer will be added. Two 6×32 silicon bolometer arrays provide background-limited imaging. The lines which are detectable by SAFIRE allow to trace the astrophysical environment in a variety of conditions.

4.4 CASIMIR

The **CA**ltech **S**ubmillimeter **I**nterstellar **M**edium **I**nvestigations **R**eceiver is a heterodyne spectrometer which covers wavelengths from 150 μm up to 600 μm. The spectrometer is based on Superconductor-Insulator-Superconductor (SIS) mixers (for wavelengths longer than 250 μm) and hot-electron-bolometric (HEB) mixers (for wavelengths shorter than 250 μm). The mixers are pumped by the amplified output of a frequency synthesizer. This signal is further multiplied up to the operation frequency. As backend spectrometers alternatively an acousto-optical spectrometer (AOS), a high resolution digital correlator and a wideband analog spectrometer can be used. Depending on the backend spectrometer a spectral resolution as high as 10^6 can be achieved. The prime objective is spectroscopical investigation of the galactic and extragalactic warm (approximately 100 K) interstellar medium.

4.5 GREAT

The **G**erman **RE**ceiver for **A**stronomy at **T**erahertz Frequencies is a modular dual-channel heterodyne spectrometer. It is focused on three scientifically selected frequency bands. The low-frequency band from 1.6 to 1.9 THz covers among other important lines the fine structure transition of CII, the mid-frequency band is centered on the cosmologically relevant 1-0 transition of HD at 2.6 THz and the rotational transition of OH, and the high frequency band covers selected frequencies between 3.5 THz and 5.2 THz, especially the fine structure transition of OI at 4.7 THz. For all channels HEB's are used as mixers. However, different types of local oscillators are foreseen: a multiplied backward wave oscillator (low frequency channel), a multiplied Gunn diode (mid frequency channel) and a far-infrared gas laser (high frequency channel). The backend spectrometers are similar to the ones used in CASIMR yielding a resolution up to 10^7.

4.6 FIFI LS

The **F**ield-**I**maging **F**ar-**I**nfrared **L**ine **S**pectrometer is an imaging, dual channel, medium resolution (R ≈ 2000) grating spectrometer which covers wavelengths from 42 – 110 μm (channel 1) and 110 – 210 μm (channel 2). The incoming signal is separated into two beams by a dichroic beam splitter. Each channel is equipped with a beam slicer and a grating. The beam slicer optically rearranges the two dimensional field into a single slit for the spectrometer. By this method the 5 × 5 pixel field of view is imaged to a 25 × 1 pixel slit. The grating disperses the light onto a 25 × 16 element detector array (Ge:Ga for channel 1, stressed Ge:Ga for channel 2). This provides good spectral coverage with medium resolution. FIFI LS is optimized for the observation of faint, extragalactic objects.

Name of Instrument	PI	Institute	Type of Instrument	Facility, PI, or Special Class
HOPI	E. Dunham	Lowell Observatory	Occultation CCD Photometer / Imager	Special Class
AIRES	**E. Erickson**	**NASA - ARC**	**Echelle Spectrometer 17 to 210 microns R=10,000**	**Facility Instr.**
HAWC	D.A. Harper	Univ. of Chicago	Far Infrared Bolometer Camera 30 - 300 microns	Facility Instr.
FORCAST	T. Herter	Cornell	Mid IR Camera 5 - 40 microns	Facility Instr.
EXES	**J. Lacy**	**Univ. of Texas**	**Echelon Spectrometer 5 - 28 microns R=1500 and R=100,000**	PI Instr.
FLITECAM	I. McLean	Univ. of California, Los Angeles	Near IR Test Camera 1 - 5 microns	Facility Instr.
CASIMIR	**J. Zmuidzinas**	**Caltech**	**Heterodyne Spectrometer 250 - 600 microns**	**PI Instr.**
SAFIRE	**S. Moseley**	**NASA-GSFC**	**Imaging Fabry-Perot 145 - 655 microns**	**PI Instr.**
GREAT	**R. Güsten**	**MPIfR, Bonn; Uni Köln; DLR, Berlin**	**Heterodyne Spectrometer 50 - 250 microns**	**PI Instr.**
FIFILS	**A. Poglitsch**	**MPE, Garching Uni Jena**	**Field Imaging Far IR Line Spectrometer 40 - 350 microns**	**PI Instr.**

Table 2. Summary of the SOFIA first light instruments (bold: spectrometer).

5. Satellite Platforms

In Fig. 6 an overview of present and future space-borne observatories, their spectral resolution and their wavelength coverage is given. Our discussion will focus on SWAS, SIRTF, ASTRO-F (IRIS), and FIRST and their spectroscopic capabilities. Since line spectroscopy is not the main task of Planck it will not be described here. The ODIN mission is described elsewhere in this volume [2].

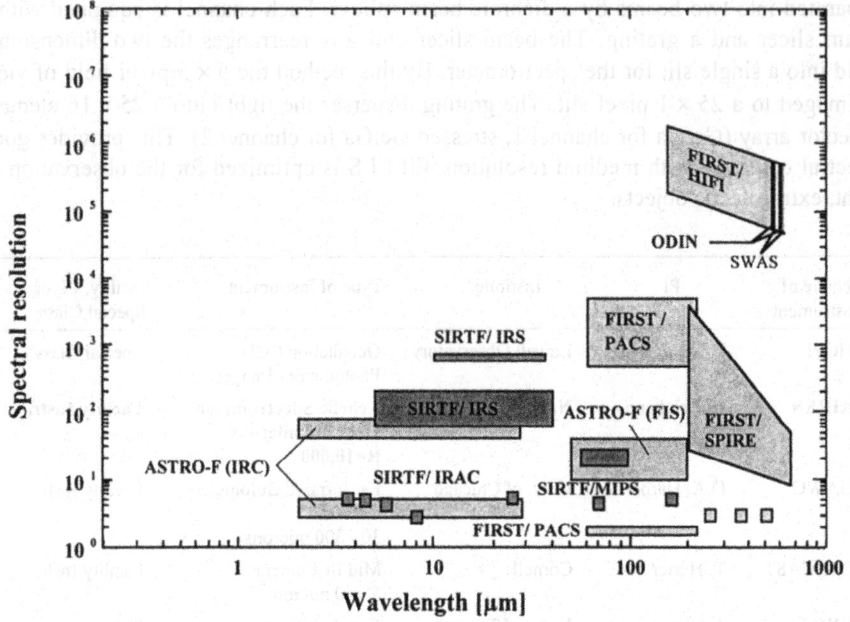

Fig. 6. Spectral resolution and wavelength coverage of FIR/submm space-borne satellite missions.

5.1. SWAS

The Submillimeter Wave Astronomy Satellite [7] was launched in December 1998. It is a NASA mission dedicated to investigate various aspects of star formation by observing five astrophysically important species which have emission lines at 490±3 GHz (O_2, CI) and 552±5 GHz. ($H_2{}^{18}O$, ^{13}CO, H_2O). By observing transitions of gaseous H_2O and O_2 whose energy above the ground state (26-27 K) corresponds well to the temperature of typical molecular clouds its composition and chemistry is studied. Especially the question of the oxygen reservoirs is addressed. The thermal balance of molecular clouds can be tested by observing these lines, because H_2O, CO and OI are predicted to be the dominant gas coolants of a cloud core at the beginning of its collapse. The UV-illuminated surfaces of molecular clouds are studied by observing the CI transition which is – due to its low critical density and low upper level energy (23.6 K) - easily excitable and therefore a good tracer of the total extent of photon dominated regions (PDR). In contrast the ^{13}CO line preferentially traces the warm, dense gas in PDRs. Up to now SWAS has detected H_2O in a variety of star forming regions, molecular clouds and solar system objects (Mars, Jupiter, Saturn and Comet Lee). For the O_2 abundance in molecular clouds an upper limit is established, which is appreciable lower than predicted by steady-state models [8].

The SWAS telescope is an off-axis Cassegrain system. It consists of a 54 ×68 cm primary mirror and secondary mirror which can be chopped by 8.5 arcmin. The absolute pointing accuracy is below 5 arcsec (1 σ). The spectrometer consists of two heterodyne receivers which are passively cooled to about 175 K. The low frequency receiver covers the band from 487-493 GHz while the high frequency receiver covers the band from 547 to 557 GHz. Each heterodyne receivers is equipped with a Schottky barrier diode mixer which is subharmonically pumped by a frequency tripled, tunable Gunn oscillator. The beams are co-aligned and the mixers operate in orthogonal linear polarizations, i.e. the signal radiation is separated into two beams by a wire grid polarization splitter. The double sideband receiver noise temperature is 2500 K (at 490 GHz) and 2200 K (at 554 GHz), respectively. This is about a factor of ten higher than state-of-the art SIS mixers can achieve. However, SIS mixer require a temperature of 4 K which is not achievable in the frame of a small satellite mission. The output of each receiver is downconverted with a bandwidth of 700 MHz centered around 1.75 GHz and 2.45 GHz, respectively. Both channels are diplexed to the 1.4-2.8 GHz input band of the acousto-optical spectrometer (AOS). The output of the AOS is forwarded to the on-board solid-state memory for later transmission to the ground.

5.2 SIRTF

The **S**pace **I**nfra**R**ed **T**elescope **F**acility [9] is a space-borne, cryogenically cooled infrared observatory. It is part of NASA's Great Observatories Program. The telescope is made from leightweight Beryllium and cooled to less than 5.5 K. It has a diameter of 0.85 m and allows diffraction limited imaging for wavelengths longer than 6.5 μm. It will be launched in 2002 and the estimated lifetime is 2.5 years (minimum) up to 5 years (goal). With SIRTF it will be possible to perform photometry in the wavelength region form 3 μm to 180 μm. Spectroscopy can be done between 5 μm and 100 μm.

From Fig. 6 the spectroscopic capabilities of SIRTF can be seen. They are primarily based on the Infrared Spectrograph (IRS) and the spectral energy distribution mode of the Multiband Imaging Photometer for SIRTF (MIPS). The Infrared Array Camera (IRAC) will be used for imaging. IRS is designed to perform diagnostic observations and redshift measurements of faint, infrared sources. High resolution ($R \approx 600$) is achieved with an echelle spectrograph. Another grating is used for the low resolution channel ($R = 60$-120). Si:As and Si:Sb detector arrays are used in both channels. In the spectroscopy mode of the MIPS instrument a low resolution grating spectrometer with a 32×32 element Ge:Ga detector array (frequency coverage 55 – 96 μm) is used. It measures source continua shapes.

5.3 ASTRO-F (IRIS)

The **I**nfra**R**ed **I**maging **S**urveyor [10] has a 0.7 m telescope which is cooled to 6 K by super-fluid liquid helium. The telescope is of Ritchey-Chretien type. Two focal-plane instruments are installed on ASTRO-F (IRIS). One is an infrared camera (IRC). This instrument consists of three independent camera systems which cover wavelengths from 2 μm to 26 μm. The IRC is equipped with grisms which add capabilities for low resolution ($R \leq 80$) spectroscopy. In this spectroscopic mode red-shifts can be estimated and sources can be classified. An InSb detector array with 512×412 pixel is used up to 5.5 μm while above two 256×256 Si:As arrays are used. The other focal-plane instrument is the far-infrared surveyor (FIS). It will survey the entire sky in the wavelength range from 50 μm to 200 μm with a resolution up to 400. To achieve this a Fourier-Transform-Spectrometer equipped with a 20 × 3 Ge:Ga detector array (up to 100 μm) and a 20 × 3 stressed Ge:Ga detector array (up to 200 μm) is used.

The main scientific objectives are the search for primeval galaxies and the investigation of galaxies at high red-shifts. IRIS will investigate star forming processes at a very early stage and the evolution of planetary systems.

5.4 FIRST

The **F**ar-**I**nfra**R**ed and **S**ubmillimetre **T**elescope [11] will perform spectroscopy and photometry in the wavelength range from 60 μm to 670 μm. The telescope is of Ritchey-Chretien design with a diameter of 3.5 m. It is radiatively cooled to an operational temperature of 80 K. The emission of black bodies with a temperature of 5 K to 50 K peak in the FIRST wavelength range. Therefore, the FIRST science objectives target at the cold universe. The FIRST science payload consists of three instruments: the photoconductor array camera and spectrometer (PACS), the spectral photometric imaging receiver (SPIRE) and the heterodyne instrument for FIRST (HIFI).

As a spectrometer PACS covers 60 – 210 μm without gaps. This is achieved with unstressed and stressed Ge:Ga detector arrays. A grating in Littrow configuration provides medium resolution (R ≈ 1500). An optical image slicer rearranges the two dimensional field of view (5×5 pixels) along a slit (1×25 pixels). PACS is optimized for the detection of weak sources.

The spectrometer part of SPIRE is an imaging Mach- Zender interferometer with two bolometer arrays. Both input ports of the interferometer are used all the time. The signal from the sky enters one input port while the signal from a calibration source enters the other port. The latter balances the power from the telescope. Two bolometer arrays optimized for 200 – 300 μm and 300 – 600 μm, respectively, will be used. The resolution will be in the range from 100 – 1000 at a wavelength of 250 μm. The field of view has a diameter of 2.6 arcmin.

HIFI is a heterodyne spectrometer which covers approximately the frequency range from 0.5 – 1.25 THz (band 1-5) and 1.41 – 1.91 THz (band 6). Each band is equipped with two orthogonally polarized mixers (band 1-5: SIS, band 6:HEB). The bands are covered without any gap. The mixers as well as the optical components that feed the mixers the signal from the telescope and combines it with the local oscillator radiation are housed in the focal plane unit inside the cryostat vessel. The local oscillator, a synthesiser which output is amplified and in several stages multiplied, is located outside of the cryostat in the spacecraft service module. Also part of this module are the backend spectrometers, an acousto-optical spectrometer and high resolution digital autocorrelators. HIFI provides one pixel per band on the sky. The lack of spatial multiplexing is compensated by its very high frequency multiplexing capabilities.

6. Comparison

It is instructive to compare the different upcoming FIR and submm observatories. For a complete comparison many factors are important. Some of them are spatial resolution, sensitivity, wavelength coverage and available instrumentation. They will be discussed here.

The spatial resolution of the FIR/ submm observatories is shown in Fig. 7. All space-borne observatories are diffraction limited above about 10 μm with a resolution which is determined by the clear aperture of the telescope. The specifications of SOFIA will allow diffraction limited observations above about 15 μm. This limit is set by air turbulence. For wavelengths longer than 70 μm FIRST offers the highest spatial resolution. SOFIA is almost as good as FIRST and three to four times better than ISO, SIRTF, and ASTRO-F (IRIS).

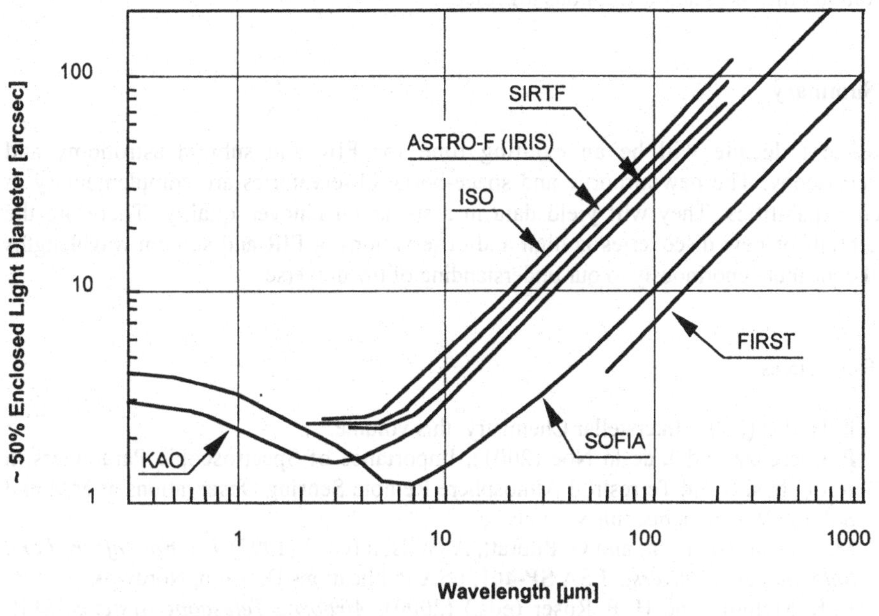

Figure 7. Spatial resolution and wavelength overage of FIR/submm observatories.

The sensitivity depends on the telescope area, the thermal background emission, and the effective emissivity and transmission. For medium (R ≈ 1000) and very high resolution (R ≈ 10^6) FIRST with a two times larger telescope area, a background temperature of 80 K (240 K for SOFIA) and no emission and absorption from the atmosphere will yield about ten times greater signal to noise ratio for a point source than SOFIA (assuming the same detector is used) [12]. The comparison of SOFIA with SIRTF and ASTRO-F (IRIS) looks similar. Although SOFIA has a much bigger telescope the low background and emissivity of the cryogenic SIRTF telescope results in an about ten times greater signal-to-noise ratio for SIRTF [6]. A limitation for SOFIA is the absorption due to a forest of telluric absorption lines of the atmosphere. This may even prohibit the observation at certain wavelengths.

A specific advantage of SOFIA compared to space-borne missions is that state-of-the-art detectors and spectrometers will be used and that the instruments can be more complex than on satellites. For example, at 100 µm the proposed imaging detector array for PACS will have 25×16 pixel while for SOFIA with its 8 arcmin field of view a 128×128 pixel array can be used. This decreases the time for spatial surveys significantly. Other examples are the heterodyne instruments. For SOFIA they will have at least two pixels while in FIRST only one pixel is available. By the year 2007 when FIRST will be launched heterodyne arrays with 16 pixels are planned for SOFIA. The advantage of more complex instruments is also reflected in Figs. 3 and 6. SOFIA with its more complex instruments will cover a wider range of wavelengths and spectral resolution than all satellite missions together.

7. Summary

The next decade will be an exciting time for FIR and submm astronomy and spectroscopy. The new airborne and space-borne observatories are complementary in their capabilities. They will yield data in a so far unachieved quality. Therefore the potential for new discoveries is high and observations at FIR and submm wavelengths will contribute enormously to our understanding of the universe.

8. References

1. E. Herbst (2001) Interstellar Chemistry, this volume.
2. P. Encrenaz and J. de la Noe (2001), Importance of Spectroscopic Parameters in Astrophysics and Terrestrial Atmosphere Remote Sensing Observations at MM and Submm Wavelenghts, this volume.
3. M. Rowan-Robinson, and G. Pilbratt, A. Wilson (eds.) (1997) *The Far Infrared and Submillimetre Universe*, ESA **SP-401**, ESA Publications Division, Nordwijk.
4. R. K. Melugin and H. P. Röser (eds.) (2000), *Airborne Telescope Systems*, SPIE **4014**, München.
5. A. Krabbe and H. P. Röser (1999) SOFIA Astronomy and Technology in the 21st Century, *Reviews of Modern Astronomy* **12**, 107-130.
6. R. Titz and H. P. Röser (eds.) (1998) *Astronomy and and Technology in the 21st Century*, Wissenschaft und Technik Verlag, Berlin.
7. G. J. Melnick et al. (2000) The Submillimeter Wave Astronomy Satellite: Science Objectives and Instrument Description, *The Astrophysical Journal* **539**, L77-L85.
8. See: All-SWAS issue (2000) *The Astrophysical Journal* **539**.
9. M. D. Bicay, C. A. Beichman, R. M. Cutri, and B. F. Madore (eds.) (1999) *Astrophysics with infrared surveys: A Prelude to SIRTF*, Astronomical Society of the Pacific (ASP) **177**.
10. H .Okuda (1997) IRTS and IRIS, ESA **SP-401**, 207-212.
11. G. L. Pilbratt (2000) The ESA FIRST Cornerstone Mission, SPIE **4013**, 142-151.
12. N. D. Whyborn (1997) The HIFI Heterodyne Instrument for FIRST: Capabilities and Performance, ESA **SP-401**, 19-24.

MILLIMETER AND SUBMILLIMETER SPECTROSCOPY IN SUPPORT OF UPPER ATMOSPHERIC RESEARCH

Edward A. Cohen
Jet Propulsion Laboratory
California Institute of Technology
4800 Oak Grove Drive
Pasadena, California 91109-8099 USA
Edward.A.Cohen@jpl.nasa.gov

Keywords: millimeter spectroscopy, upper atmosphere, limb sounding

Abstract This paper describes the laboratory program in millimeter and submillimeter spectroscopy at the Jet Propulsion Laboratory (JPL). The program directly supports ongoing and developing field measurement programs; in particular, the Microwave Limb Sounder (MLS) now in orbit aboard the Upper Atmospheric Research Satellite (UARS), the balloon borne Submillimeter Limb Sounder (SLS), the Far Infrared Limb Observing Spectrometer (FILOS), the 2.5 THz balloon OH spectrometer (BOH), and the MLS to be flown in 2002 on the Earth Observing System (EOS) AQUA platform. This program provides frequencies, linewidths and transition moments for species sought by the observing program, which have promise of being observed by instrumentation being developed by NASA, and which improved atmospheric models indicate are important. These data along with pertinent data available from other sources are analyzed and the results made available to the user community via the JPL "Microwave, Millimeter, and Submillimeter Spectral Line Catalog." From these data, accurate brightness temperatures and line profiles can be obtained. In addition, this program supplies necessary microwave through far infrared data that aid in the interpretation and assignment of infrared and optical spectra of importance to stratospheric research. Particular attention is paid to molecules whose high resolution infrared spectra are contained in the JPL and HITRAN infrared databases and to transient species relevant to research being done by the JPL kinetics group. This program has also been concerned with the determination of molecular structures and other molecular properties of halogen oxides and oxo-acids which affect chemical reactivity.

Introduction

NASA has a large ongoing stratospheric research program. In this program JPL has a leading role in the observation of millimeter wave thermal emis-

J. Demaison et al. (eds.), Spectroscopy from Space, 59–71.

sion of stratospheric species and the development of instrumentation to perform those measurements at increasing sensitivity and at higher frequencies. Millimeter wave radiometers have been successfully flown by JPL on aircraft, balloons, and UARS. A submillimeter radiometer has made a number of successful balloon flights and submillimeter channels will be used on the MLS to be flown on the 2002 AQUA platform of EOS. The FILOS far infrared radiometer has measured OH and other species near 3.1 THz and a heterodyne channel near 2.5 THz will be part of the EOS MLS. JPL was responsible for the high resolution infrared Atmospheric Trace Molecule Spectroscopy Experiment (ATMOS) experiments and is now responsible for the planned Tropospheric Emission Spectrometer (TES) to be flown on the AQUA platform in 2002. The JPL infrared spectroscopists who provide the supporting spectroscopic studies and analyses are highly regarded by the international community. The JPL Kinetics Group has played a prominent role in characterizing many of the major stratospheric chemical processes. We have worked closely with the investigators responsible for both field measurements and other laboratory programs on a wide variety of problems in upper atmospheric research.

The millimeter spectroscopy laboratory benefits from having a strong technology development program at JPL, and from regular interaction with the MLS team, the infrared spectroscopists, and the chemical kinetics group. Throughout the course of this program we have collaborated with a number of workers on a variety of problems requiring the application of spectroscopic techniques and analysis to upper atmospheric research. Under other tasks the millimeter spectroscopy laboratory has supported observational programs in interstellar and planetary radio astronomy. In this way a rather modestly funded task has been able to provide essential and efficient support to the NASA Upper Atmospheric Research Program.

In this paper some of the past accomplishments of the laboratory program will be described in terms of its contributions to MLS support, infrared analysis, and spectroscopy of reactive species. Some of the projects currently underway will be itemized. It is hoped that the information will suggest productive areas of research for those interested in carrying out laboratory studies.

MLS Support

The JPL Submillimeter–Microwave Spectroscopy Laboratory has supported the field measurements program closely in supplying the basic spectroscopic parameters necessary for the interpretation of Earth's atmospheric spectra since 1976. The interpretation of MLS data has relied heavily on measurements and calculations of molecular spectra carried out under this program. The MLS web page (http://mls.jpl.nasa.gov) contains links to several documents which describe the goals of MLS program as well as some of the requirements for laboratory data. Among the most important contributions have been a number of studies of ClO,[1–4] and O_3[5–10]. The ClO work has provided both the positions and air broadened linewidths that have been used for observations at both 204 and 649 GHz. The ozone data in particular have been essential to the observing community as ozone is the principal source of interfering features throughout the millimeter region. This body of data has continued to be a principal driver in the selection of observing objectives and design parameters for the field measurement program.

This program has made significant contributions to the JPL "Microwave, Millimeter and Submillimeter Spectral Line Catalog"[11,12] which is a principal source of line position and strength information for the millimeter and far infrared field measurement community. This involves not only original measurements in our laboratory, but critical evaluation of literature data. Air broadened linewidths, although very important for analyses of atmospheric data, are not included. There are relatively few reliable linewidth measurements in the literature in the microwave and infrared regions. Significant developments in linewidth measurement technique are described by Belov and Tretyakov elsewhere in this volume. The importance of the catalog is so well recognized that our colleagues have frequently made their measurements available to us prior to publication. Waters has utilized the catalog extensively in his chapter on microwave limb sounding[13] and has described in some detail the importance of a program in molecular spectroscopy to the field.

The catalog was initiated more than twenty five years ago as an astrophysics supported line list by R. L. Poynter. During the definition phases of the UARS Microwave Limb Sounder experiment, R. K. Kakar was supported to develop line lists for molecules of atmospheric importance. This work was continued by H. M. Pickett. Since there was some overlap of the two line lists, Poynter and Pickett merged the lists into a single catalog that was made available to other groups. The catalog has been of great use to those studying the millimeter and far infrared regions, and portions of it have been incorporated into the HITRAN compilations.[14] The ongoing task in support on upper atmospheric research has devoted a considerable portion of its resources to providing data and computations for the catalog although it does not directly support distribution or software development. At present, continuously updated versions as well as much of the software used for performing the spectroscopic analyses[15] are available via anonymous FTP and on the internet directly from our laboratory (http://spec.jpl.nasa.gov). A paper describing the newest version of the JPL Submillimeter, Millimeter, and Microwave Spectral Line Catalog was published recently[12].

We have provided important original contributions for a number of molecules of atmospheric importance and of importance to our understanding of halogen oxide chemistry. These include O_3[5–10,16], ClO[1–4], ClOOCl[17],

$ClOClO_2$[18], $ClClO_2$[19,20], HOCl[21], $HOClO_3$[22], OClO[23], BrO[24,25], BrOBr[26,27], OBrO[26,28], HOBr[29–31], OH[32], H_2O_2[33], $a^1\Delta$ O_2[34,35], O_2[36], CO[36], HNO_3[37], $ClONO_2$[38], HO_2NO_2[39], COF_2[40,41], SO_2[42], IO[43], OIO[44], and FClO[45]. Spectra of all those species that may be observed *in situ* have been placed in the JPL catalog. Others are being prepared for entry in order to generate a more comprehensive database for those engaged in laboratory studies. In addition a number of new submillimeter and far infrared transitions have been measured for a variety of molecules in order to improve the reliability of frequency predictions in the far infrared region for cataloging purposes. For many of the species, the original data files used in the generation of the catalog entries are available from our web site.

Molecules for which calculations have been recently entered or improved in the catalog include $O^{35}ClO$, $O^{37}ClO$, $O^{79}BrO$, $O^{81}BrO$, CO, ^{13}CO, $C^{17}O$, $C^{18}O$, H_2O (ground state and ν_2=1), ^{79}BrO (ground state and v=1), and ^{81}BrO (ground state and v=1). Calculations for IO and OIO have been prepared, but are not yet available for distribution. $ClClO_2$ (chloral chloride) and $HOClO_3$ (perchloric acid) calculations have been performed, but have not been placed in the catalog. In addition, we have observed a number of previously unmeasured transitions of the rarer isotopes of water in the 550 - 700 GHz region. These include HDO (ground state and ν_2=1), D_2O (ground state and ν_2=1), $H_2{}^{17}O$ (ground state and ν_2=1), $HD^{18}O$, and $H_2{}^{18}O$. Improved catalog entries are in preparation. In collaboration with other laboratories, a paper on the rotational spectra of SO_2, $^{33}SO_2$ and $SO^{17}O$ has been published[41] which extends the range of observations and improves the molecular parameters.

The spectrometer used for most of this work has been described in a number of the above referenced papers. The basic spectrometer is described in Refs. 17 and 18. THz laser sideband measurements are reported in Refs. 32 and 46. The new difference frequency spectrometer with capability from several hundred GHz to the THz region is described in Refs. 47 - 49. Recently, we have incorporated 110 - 270 GHz and 570 - 720 GHz, phase locked backward wave oscillators (BWO) for use with our submillimeter spectrometer. We have made appropriate modifications in the modulation schemes and sweeping software so that the tube can be used for precise frequency and linewidth measurements. Similar systems are described by Belov and Tretyakov in this volume. In addition, we have successfully used planar antiparallel pair Schottky diode multipliers designed and built by P. H. Siegel of the Submillimeter Wave Advanced Technology Group to observe spectra near 919.31 GHz. The multiplier generated thirteenth harmonic of $\approx$70.716 GHz using $\approx$10 mw of fundamental klystron power. These planar multipliers were also used to follow commercially available multipliers and frequency synthesizers to generate a wide range of submillimeter frequencies. Recently we have used them in an astrophysics supported program to produce 10 GHz long scans in the 525 GHz region. In this case the synthesizer operated near 12.5 GHz and was followed by a commercial times six multiplier. The JPL multiplier was then tuned to optimize seventh harmonic. The result was forty-second harmonic of a synthesized frequency. The reproducibility of scans taken with these devices and synthesized frequencies will be particularly useful for linewidth measurements. For frequency measurements, scans of $\approx$20 GHz can be carried out unattended with subsequent automated peak measurement.

Linewidth measurements in our laboratory are usually done by using the convolution method developed by Pickett[50]. Application of this method to the atmospheric molecules O_3 and ClO may be found in Refs. 4 and 9. More recently, in a collaborative effort with the De Lucia group at Ohio State University we studied the N_2 and O_2 broadening of several transitions of HNO_3 which contribute to the background emission observed by MLS[37]. The MLS team has used these measurements interpret the background emission in terms of upper atmospheric HNO_3 distribution.[51,52] The joint study represents one of the few examples for which different investigators using different techniques have measured identical transitions in order to validate the results[53]. The comparison of results showed an absolute accuracy of $\approx 3\%$ can be readily obtained.

Since the strength of a rotational transition is dependent upon the individual components of the permanent molecular dipole moment, it is important to measure the dipole moment of new molecules. The techniques for doing this in the microwave region at low frequencies and for well resolved Stark effect are extensively documented in the literature. In the millimeter and submillimeter wavelength regions the measurements are complicated by the fact that the transitions are often high J transitions with unresolvable Stark effect. Moreover, the type of large diameter cells which are frequently used for observations of highly reactive species are not compatible with the application of a strong homogeneous electric field. The polarization of the high frequency radiation is not usually exactly known with respect to the applied electric field.

We have developed a technique for measuring dipole moments under weak field conditions for which the Stark effect is unresolved and the microwave polarization is not well known. In one conducts an experiment using a double pass cell, the rooftop reflector used to obtain two passes through the sample also rotates the reflected radiation by 90°. Therefore, it is reasonable to assume that the Stark shifted transition contains equal contributions from the $\Delta M = 0$ and $\Delta M = \pm 1$ components. For second order Stark shifts of the energy levels of the form

$$\Delta E/\mathcal{E}^2 = A + BM^2$$

the shift in frequency can be expressed in closed form as the intensity weighted average of all the Stark components. Here A and B are the Stark coefficients and $\mathcal{E}$ is the applied electric field. For the present case of equal parallel and perpendicular polarization, one obtains

$$\Delta\nu/\mathcal{E}^2 = \Delta A + \Delta B[8J(J+1) - 1]/20$$

for $\Delta J = 0$ with ΔA and ΔB referring to the differences between upper and lower states and

$$\Delta\nu/\mathcal{E}^2 = \pm[\Delta A + \Delta BJ(6J+7)/20 + B_{J+1}(J+1)/2]$$

for a $J+1 \leftarrow J$ or $J \leftarrow J+1$ transition where $\Delta A = A_{J+1} - A_J$ and $\Delta B = B_{J+1} - B_J$. The upper sign is for $J+1 \leftarrow J$ and the lower for $J \leftarrow J+1$. For individual hyperfine transitions, F is to be used instead of J.

Because the distribution of microwave power in the sample cell is a function of frequency and not well controlled and because the applied field is not very homogeneous, the transitions which are used for calibration should be preferably in the same frequency region as those used for the dipole determination. While this method is not as accurate or precise as measurements of individual components, it is quite useful when more conventional techniques are unavailable. Applications of the method can be found in Refs. 20 and 28.

Infrared Analysis

In general, the contributions of rotational spectroscopy to infrared analysis are precise rotational and centrifugal distortion constants as well as very accurate rotational energies for the ground vibrational states and often excited states as well. These can be extremely useful in assigning complex infrared spectra. Merged fits of infrared and rotational data usually provide excellent molecular constants. Infrared analyses in the literature which utilize rotational data, energy levels, or molecular constants from this laboratory include works on O_3 , COF_2[40,54,55], ClO[56], HOBr[29,30] and $HOClO_3$[57]. The large body of HNO_3 rotational data obtained by the De Lucia group[58–65] has also been used extensively. Analyses of O_3 and HNO_3 have been recently reviewed by Flaud and Bacis[66] and Perrin,[67] respectively. Currently, we are re-examining the line strengths of the ν_9 and $2\,\nu_5$ bands on HNO_3 using continuously flowing anhydrous samples[68].

Recently we have provided G. C. Toon with new calculations of the ν_5 band of phosgene, $COCl_2$. These calculations which utilized earlier rotational data obtained from the literature have allowed the definitive identification of the $COCl_2$ molecular signature in high-resolution spectra taken by the MKIV balloon borne FT spectrometer[69].

Spectroscopy of Reactive Species

In addition to supporting the observing programs, we are continuing to collaborate with the kinetics group at JPL in order to unambiguously identify and characterize reactive and transient species that play important roles in upper atmospheric chemistry. Our early efforts led to the first assigned high resolution spectra and structural determination of ClOOCl [17] and $ClOClO_2$ [18]. These remain the only high resolution studies of these compounds. Subsequently we were able to synthesize the second most stable form of a Cl_2O_2 molecule, chloral chloride and determine its structure and many of its molecular properties.[19,20] We were also able to re-examine spectra taken in our studies of ClOOCl and $ClOClO_2$ and show the $ClClO_2$ did not form in observable quantities from the ClO self reaction or from Cl + OClO under the conditions of temperature and pressure described in our earlier work[17,18]. The determination of the ClOOCl structure was an important piece of confirming evidence that the peroxide is the primary product of the ClO self reaction and an important contributor to polar ozone depletion processes as proposed by Molina and Molina[70].

Most of the unstable molecules that we have studied have been referred to above. These are primarily the halogen oxides and oxoacids. The analyses of these spectra have revealed significantly more detail about the molecular structure and chemical bonding in these compounds than was previously known. In addition, spectra of the starting materials ClF_3 and $FClO_2$[71], taken during studies of FClO and $ClClO_2$ were analyzed in great detail by H. S. P. Müller at Universität zu Köln after he left JPL.

Recent work on unstable molecules has included the following:

1. A thorough analysis of the rotational spectra of the four main isotopic species of chloryl chloride, $ClClO_2$[20].
2. The analysis of the microwave spectra of the Coriolis coupled ν_4 and ν_6 states of $ClClO_2$
3. Observations of the microwave spectrum and Stark effect of FClO[45] and determination the dipole moment.
4. Observation of the rotational spectra of the $X_1\,^2\Pi_{3/2}$ state of IO up to $v = 13$ and extension of the region of observations to $\approx$ 700 GHz[43].
5. The first observation of the rotational spectrum of the $X_2\,^2\Pi_{1/2}$ state of IO and its excited vibrational states up to $v = 9$ and the first rotational spectra of $I^{18}O$ in both the X_1 and X_2 states[43].
6. Observation and analysis the first rotational spectra of OIO in the ground vibrational state and first excited bending state. This is the only high-resolution study of the molecule[44].
7. Observation of BrO rotational spectra up to $v = 8$ and $v = 7$ in the X_1 and X_2 states respectively[25].
8. Observation of the rotational spectrum of ClO up to $v = 2$ as well as $Cl^{18}O$ in natural abundance[71].

The chloral chloride paper reports the synthesis, rotational spectra of the four major isotopic species, molecular structure, harmonic force field, quadrupole coupling constants, and dipole moment. A paper is in preparation in collaboration with H. S. P. Müller on rotational spectra of the Coriolis coupled ν_4 and ν_6 states. $ClOClO_2$, $ClClO_2$, and $FClO_2$[71,73,74] are all characterized by a long weak bond to the ClO_2 group. FClO also has a weaker than normal FCl bond.

The FClO work was done in collaboration with Müller, who has recently reported an analysis of the ν_1 and ν_2 infrared bands[75]. The primary contribution of the JPL work was a determination of the molecular dipole moment. The spectra taken at JPL have been analyzed along with those taken at Köln, infrared spectra taken at Wuppertal, and molecular beam spectra taken at Zürich. Together these give a rather complete picture of the molecule. Dr. Müller has continued work on the molecule at Universität zu Köln. The primary objective of the present work is determination of molecular parameters for comparison with recent *ab initio* studies[76–78].

The work on the halogen monoxides was not intended to be extensive, but we discovered during our studies of the IO rotational spectrum that very highly excited vibrational states were being formed in a chemiluminescent reaction of molecular iodine with oxygen atoms. We were initially able to follow the transitions up to $v = 12$ in the $X_1\,{}^2\Pi_{3/2}$ electronic state using O atoms produced in an external microwave discharge. It is interesting the $v = 12$ state is approximately the same energy as the $a\,{}^1\Delta_g$ state of O_2 and the ${}^2P_{1/2}$ state of atomic I. It is well known that energy exchange between $O_2\ a\,{}^1\Delta_g$ and I atom is the mechanism that drives the high power chemical oxygen iodine lasers (COIL). The reaction $O + I_2 \rightarrow IO + I$ is not exothermic enough to populate states higher than $v = 10$. Earlier experiments with ClO and BrO using the products of an O_2 discharge plus the halogen molecules did not appear to give anomalous vibrational distributions. Using a dc discharge through a flowing mixture of I_2 and O_2 in the sample cell we were able to make the first observations of rotational spectra of the $X_2\,{}^2\Pi_{1/2}$ electronic state. At present, we have observed vibrational levels up to $v = 13$ for $X_1\,{}^2\Pi_{3/2}$ and $v = 9$ for $X_2\,{}^2\Pi_{1/2}$. In addition, $I^{18}O$ has been observed for both the X_1 and X_2 states up to $v = 5$.

Using the dc discharge cell that was so effective for vibrational excitation of IO, measurements of BrO have now been extended up to $v = 8$ and $v = 7$ for the X_1 and X_2 states, respectively. Excited vibrational levels of ClO were not obtained with these methods, however good signal to noise at thermal populations allowed measurements for the $v = 2$ levels of both the X_1 and X_2 states as well as ${}^{18}O$ in natural abundance.

The Hamiltonian of Brown, *et. al.* [79], with explicit isotope dependencies for each parameter, has provided a set of mass and nuclear moment independent parameters for each of the halogen monoxide species, IO, BrO and ClO. The electron spin-rotation constant, γ, and the centrifugal distortion of the spin-orbit splitting, A_D, which are normally correlated, have been separately determined by the isotope dependence of their contributions to the spectrum. Interatomic potentials have been derived from the mass-independent parameters that are accurate up to the observed excitation energies for each molecule. The hyperfine parameters have been compared with the literature values[80–82] of the appropriate calculated relativistic radial integrals of the halogens.

During the IO investigation, a large number of unidentified lines appeared which we suspected might be OIO. We pursued this investigation and were successful in obtaining the first rotational and only high-resolution spectrum of OIO. OIO is an asymmetric prolate rotor (κ = -0.690) with a $^2B_1$ electronic ground state. It was formed initially from the products of a microwave discharge in O_2 passing over molecular iodine and later with greater yield in a DC discharge through a mixture of O_2 and I_2 vapor. Although the experiment was hampered by copious solid deposits and apparently inefficient production of OIO, the rotational spectrum was quite prominent in survey scans in the millimeter and submillimeter regions. Facilitated by predictions of fine and hyperfine patterns from ESR results[83], it was eventually possible to identify high J, R branch transitions with $K_c \approx J$. Over 550 ground state transitions and over 160 transitions of the excited bending state have been included in the fits. The transitions cover a wide range of rotational quantum numbers and permit the accurate determination of an extensive set of molecular parameters. The parameters have been interpreted in terms of the molecular geometry, harmonic force field, and electronic structure.

Continuing Work

During the last year the MLS science team listed its highest priorities for laboratory measurements of air broadened linewidths. These are as follows:

2.5 THz lines of OH, O_2, and H_2O to improve accuracy of OH measurement (OH: 2.510 THz and 2.514 THz; O2: 2.502 THz; H_2O: 2.532 THz).

HCl line at 625.9 GHz to improve accuracy of chlorine loading measurement.

$^{18}O^{16}O$ (isotopic O_2) line at 233.9 GHz to improve accuracy of temperature measurement.

O_3 lines to improve accuracy of ozone measurement. (235.7 GHz, 237.1 GHz, 242.3 GHz, 243.5 GHz, 625.4 GHz, 2.51 THz, 2.54 THz)

HO_2 lines at 649.7 and 660.5 GHz to improve accuracy of HO_2 measurement.

BrO lines at 624.8 and 650.2 GHz to improve accuracy of bromine measurement.

They also listed several other features whose linewidth parameters including temperature dependence are somewhat less critical but still of high priority.

HCN line at 177.3 GHz

CO line at 230.5 GHz

HOCl line at 635.9 GHz

CH_3CN lines at 183.9, 202.3, 624.8, 626.4, and 660.7 GHz

O_3 lines at 239.1, 231.3, 248.2, 249.8, and 250.0 GHz to improve CO retrievals

While these required measurements are for only those transitions which are to be observed from space or which may interfere with observations, additional measurements of these and other transitions for the same molecules would be extremely helpful in extending the very sparse database of linewidths of rotational transitions. In addition, the MLS team needs improved measurements of the non-resonant absorption of water vapor and air throughout these spectral regions. There are measurement programs in progress at our institution and elsewhere which address these requirements, but there remains much to be done in the area of linewidth measurements.

In conjunction with work being done in the kinetics group, we are attempting obtain definitive identification of the volatile compounds evolved from the solid condensate of the ClO self reaction. This material is trapped at 77K and slowly warmed. Our previous work on the chlorine oxides and oxo-acids provides precise predictions of the rotational spectra of ClOOCl, OClO, ClO, $ClClO_2$, $ClOClO_2$, HOCl and $HOClO_3$. It not only will allow identification of the species, but also will give a moderately good indication of their relative abundances. In our earlier experiments in this laboratory we found only ClOOCl, but those experiments did not involve condensation of products. In our earlier work on OBrO we observed BrO and Br_2O as well as OBrO over a condensate from the reaction of oxygen atoms with Br_2.

An example of an another area in which rotational spectroscopy can make a contribution to atmospheric measurements is in the comparison of absorption cross sections in other spectral regions with those of rotational transitions. A recent example has recently been reported by Vander Auwera, *et* al[84]. We are planning simultaneous measurements of the rotational and UV spectra of OBrO in order to obtain an independent measurement of the UV cross section. We have already measured the OBrO permanent dipole moment and can calculate rotational line strength with good accuracy. The amount of OBrO in the optical path can then be determined. Renard, *et al.*[85,86] have reported detection of stratospheric OBrO in amounts much larger than model calculations would allow[87]. An independent measurement of OBrO cross sections will provide a basis for interpreting atmospheric spectra.

A similar set of measurements would also be useful for OIO. In a paper by Ingham, *et al.*[88] on the photodissociation of IO and OIO, the uncertainty of OIO absorption cross section of more than a factor of two is a limiting factor in the precision of the results. Unlike OBrO, our present method of OIO production does not lend itself to dipole moment measurement. We are investigating cleaner and more efficient methods of OIO production. If successful, we shall measure the dipole moment, determine the rotational line strengths, and attempt simultaneous measurements of rotational and UV spectra.

Conclusion

Examples have been presented of studies of rotational spectra that have application to upper atmospheric research. These have been not only applications for interpretation of atmospheric spectra, but also structural studies of molecules which have a role in atmospheric chemistry. It is hoped that the examples and the associated references will be helpful to those wishing to participate in this field of study.

Acknowledgment

This work was performed at the Jet Propulsion Laboratory, California Institute of Technology, under a contract with the National Aeronautics and Space Administration.

References

1. R. K. Kakar, E. A. Cohen and M. Geller, *J. Mol. Spectrosc.* **70**, 243 (1978).
2. E. A. Cohen, H. M. Pickett, and M. Geller, *J. Mol. Spectrosc.* **106**, 430 (1984).
3. H. M. Pickett, D. E. Brinza and E. A. Cohen, *J. Geophys. Res.* **86C**, 7279 (1981).
4. J. J. Oh and E. A. Cohen, *J. Quant. Spectrosc. Radiat. Transf.* **52**, 151 (1994).
5. E. A. Cohen and H. M. Pickett, *J. Mol. Struct.* **97**, 97 (1983).
6. C. Chiu and E. A. Cohen, *J. Molec. Spectrosc.* **109**, 239 (1985).
7. H. M. Pickett, E. A. Cohen, and J. S. Margolis, *J. Molec. Spectrosc.* **110**, 186 (1985).
8. H. M. Pickett, E. A. Cohen, L. R. Brown, C. P. Rinsland, M. A. H. Smith, V. Malathy Devi, A. Goldman, A. Barbe, B. Carli and M. Carlotti, *J. Mol. Spectrosc.* **128**, 151 (1988).
9. J. J. Oh and E. A. Cohen, *J. Quant. Spectrosc. Radiat. Transf.* **48**, 405 (1992).
10. E. A. Cohen, K. W. Hillig, II, and H. M. Pickett, *J. Mol. Struct.* **352/353,** 273 (1995).
11. R. L. Poynter, and H. M. Pickett, *Appl. Optics,* **24**, 2235 (1985).
12. H. M. Pickett, R. L. Poynter, E. A. Cohen, M. L. Delitsky, J. C. Pearson and H. S. P. Müller, "Submillimeter, Millimeter and Microwave Spectral Line Catalogue," JPL Publication 80-23, Rev. 4 (1996), and *J. Quant. Spectrosc. Radiat. Trans.* **60,** 883(1998).
13. J. W. Waters, "Microwave Limb Sounding," Chapter 8 in *Atmospheric Remote Sensing by Microwave Radiometry,* M. A. Janssen, Editor, John Wiley & Sons, Inc. 1993.
14. L. S. Rothman, C. P. Rinsland, A. Goldman, S. T. Massie, D. P. Edwards, J.-M. Flaud, A. Perrin, C. Camy-Peyret, V. Dana, J. Y. Mandin, J. Schroeder, A. McCann, R. R. Gamache, R. B. Wattson, K. Yoshino, K. V. Chance, K. W. Jucks, L. R. Brown, V. Nemtchinov, and P. Varanasi, *J. Quant. Spectrosc. Radiat. Transf.* **60**, 665 (1998).
15. H. M. Pickett, *J. Mol. Spectrosc.* **148**, 371 (1991).
16. H. M. Pickett, D. B. Peterson, and J. S. Margolis, *J. Geophys. Res.* **97**, 20787 (1992).
17. M. Birk, R.R. Friedl, E. A. Cohen, H.M. Pickett, and S.P. Sander, *J. Chem. Phys.* **91**, 6588 (1989).
18. R. R. Friedl, M. Birk, J. J. Oh, E. A. Cohen, *J. Mol. Spectrosc.* **170**, 383 (1995).
19. H. S. P. Müller and E. A. Cohen, *J. Phys. Chem.* **101,** 3049 (1997).
20. H. S. P. Müller, E. A. Cohen, and D. Christen, *J. Chem. Phys.* **110,** 11865 (1999).
21. H. E. Gillis Singbeil, W. E. Anderson, R. Wellington Davis, M. C. L. Gerry, E. A. Cohen, H. M. Pickett, F. J. Lovas and R. D. Suenram, *J. Mol. Spectrosc.* **103**, 466 (1984).
22. J. J. Oh and E. A. Cohen, Molecular Spectroscopy Symposium, Ohio State University, June 1992.
23. H. S. P. Müller, G. O. Sørensen, M. Birk, and R. R. Friedl, *J. Molec. Spectrosc.* **186,** 177 (1997).
24. E. A. Cohen, H. M. Pickett and M. Geller, *J. Mol. Spectrosc.* **87**, 459 (1981).

25. B. J. Drouin, C. E. Miller, H. S. P. Müller, and E. A. Cohen, *J. Mol. Spectrosc.* **205,** 128 (2001).
26. H. S. P. Müller, C. E. Miller, and E. A. Cohen, *Agew. Chemie,* **108**, 2285 (1996), *Agew. Chemie Int. Ed. (Eng.)* **35**, 2129 (1996).
27. H. S. P. Müller and E. A. Cohen, *J. Chem. Phys.* **106,** 8344 (1997).
28. H. S. P. Müller, C. E. Miller, and E. A. Cohen, *J. Chem. Phys.* **107,** 8292 (1997).
29. E. A. Cohen, G. A. McRae, T. L. Tan, R. R. Friedl, J. W. C. Johns, and M. Noël, *J. Mol. Spectrosc.* **173,** 55 (1995).
30. G.A. McRae and E.A. Cohen, *J. Mol. Spectrosc.* **139,** 369 (1990).
31. Y. Koga, H. Takeo, S. Kondo, M. Sugie, C. Matsumura, E. A. Cohen, and G.A. McRae, *J. Mol. Spectrosc.* **138,** 467 (1989).
32. G. A. Blake, J. Farhoomand, and H. M. Pickett, *J. Mol. Spectrosc.* **115,** 226 (1986).
33. E. A. Cohen and H. M. Pickett, *J. Mol. Spectrosc.* **87,** 582 (1981).
34. E. A. Cohen, M. Okunishi, and J. J. Oh, *J. Mol. Struct.* **352/353,** 283 (1995).
35. K. W. Hillig II, C. C. W. Chiu, W. G. Read, and E. A. Cohen, *J. Mol. Spectrosc.* **109,** 205 (1985).
36. W. G. Read, K. W. Hillig, II, E. A. Cohen, and H. M. Pickett, *IEEE Trans. Ant. and Prop.* **36,** 1136-1143 (1988).
37. T. M. Goyette, F. C. De Lucia, and E. A. Cohen, *J. Quant. Spectroscopy & Rad. Transf.* **60,** 77 (1998).
38. H. S. P. Müller, P. Helminger, and S. H. Young, *J. Mol. Spectrosc.* **181,** 363 (1997).
39. R. D. Suenram, F. J. Lovas, and H. M. Pickett, *J. Mol. Spectrosc.* **116,** 406 (1986).
40. E. A. Cohen and W. Lewis-Bevan, *J. Mol. Spectrosc.* **148,** 378 (1991).
41. E. A. Cohen, J. J. Oh, L. R. Brown, D. B. Peterson, and R. D. May, Symposium on Molecular Spectroscopy, Ohio State University, June, 1992.
42. H. S. P. Müller HSP, J. Farhoomand, E. A. Cohen, B. Brupbacher- Gatehouse, M. Schäfer, A. Bauder, and G. Winnewisser,*J. Mol. Spectrosc.* **201,** 1 (2000).
43. C. E. Miller and E. A. Cohen, International Symposium on Molecular Spectroscopy, Paper TJ06, Ohio State University, June, 1999.
44. C. E. Miller and E. A. Cohen, International Symposium on Molecular Spectroscopy, Paper TJ07, Ohio State University, June, 1999.
45. H. S. P. Müller and E. A. Cohen, International Symposium on Molecular Spectroscopy, Paper TJ08, Ohio State University, June, 1999.
46. J. Farhoomand, G. A. Blake, M. A. Frerking and H. M. Pickett, *J. Appl. Phys.* **57,** 1763 (1985)
47 P. Chen P, J. C. Pearson JC, H. M. Pickett HM, S. Matsuura S,and G. A. Blake, *Astrophys. J. SUPPL. S.* **128,** 371 (2000).
48. S. Matsuura, P. Chen, G. A. Blake, J. C. Pearson, and H. M. Pickett, *IEEE T. Microw. Theory.* **48,** 380 (2000).
49. S. Matsuura, P. Chen, G. A. Blake, J. C. Pearson, and H. M. Pickett, *Int. J. Infrared Milli.* **19,** 849 (1998).
50. H.M. Pickett, *Appl. Opt.* **19,** 2745 (1981).
51. M. L. Santee, G. L. Manney, W. G. Read, L. Froidevaux. and J. W. Waters, *Geophys. Res. Let.* **23,** 3207 (1996).
52. M. L. Santee, W. G. Read, J. W. Waters, L. Froidevaux, G. L. Manney, D. A. Flower, R. F. Jarnot, R. S. Harwood, and G. E. Peckham, *Science,* **267,** 849 (1995).
53. R. R. Gamache, J. M. Hartmann, and L. Rosenmann, *J. Quant. Spectroscopy and Rad. Transf.* **52,** 481 (1994).
54. R. D'Cunha, V. A. Job, G. Rajappan, V. Malathy Devi, W. J. Lafferty, A. Weberi, *J. Mol. Spectrosc.* **186,** 363 (1997).
55. M. J. W. McPhail, G. Duxbury, and R. D. May, *J. Mol. Spectrosc.* **182,**118 (1997).
56. J. B. Burkholder,P. D. Hammer,C. J. Howard, A. G. Maki, G. Thompson, and C. Chackerian, *J. Mol. Spectrosc.* **124,** 139 (1987).
57. M. S. Johnson, F. Hegelund, and B. Nelander, *J. Mol. Spectrosc.* **190,** 269 (1998).
58. J. K. Messer, F. C. De Lucia, and P. Helminger, *J. Mol. Spectrosc.* **104,** 417 (1984).
59. R. A. Booker, R. L. Crownover, F. C. De Lucia, and P. Helminger, *J. Mol. Spectrosc.* **128,** 62 (1988).
60. R. A. Booker, R. L. Crownover, and F. C. De Lucia, *J. Mol. Spectrosc.* **128,** 306 (1988).
61. R. L. Crownover, R. A. Booker, F. C. De Lucia, and P. Helminger, *J. Quant. Spectrosc. Radiat. Transfer.* **40,** 39 (1988).
62. T. M. Goyette and F. C. De Lucia, *J. Mol. Spectrosc.* **139,** 241 (1990).

63. T. M. Goyette, C. D. Paulse, L. C. Oesterling, F. C. De Lucia, and P. Helminger, *J. Mol. Spectrosc.* **167,** 365 (1994)
64. C. D. Paulse, L. H. Coudert, T. M. Goyette, R. L. Crownover, P. Helminger, and F. C. De Lucia, *J. Mol. Spectrosc.* **177,** 9(1996)
65. T. M. Goyette, L. C. Oesterling, D. T. Petkie, R. A. Booker, P. Helminger, and F. C. De Lucia, *J. Mol. Spectrosc.* **175,** 395 (1996).
66. J.-M. Flaud and R. Bacis, *Spectrochim. Acta A,* **54,** 3 (1998).
67. A. Perrin, *Spectrochim. Acta A,* **54,** 375 (1998).
68. R. A. Toth, E. A. Cohen, and L. R. Brown, *J. Mol. Spectrosc.* in preparation.
69. G. C. Toon, J.-F. Blavier, B. Sen, and B. J. Drouin, *Geophys. Res. Lett.* submitted.
70. L. T. Molina and M. J. Molina, *J. Phys. Chem.* **91,** 433 (1987).
71. H. S. P. Müller, *J. Mol. Struct.* **517-518,** 335 (2000).
72. B. J. Drouin, E. A. Cohen, C. E. Miller, and H. S. P. Müller, Symposium on Molecular Spectroscopy, Ohio State University, June 2000.
73. C. R. Parent and M. C. L. Gerry, *J. Mol. Spectrosc.* **49,** 343 (1974).
74. A. G. Robiette, C. R. Parent, and M. C. L. Gerry, *J. Mol. Spectrosc.* **86,** 455 (1981).
75. H. S. P. Müller, *Chem. Phys. Lett.* **314,** 396 (1999).
76. J. S. Francisco and S. P. Sander, *Chem. Phys. Lett.* **241,** 33(1995).
77. Y. M. Li and J. S. Francisco *J. Chem. Phys.* **110,** 2404 (1999).
78. T. J. Lee, *J. Phys. Chem.* **98,** 3697 (1994).
79. J. M. Brown and J. K. G. Watson, *J. Mol. Spectrosc.* **148,** 371 (1991).
80. P. Pyykkö and M. Seth, *Theor. Chem. Acc.* **96,** 92 (1997).
81. I. Lindgren and A. Rosén, Case Studies in Atomic Physics, **4,** 197 (1974).
82. P. Pyykkö and L. Wiesenfeld, *Mol. Phys.* **43,** 557 (1981).
83. J.R. Byberg, *J. Chem. Phys.* **85,** 4790 (1986).
84. J. Vander Auwera, J. Kleffmann, J.-M. Flaud, G. Pawelke H. Burger D. Hurtmans and R. Petrisse, *J. Mol. Spectrosc.* **204,** 36 (2000).
85. J. B. Renard, M. Chartier, C. Robert, G. Chalumeau, G. Berthet, M. Pirre, J. P. Pommereau, F. Goutail, *Appl. Optics,* **39,** 386 (2000).
86. J. B. Renard, M. Pirre, C. Robert, D. Huguenin, it J. Geophys. Res.-Atmos. **103,** 25383 (1998).
87. M. P. Chipperfield, T. Glassup, I. Pundt, O. V. Rattigan, *Geophys. Res. Lett.* **25,** 3575 (1998).
88. T. Ingham, M. Cameron, and J. N. Crowley, *J. Phys. Chem. A* **104,** 8001 (2000).

LABORATORY SUBMILLIMETER-WAVE SPECTROSCOPY

S.P. BELOV, M.YU. TRETYAKOV

Institute of Applied Physics of Russian Academy of Sciences
Uljanova St. 46, Nizhny Novgorod 603600, Russia
Fax. +7-(8312)-36-37-92, e-mail: belov@appl.sci-nnov.ru
http://www.appl.sci-nnov.ru/mwl/index.htm

1. Introduction

High-resolution submillimeter-wave (HRSMW) laboratory spectroscopy, as well as MW and IR spectroscopy, provides important information for both interstellar space and earth atmospheric research.

In the last 6–7 years a number of the modern submm-wave spectrometers have been developed in Russia [1], Germany [2], USA [3], France [4], Japan [5] and other countries. Line position, line profile, pressure broadening and pressure shift parameters, their temperature dependencies, absolute intensities of the lines and absorption in continuum can be measured now with these spectrometers up to 1.2 THz as easily as in the centimeter-wave region. Impressive progress in the development of HRSMW spectroscopy is based largely on the successful development of broadband digitally tunable sources of radiation with narrow linewidths (below 10 kHz at 1 THz) and a relatively high power (1–50 mW CW), which can cover the entire frequency region from several tens of GHz to well over 1 THz.

2. Key Elements of BWO-Based Broadband Digitally Tunable Source of Submillimeter-Wave Radiation

Nearly ideal sources for HRSMW laboratory spectroscopy are the "ISTOK"-made backward wave oscillator (BWO) tubes [http://www.istok.com] stabilized in frequency and phase against harmonics of a RF synthesizer. A set of the ISTOK's BWOs covers continuously the frequency region from 36 GHz up to 1250 GHz.

In general, a BWO is a broadband voltage tunable source of mm and submm coherent radiation. One BWO can cover continuously 100–150 GHz with a voltage variation of a 2–3 kV and a tuning rate of 30–70 MHz/Volt. Because a BWO is an electron beam device, its frequency can be tuned over its entire frequency region at a high rate: 10^6–10^7 MHz/s and faster. The linewidth of a phase-locked BWO is below 1 kHz at 100 GHz and depends upon the spectral purity of the reference RF synthesizer signal, the harmonic used, and the phase noise of the phase-lock loop (PLL) electronic circuits. The BWO radiation linewidth increases with increasing frequency, but at 1 THz it still can be below 10 kHz. The natural linewidth of ISTOK BWO is probably below 100 Hz. BWOs output power ranges from tens of mW at 200–500 GHz to about 1 mW at 1.2 THz. The output power variations of a single BWO with a frequency have

J. Demaison et al. (eds.), Spectroscopy from Space, 73–90.

a typical period of ~ 1–1.5 GHz and a depth of the tens of per cents. All of ISTOK's tubes have rectangular waveguide outputs which are slightly oversized beginning with the 200 GHz tube. To reduce the losses, 3.6×1.8 mm^2 waveguides are used for the submm-wave tubes beginning with 600 GHz. With appropriate care a submm-wave BWO can operate 2–3 years (~ 1000 h) without a degradation of performance.

The highest spectral purity and broadband frequency tunability of a phase-locked BWO can be achieved only when the elements of its phase-lock system have complimentary properties. The first and most important requirement is that the mixer-multiplier of such system must operate in the entire frequency region covered by BWOs up to 1.2 THz. A quasi-optical broadband submm-wave mixer-multiplier [1] is the first key element of that system. The second one is a low noise HEMT (high electron mobility transistor) amplifier [2] for the intermediate frequency (IF) signal of the mixer-multiplier output. Both of these key elements were successfully designed for the first time for the Cologne THz spectrometer.

The quasi-optical broadband mixer-multiplier consists of a parabolic mirror, a short tapered ridged waveguide for the 78–118 GHz reference signal, a planar filter for the 350 MHz IF signal output, and a submm-wave GaAs planar Schottky diode with a cutoff frequency of ~ 6 THz. The diode is mounted in the focus of the parabolic mirror illuminated by the BWO radiation. It is glued to a tip of the waveguide ridge by one side and to the IF signal filter by another one. The construction of this mixer-multiplier effectively couples both the reference signal and the submm-wave radiation to the planar Schottky diode. Once the mixer-multiplier is properly adjusted, no further tuning is required. Additionally it has a high reliability because of the planar Schottky diode is very stable and reliable.

At frequencies above 850 GHz, the BWO phase stabilization becomes more and more difficult due to several factors: 1. output power of the BWO decreases with frequency and spreads into many radiation modes of its oversized output waveguide, 2. high harmonics of the reference signal are required, 3. absorption of radiation due to atmospheric water decreases power coming to the mixer, 4. the diode performance deteriorates. In order to provide a stable lock-in signal at THz frequencies, the BWO PLL needs an extremely low noise amplification of the IF signal. A HEMT amplifier provides a flat gain of 21 dB in 310–390 MHz IF band with an average noise figure of ~ 0.5 dB. This corresponds to a 36 K noise temperature. It maintains a signal-to-noise ratio of PLL signal above 36 dB, which is required for a stable operation of the PLL, and stabilize the BWOs against harmonics of the KVARZ mm-wave synthesizer in all frequency region up to 1.25 THz [6].

The KVARZ synthesizer [1] includes two blocks: a basis unit and BWO-based mm-wave unit. It can operate in 53–78, 78–118 and 118–178 GHz frequency regions with the same basis unit and has a regular GPIB interface. The linewidth of KVARZ synthesizer output signal is below 1 kHz; minimal frequency step size is 100 Hz.

Instead of a KVARZ synthesizer, any MW synthesizer (Hewlett Packard, Rhode & Schwarz, Marconi, etc.) that can operate above 2 GHz can be used in a combination with a BWO-based mm-wave generator unit developed in the Institute of Physics of Microstructures of RAS, Nizhny Novgorod, or with a phase-stabilized Gunn oscillator [5].

BWO-based broadband digitally tunable sources of submm-wave radiation have overcome many technological difficulties previously associated with the submm-wave region and extended precise microwave broadband scanning spectroscopy to THz frequencies. These sources can be used for spectroscopy with any absorption cell (discharge cell, high-temperature cell, multipass cell etc.), with a molecular beam [7], with a frequency multiplier [8, 9], and even for a FIR side-band spectroscopy [10]. Free running BWOs can also be successfully used [3]. The advantages of the BWO-based sources are illustrated by the results presented below.

3. Broadband Doppler-Limited Spectroscopy with the Cologne THz Spectrometer

The Cologne THz spectrometer has been described in detail in refs.[1, 2, 6]. It is a computer controlled video-type spectrometer with the precise broadband tunable BWO-based radiation source described above, free-space absorption cell, magnetically tuned liquid-He cooled hot electron InSb bolometer, rubidium frequency standard, digital lock-in amplifier and data acquisition system. In the Doppler limited mode, the spectrometer can operate up to 1.25 THz with a sensitivity up to $\sim 10^{-8}$ cm^{-1} in favorable cases. Tuning of the frequency over the region of interest is stepwise.

The size of a frequency step in the broad scanning mode can be as small as the minimum frequency step size of the RF synthesizer multiplied by the harmonic number. For the methanol *K*-doublet presented in Fig. 1, the minimum frequency step can be 1.1 kHz. The absolute frequency of each point of the spectrum is known with an uncertainty of a $\sim 1 \cdot 10^{-11}$ (the short-term stability of the Rb frequency standard which fed KVARZ synthesizer and PLL phase-sensitive detector) multiplied by $N^{0.5}$, where N is a number of the frequency steps per second.

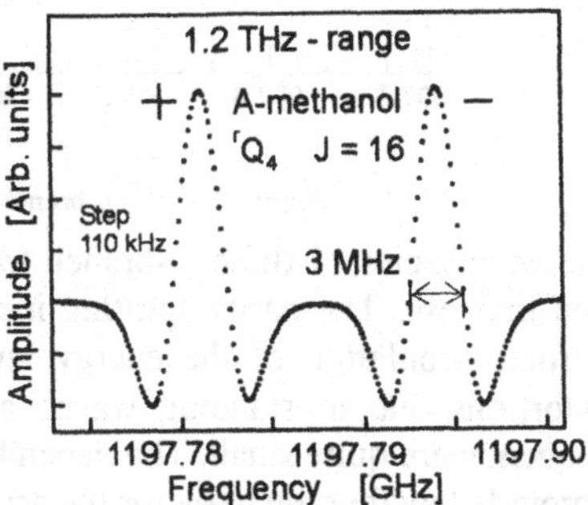

Figure 1. Step-by-step record of *K*-doublet of *A*-methanol at 1.2 THz .

For strong isolated lines without internal structure, the precision of the frequency measurements with the Cologne THz spectrometer in the Doppler limited mode is about ±5 kHz. This value was confirmed by studying the $^{12}C^{16}O$ rotational spectrum and by comparing the results of that study with Lamb-dip measurements which are known to better 1 kHz [11]. The absorption lines (Fig. 1) are recorded in the second derivative form with the 2*f*-detection mode of the lock-in amplifier. In the frequency region above 1 THz, continuous frequency scans of a few GHz can be achieved. Below 700 GHz scans can be done over the entire frequency band of a BWO excluding regions of strong absorption by atmospheric water lines

The advantages of the broadband tuning capabilities of the spectrometer and its high sensitivity have been very impressively demonstrated by a methanol spectrum study [12]. Methanol is a molecule with internal rotation and it is very abundant in interstellar space. Its rotational-torsional spectrum was for many years, and still remains, an area of very active research. Prior to our high-resolution submm-wave study, the methanol spectrum had been investigated up to 6 THz. "Atlas of Assigned Lines from 0 to 1258 cm^{-1}" [13] listed about 35 000 methanol transitions. However, between 600 GHz and 1250 GHz only a few lines were measured and assigned. With the Cologne THz spectrometer, three *R*-branches, *J*=12–11, 18–17, 22-21, and fourteen *Q*-branches (all in this frequency region) of methanol in torsional states V_t = 0, 1, 2 were measured in about one month. The *Q*-branches in V_t=2 state were observed for the first time. One of the most compact of them is presented in Fig. 2.

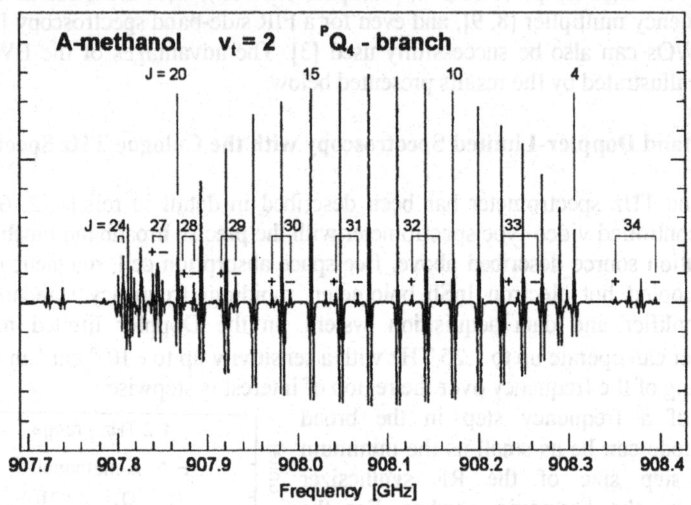

Figure 2. The $^PQ_4$-branch of *A*-methanol in V_t=2 torsional state .

The frequencies of these *Q*-branch transitions decrease with *J*, turn back at *J*=24 and then increase. The parity splitting increases with *J* and becomes visible at *J*=21. The thermal population of the energy levels is beautifully presented. The line intensity distortions due to standing waves and variations of BWO output power with the frequency are quite small. The dependencies of the line intensities and *K*-splittings with *J* provide together the basis for the assignment of all *K*-doublets up to *J*=34. These lines were not included in the fit [12] because of a large uncertainty of the predictions and because of the high *J*. Computing time increases drastically with *J*.

It is necessary to mention here that the predicted position of the *A*-type, V_t=2, $^PQ_4$-branch was about 5 GHz from where it was actually observed. The V_t=2 state is above the torsional barrier and predictions due is difficult, especially for the methanol molecule. All others newly measured *Q*-branches (except V_t=0) were shifted from predicted frequencies by a few GHz also. However, there were no problems finding and measuring these *Q*-branches with the Cologne THz spectrometer because of its broadband tunability and high sensitivity. Lack of such features have complicated the study of THz methanol spectra by other techniques.

The new data containing transitions with $J \leq 24$ and $|K| \leq 8$ have been added to the previously measured lines to form a global data set containing lines with the frequencies below 1.2 THz. The fit of that data set included 790 *A*–methanol lines and 753 *E*–methanol lines. It has provided the new set of spectroscopic parameters for methanol listed in [12]. In general, the low-order constants were not changed too much but the higher-order constants have changed considerably in a comparison with those previously known. The reason for this is that the previous data set did not include the lines that could not be well predicted. Measurements of such lines were restricted at THz frequencies both with a high-resolution technique because of limited scanning ability, and with a FTS because of inadequate sensitivity. The BWO-based source has overcome both of these problems and opened THz frequency region for precise broadband scanning high-resolution spectroscopy with high sensitivity.

4. Broadband sub-Doppler Saturation Spectroscopy with the Cologne THz Spectrometer

It is well known that Lamb-dip saturation spectroscopy greatly enhances both the resolution and the precision of the molecular transition measurements.

The Cologne THz spectrometer has really opened the submm-wave region for Lamb-dip measurements. Moreover, it can be used without any changes in its hard- and software for sub-Doppler saturation spectroscopy up to about 1 THz (Fig. 3).

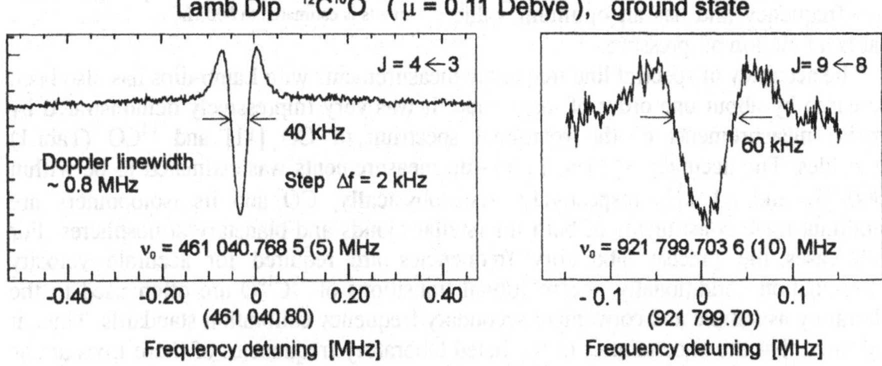

Figure 3. The Lamb-dips of the rotational transitions $J = 4 - 3$ and $J = 9 - 8$ of CO molecule recorded with the Cologne THz spectrometer equipped with the free-space absorption cell.

A small portion of the radiation which is reflected back to the absorption cell mainly from the bolometer is enough to saturate a strong molecular transition and to induce a Lamb-dip phenomena because of the large output power of the BWO and a high spectral purity of its radiation. Of course, it is necessary to reduce the pressure of a gas in the absorption cell, to decrease considerably the amplitude of the source frequency modulation, and to adjust carefully the optics and the beam of the BWO radiation. Apart from that, there are no other special requirements. Moreover, with a free-space absorption cell (which is not the best one for the induction of the Lamb-dip phenomena), the saturation dips have been observed above 900 GHz even for the CO molecule which has a small dipole moment of 0.11 D (Fig. 3). This means that spectra of a large number of molecules that have strong lines in the mm and submm-wave regions can now be studied with sub-Doppler resolution.

The resolution of the Cologne THz spectrometer in the sub-Doppler saturation mode is between one and two orders of magnitude better than in the Doppler limited mode. The experimental linewidth of the saturation dips can be varied from ~100 kHz to about 20 kHz.

The narrowest Lamb-dip with a linewidth of only 15 kHz was observed for the J=2–1 transition of CO at 230 GHz [11]. The increased resolution is demonstrated in Fig. 4 where the magnetic hyperfine structure of J=1–0, F_1=5/2–3/2 transition of $H^{35}Cl$ due to the hydrogen nuclear spin-rotation interaction is presented [14]. The linewidth of these Lamb-dips is about 25 kHz at 626 GHz, whereas the Doppler width is estimated as 1.3 MHz. A signal-to-noise ratio for the saturation dips depends upon both the gas pressure in the absorption cell and the frequency of the molecular transition (Fig. 3). Typically it decreases with frequency and has an optimum value that is a function of pressure.

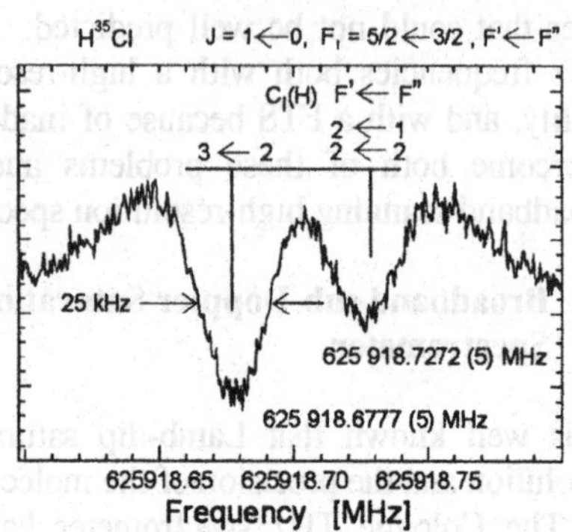

Figure 4. Magnetic hyperfine structure of J=1-0, F_1=5/2-3/2 transition of $H^{35}Cl$ molecule. The linewidth of the observed Lamb-dips is about 25 kHz. The accuracy of frequency measurements is estimated to 500 Hz.

The accuracy of spectral line frequency measurements with Lamb-dips has also been increased by about one order of magnitude. It was very impressively demonstrated by precise measurements of the rotational spectrum of CO [11] and ^{13}CO (Tabl.1) molecules. The accuracy of these Lamb-dip measurements was estimated to be within ±500 Hz and ±1 kHz respectively. Astrophysically, CO and its isotopomers are important trace constituents of both interstellar clouds and planetary atmospheres. For both cases the precise laboratory frequencies are required for accurate velocity determination. Additionally, the rotational transitions of $^{12}C^{16}O$ are often used in the laboratory as simple and convenient secondary frequency calibration standards. Thus, it is desired that the uncertainties of the listed laboratory frequencies of these lines are as small as possible and reflect the most recent technological advances. It is hard to believe, but the precise Lamb-dip frequencies of the first six rotational transitions of $^{12}C^{16}O$ fitted together with all other available data up to J=38–37 reduce the 1σ uncertainty of calculated frequency of J=1–0 transition down to 60 Hz [11]. It is an extremely high accuracy that was not available until that study. The uncertainties of the other listed frequencies increases with J but still does not exceed 1 kHz for transitions below J=12–11. The recent Lamb-dip measurements of J=9–8 transition with the Cologne THz spectrometer has provided an additional check of the accuracy of the listed frequencies. The measured frequency 921 799.703 6(10) MHz (Fig. 3) is in excellent agreement with 921 799.704 2(4) MHz listed in [11].

The precise measured frequencies of ^{13}CO transitions are given in Tab.1. The magnetic hyperfine structure due to the ^{13}C nuclear spin-rotational coupling is resolved for the five low-J transitions. The strongest components of the structure are separated by only ~35 kHz. The details of the measurements are given in [15]. It is interesting to mention that some ^{13}CO lines beyond 1.2 THz have been measured with the BWO frequency multiplier and BWO sideband spectrometer [10]. As it can be seen from the "Obs.-Calc." values (Tab.1), the accuracy of measurements with the last one is very high.

TABLE 1. Sub-Doppler and Doppler-Resolved Rotational Transitions of $^{13}C^{16}O$ [15].

J'	F'	←	J''	F''	Obs. Frequencies [a] [MHz]	Obs.-Calc. [kHz]	Rel. Int.
1		←	0		110 201.3541(51)[b]	0.2	
2	1.5	←	1	1.5	220 398.619(10)[c]	5.2	0.067
2	1.5	←	1	0.5	220 398.6635(20)[c]	-2.5	0.333
2	2.5	←	1	1.5	220 398.6998(20)[c]	-1.1	0.600
3	2.5	←	2	1.5	330 587.9470(20)[c]	0.1	0.400
3	3.5	←	2	2.5	330 587.9802(20)[c]	-1.5	0.571
4	3.5	←	3	2.5	440 765.1556(20)[c]	0.1	0.429
4	4.5	←	3	3.5	440 765.1903(20)[c]	0.6	0.556
5	4.5	←	4	3.5	550 926.2663(20)[c]	-0.1	0.444
5	5.5	←	4	4.5	550 926.3016(20)[c]	0.4	0.546
6	5.5	←	5	4.5	661 067.2586(20)[c]	0.6	0.455
6	6.5	←	5	5.5	661 067.2936(20)[c]	0.8	0.539
7		←	6		771 184.125(5)[c]	0.6	
8		←	7		881 272.808(5)[c]	-0.6	
9		←	8		991 329.305(5)[c]	-1.6	
11		←	10		1 211 329.636(50)[d]	-26.9	
12		←	11		1 321 265.42(10)[e]	-63.2	
14		←	13		1 540 988.23(10)[e]	-91.5	
15		←	14		1 650 767.344(55)[d]	35.0	
17		←	16		1 870 140.359(25)[f]	3.7	
18		←	17		1 979 726.393(15)[f]	0.2	
19		←	18		2 089 240.033(55)[d]	-62.4	
25		←	24		2 744 579.059(60)[d]	-32.4	
26		←	25		2 853 474.444(60)[d]	40.1	
28		←	27		3 070 948.140(70)[d]	56.2	
30		←	29		3 287 972.525(100)[d]	-82.4	

[a] For unresolved hyperfine structure the F values and relative intensities are omitted. In these cases the calculated frequencies were obtained by using intensity weighted averages of individual hyperfine structure components.
[b] Observed frequency taken from Winnewisser *et al.*
[c] Frequencies measured with the Cologne terahertz spectrometer.
[d] Frequencies taken from Zink *et al.*
[e] Frequencies were recorded by multiplying the BWO output.
[f] Frequencies measured with the Cologne sideband spectrometer.

5. Broadband High-Resolution Spectroscopy with the FASSST Spectrometer

Fast Scan Submillimeter Spectroscopy Technique (FASSST) has been designed primarily for the study of mm and submm high-resolution molecular spectra containing many thousands of lines. It allows scanning over tens of GHz in a few seconds and obtaining spectra with Doppler limited resolution and very good sensitivity (Fig. 5, time constant 1 μs). FASSST is a fundamentally simple and powerful spectrometer, similar in some aspects to FTIR spectrometer, which has had an enormously wide area of application.

It is a video spectrometer with a free-running BWO, liquid-He cooled hot electron InSb bolometer and "optical" scheme of calibration of the frequency scale. This scheme includes a non tunable long-path Fabry-Perot (FP) cavity with ~5 MHz mode spacing, a known reference spectrum and sophisticated software. During the BWO frequency scan, the computer samples four different signals at a 1 MHz rate. These are the investigated and reference spectra, FP cavity fringes and BWO cathode voltage. The scan can be made over the entire frequency range of the BWO or any part of it. When the scan is finished, the software automatically does the following: 1) finds all lines exceeding a set amplitude level in the both of spectra; 2) recognizes and assigns frequencies of the reference spectrum lines; 3) determines the frequency spacing of the FP cavity fringes from the absolute frequency of the reference lines and number of the fringes; 4) calculates the absolute frequency of each observed line of the investigated spectrum by linear interpolation of the frequency between two adjoining fringes, and 5) lists the frequency and amplitude of each measured line [3]. Thus, the spectrum can be measured and listed practically automatically.

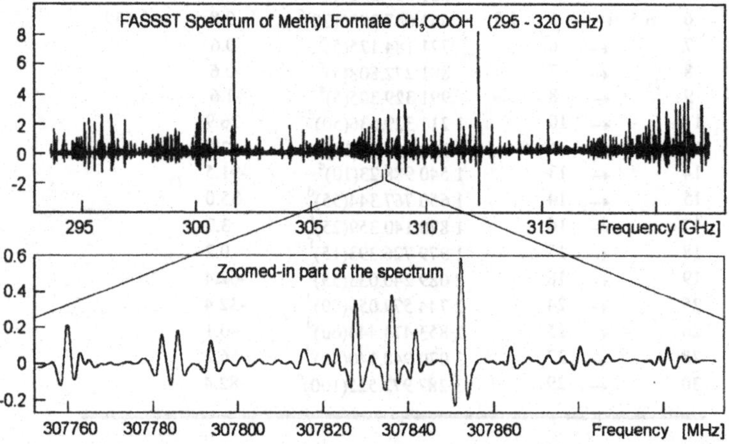

Figure 5. **An example of the high-resolution FASSST spectrum. A single 30 GHz frequency scan of the Methyl Formate spectrum has been taken in about 3 sec.**

A small fraction of the methyl formate ($C_2H_4O_2$) FASSST spectrum is presented in Fig. 5. The upper trace includes 1950 lines that have been taken with a single 30 GHz scan in ~ 3 seconds. As can be seen from the lower trace (0.01 part of the upper spectrum) it is a high-resolution spectrum. The noise is not seen, there are only lines presented in a second derivative form. For strong absorption lines signal-to-noise ratios up to 10^4 can be obtained with an integration time of 1 μs (1 MHz of bandwith), which is typical for FASSST. The accuracy of the line position determination depends mainly on the quality of the FP fringes and linearity of the frequency scan from one FP mode to the next one. Because of BWO is very sensitive to voltage variations (~ 50–70 MHz/V) the linearity of the scan can deteriorate. This may be caused by the variations of the BWO cathode voltage due to the ripples and switching effects of the BWO power supply, the AC filament current, picked up voltages and so on. These variations have to be effectively "frozen" by the fast scan [3]. With a tuning rate of 10 GHz/s and a FP mode spacing of 5 MHz, the accuracy of the frequency measurements with the FASSST ranges typically within 50–100 kHz for well resolved lines in the frequency region below ~ 600 GHz. This accuracy can serve very well for many practical applications of the FASSST [16].

FASSST provides full utilization of the properties of a BWO as an electronic generator, especially its tunability, bandwidth, high spectral purity and brightness. The ability of FASSST to survey high-resolution broadband spectra in a short time and with a high sensitivity is very useful for laboratory spectroscopy in the mm and submm-wave regions. FASSST reduces considerably the time spent searching for new lines and groups of the lines which cannot be predicted very well as was the case for the methanol spectrum, Fig. 2, and/or depend critically on the conditions of the experiments (discharge, photolysis, chemical reaction etc.). Scanning a phase-locked BWO over its entire frequency band requires too much time, while the accuracy of the line position determinations with a uncalibrated, free-running BWO is not accurate enough for the real needs of spectroscopy (loop analysis, for example). FASSST is a suitable device for the study of spectra of the large molecules with many thermally populated modes, internal rotors or large amplitude motions. It can be adapted for fast measurements of low-resolution spectra and applied to atmospheric research as well.

6. Spectroscopy With Radio Acoustic Detector: Lines Shape, Line Shift, Line Broadening

Precise and reliable laboratory measurements of the spectral parameters (line positions, linewidths, intensities) of minor atmospheric constituents and dependencies of these parameters on pressure, temperature, and other components of the gas mixture are very important for accurate retrievals of the atmospheric profiles (temperature, pressure, concentrations) from the remote sensing data.

The spectroscopic techniques described above allow one to measure all of these parameters in submm-wave region. An excellent example of a lineshape study is that of the $J = 3 - 2$ rotational transition of CO in collision with N_2 and O_2 at different temperatures from 273 K to 393 K [17]. In this study a computer controlled video-type spectrometer with a phase-locked BWO was employed. The true line profile measurements undertaken with that spectrometer (which is very similar to the Cologne THz spectrometer) has shown that its sensitivity and accuracy are high enough to see clearly difference between observed line shape and calculated one with a Voigt profile. Thus, different lineshape theories can be tested and compared to each other by fitting the experimental data. It means that experimental accuracy in favorable cases is not the limiting factor for lineshape studies with the video-type spectrometers, but theory. On the other hand, the baseline distortions due to reflections of radiation (mainly from bolometer and windows of the absorption cell) result in an important limitation. Accurate and reliable data can be obtained with a video-type spectrometer only in the low-pressure range, typically below a few tenths of Torr [17]. It is a very difficult task to avoid both the reflections from the bolometer and the baseline distortions in submm-wave region.

The pressure range of lineshape measurements can be extended to about 10 Torr with an another type of spectrometer, called RAD (Radio Acoustical Detection), which is described in Refs. [18, 19]. The absorption cell of the RAD spectrometer is equipped with a highly sensitive microphone. A resonance absorption by the gas of the chopped (or modulated in frequency) BWO radiation induces a periodic pressure variation (acoustic signal) inside the cell, which is registered by microphone. If the radiation power is not absorbed by the gas, there is no signal to the microphone. It means that there is no baseline in RAD spectrometer.

RAD has successfully been used for study of pressure shifts [20, 21] and pressure broadenings [22, 23] of spectral lines in submm-wave region. These observed with RAD pressure shifts for the $s(1,0) \leftarrow a(0,0)$ transition of $^{14}NH_3$ at 572.5 GHz by different molecules (H_2O, N_2O, PH_3, OCS, NH_3, C_6H_5Cl, C_3H_7ON) are presented in Fig. 6. The results of these measurements demonstrate not only existence of the large pressure shifts (up to 10 MHz/Torr) of this spectral line, but also the fact that the value of the shift may vary strongly or even change sign depending on the perturbing molecule.

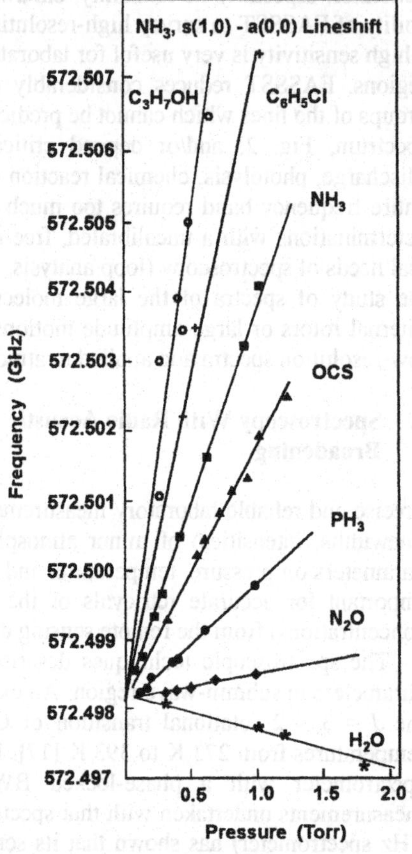

Figure 6. The frequency dependence of the $s(1.0) \leftarrow a(0.0)$ transition of ammonia $^{14}NH_3$ molecule on the pressure of various perturbing gases.

At the present time, the RAD is equipped with a Baratron pressure gauge for precise pressure broadening and line shape measurements [24]. As an example, the true line profile measurements of the rotational transition J= 5–4 of CO with the acoustic cell and phase-locked BWO are presented in Fig. 7. The measured line shape is fitted to Voigt profile. The maximum value of the residual (*obs.– calc.*) shown in the lower part of Fig. 7 is only ~0.45% of the peak amplitude of the line, but still clearly demonstrates an inadequacy of the fitting model to the observed line profile. The linewidths of the J=5–4 transition of CO retrieved from true line shape measurements at 15 different pressures of pure CO are presented in Fig. 8. Very small deviations of the data from the linear pressure broadening dependence confirm that the accuracy of the lineshape measurements with RAD is very high. Both the pressure shift and pressure broadening parameters of mm and submm lines can now be measured with RAD to an accuracy of a few kHz per Torr [23, 24].

The sensitivity of the RAD spectrometer is quite high, typically ~10^{-8} cm^{-1} (time constant 1 s). It is important to mention that the RAD's sensitivity depends on the radiation power and sensitivity of absorption cell's built in microphone, but it does not necessarily depend from the length of the cell. So, RAD absorption cell can be only a few centimeters long. The period of a standing wave into this cell will be a few GHz. Additionally, the amplitude of that standing wave can be considerably suppressed by a special shape of the cell windows (there is no bolometer or any other detector behind the cell). In comparison with a common video-type spectrometer these factors together allow the RAD spectrometer to reduce the influence of the standing wave on the line shape, to extend the range of available pressures of lineshape measurements by an order of the magnitude, and to increase accuracy of the pressure broadening and especially of the pressure shift measurements. The gas temperature in an acoustic cell can be varied and temperature dependence of the lineshape parameters can be measured as well [25]. A new design of the acoustic cell and the microphone allows one to vary temperature from liqud-N_2 temperature up to 400–450 K [26].

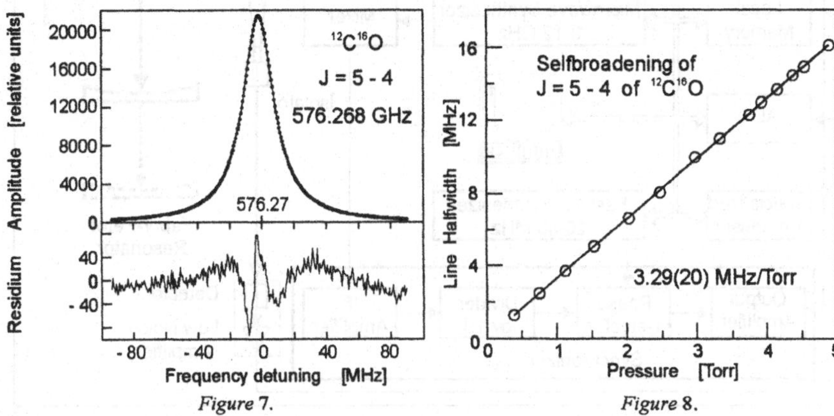

Figure 7. *Figure* 8.

Fig. 7. An example of spectral line recorded with the acoustic cell in the true line profile. The residual of the data fit to Voigt profile (solid line) is presented below. The line is observed in a mixture of carbon monoxide with nitrogen gases under total pressure of 4.238 Torr.

Fig. 8. The pressure broadening of spectral line J =5 – 4 of CO at 576.268 GHz studied with acoustic detector. Broadening parameter of the transition due to the CO – CO collisions at room temperature is determined as 3.29 (20) MHz/Torr [27].

7. Wide Range High Sensitive Resonator Spectroscopy of Broad Lines

The capabilities of high and super-high resolution microwave spectrometers operating with gases at low pressures are complemented by resonator spectrometers with which spectral lines are detected through variation of width of "probe resonance" over the line profile [28, 29]. The principle of resonator spectrometer operation can be explained as follows: a gas sample placed in the resonator decreases its quality. The measurement of radiation absorption by the sample becomes a measurement of the quality factor of the resonator that can be done by measurement of width of resonance response. The sensitivity of such a spectrometer improves with both the accuracy of the resonance width measurement and the quality factor of the resonator. Due to indirect measurement of spectral line amplitude, resonator spectrometers do not have the problems associated with traditional microwave video-spectrometers resulting from baseline variations which limit the range of maximum working pressures of the sample to the level of below 1 Torr. Use of a resonator spectrometer allows the study molecular lines in gases at pressures up to one atmosphere and higher. Another advantage of a resonator spectrometer of this type is that the measurement of radiation absorption can be made in absolute units including non-resonant absorption measurement at any frequency point.

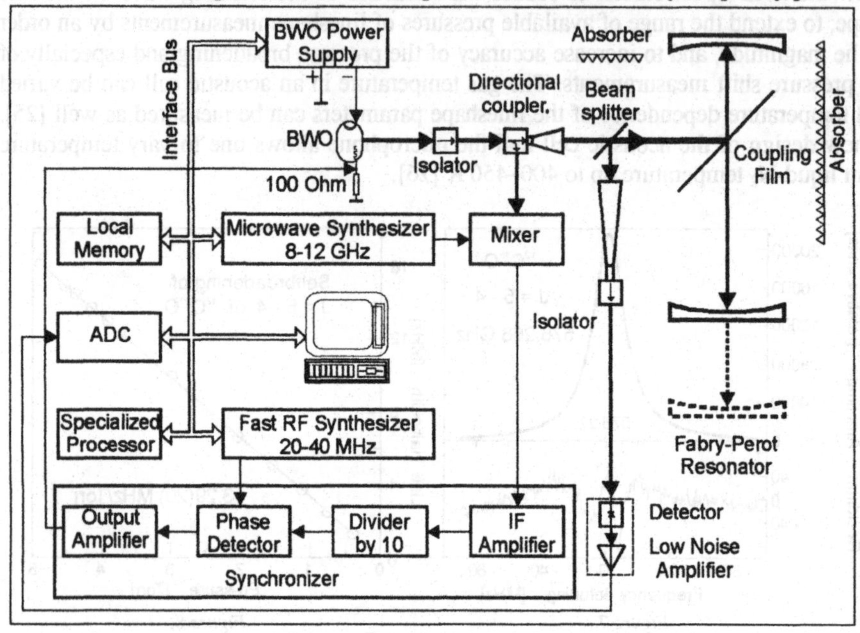

Figure 9. Block-diagram of resonator spectrometer.

A block diagram of the resonator spectrometer [29] is presented in Fig. 9. The synthesized frequency radiation source is based on a backward-wave oscillator (BWO) tube phase locked to a harmonic of a microwave (8 – 12 GHz) synthesizer. An essential property in comparison with radiation sources of the other aforementioned spectrometers is the introduction here of the possibility of really fast digital frequency change by any requested law within 0.2 GHz range. This advantage was achieved of by use of a specially developed synthesized radio frequency source with discrete steps of 0.03 Hz and a switching time of about 200 ns without phase jumps. The source is based on 0 – 50 MHz direct digital synthesizer microchip. The signal from the source is used as a reference for the phase detector of the BWO PLL, so that the tube output radiation frequency follows the function of the radio source frequency change. Fast digital back and forth frequency sweeping of radiation around the frequency of the peak frequency of the resonator response is very important to minimize resonance width measurement error associated with a drift of the resonance central frequency during the time of measurement. The precision radiation frequency control, signal acquisition and processing were done by computer. Results of each fast frequency scan were recorded and processed separately.

A Fabry-Perot resonator is the sensor for the spectrometer. Special efforts were undertaken to achieve a quality factor as high as possible. Diffraction, reflection and coupling losses were minimized. A quality factor of about 600 000 was achieved for a cavity length of about 30 cm.

This set-up combines high-enough power, fast sweeping, precisely frequency controlled and broad-range a radiation source with a resonator having the highest possible quality factor for a cavity with non-cooled mirrors.

The basic procedure in a resonator spectrometer is measurement of the width of Fabry-Perot resonance. An experimentally observed resonance curve of the Fabry-Perot resonator near 85 GHz is presented in Fig. 10. The curve is an accomulation of 500 fast scans separately recorded one after another, then centered and coadded. The resulting curve was fitted then to a Lorentzian profile. The residuals of the fit, multiplied by 100, are presented in lower part of the figure. They indicates the high quality of the data and adequacy of the resonance model. The width of the resonance was defined from the fit as 164 728 (20) Hz. A resonance width measurement accuracy of 20 Hz was reached for the first time. It corresponds to a radiation absorption coefficient sensitivity limit of the spectrometer of $4{\cdot}10^{-9}cm^{-1}$ or 0.002 dB/km which exceeds parameters of other previously known analogs in more than order of magnitude.

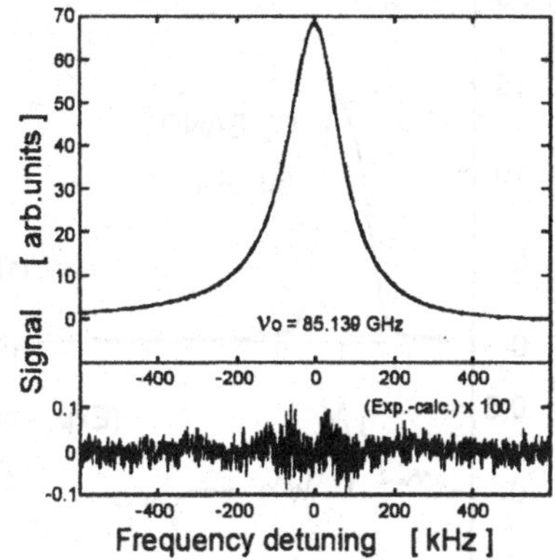

Figure 10. Observed curve of Fabry-Perot resonator and result of its fitting to Lorentzian profile.

The capabilities of the spectrometer were demonstrated with *in situ* analysis of laboratory air. Scans covering 44 – 98 GHz and 113 – 200 GHz were recorded in one experiment for each range. The records are presented in Fig. 11. Broad band frequency scanning in these experiments was produced without mechanical tuning of the resonator. Changing the source frequency from one longitudinal mode of the resonator to another was only necessary. The circles represent experimental points, and the solid line in lower frequency range corresponds to calculation according to an atmosphere model MPM-92 [30]. The solid line in the higher range is the Van Vleck-Weisskopf shape with the addition of linear and quadratic with frequency terms fitted to the experimental points. Differences between observed and calculated spectra multiplied by ten are presented below.

An analysis of the lines of atmospheric oxygen and atmospheric water observed in the experiment was performed. It was found that the water line at 183 GHz at atmosphere pressure fits the Van Vleck-Weisskopf shape over twenty halfwidths down and six halfwidths up from the line center within experimental accuracy. The experiment yielded a value of 3.985(40) MHz/Torr for the dry air broadening parameter for this water line, which is more accurate than other previously known values.

The experimentally measured integral absorption coefficient for the water line coincides within 2% of the theoretically calculated integral absorption for amount of water in atmosphere under the experimental conditions. This proves the ability of the method to measure spectral line absorption coefficients.

The oxygen line at 119 GHz was for the first time experimentally investigated by microwave method directly in an air at the atmospheric pressure. It also fits well to the

Van Vleck-Weisskopf shape within 10 halfwidths. The dry air broadening parameter of the line was determined to be 2.14(9) MHz/Torr [31].

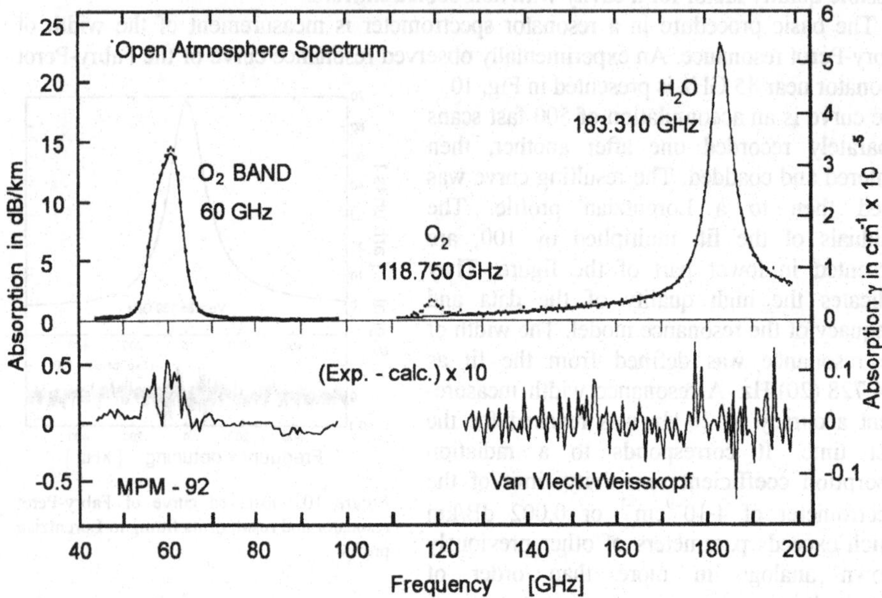

Figure 11. Experimental and calculated spectra of laboratory atmosphere inside open Fabry-Perot resonator.

In addition to broad-band spectral records, the spectrometer can perform the absolute measurement of radiation absorption at any frequency point within the working range by use of the method of resonator length variation. The essence of the method is the measurement of widths of two consequential modes of the resonator corresponding to two different resonator lengths (or measurement of widths of a series of resonances obtained by varying the resonator length), keeping the central frequency of the radiation constant [29]. Then the radiation absorption coefficient can be unambiguously calculated from the width values and known mode numbers, the resonator length, the radiation frequency and the radius of curvature of the resonator mirrors. Absorption in the 140 GHz atmospheric window was studied. A measurement accuracy of 0.0027 dB/km is reached. The parameters achieved are sufficient to allow the system to be used for study of the old question of the nature of the so-called nonresonant atmospheric absorption.

The system could be applied for real time monitoring of the atmosphere and technological processes monitoring. In particular it is possible to develop, on the basis of the spectrometer and using existing solid state radiation source technology, commercial devices such as absolute humidity meter and oxygen (or other gas of interest) meter with absolute self calibration.

Although the experiments presented in this paragraph were done in the millimeter-wave range, similar techniques exists up to frequencies exceeding 1 THz, so similar measurements are affordable in the whole millimeter and submillimeter wave bands.

8. Further Developments

The systems described above have been developed as general purpose laboratory spectrometers for study of high resolution spectra in the mm and submm wave regions. They have an excellent resolution, an extremely high accuracy of frequency measurements, a very good sensitivity and can operate in the frequency region up to 1.2 THz. Nevertheless, some parameters of these spectrometers can be further improved, and their applications can be expanded. Some future developments are pointed out, for example, in [1] and [3]. A few more technical improvements, which can be done in a short time if granted, are listed below:

8.1 None Reflection Bolometer

A magnetically tuned liquid-He cooled InSb hot electron bolometer is widely used as a highly sensitive, broadband and quite fast detector of a submm-wave radiation. Despite its excellent haracteristics, it has a drawback: reflection. A part of the submm-wave radiation focused to the bolometer is reflected back to the absorption cell and produces a standing wave. This has been used for a sub-Doppler saturation spectroscopy (section 3). However in the most cases this standing wave produces negative effects. It results in an important limitation for the lineshape measurements (section 6), decreases accuracy of the line center frequency measurements and sensitivity of the video-type spectrometers. The development of a bolometer without (or with a considerably lower level) of the reflection back to the absorption cell, will allow to improve performance. A preliminary discussion has shown that it is possible and that an existed bolometer can be improved and adapted to the spectroscopy needs [32].

8.2 Absolute Intensity Calibration of the FASSST Spectra

As it was pointed out in [3], amplitude calibration of the FASSST spectra is achieved by linear interpolation between reference spectral lines of known intensity. This procedure does not provide a good accuracy and is not reliable because of the assumption that the power of a BWO radiation is a constant. A calibrated scale for each spectral element of the FASSST spectrum will improve the performance of this spectrometer: i) the accuracy and especially reliability of its absolute intensity measurements for spectral lines will be improved considerably; ii) the area of its application can be expanded from the study of the high-resolution spectra to include the low-resolution one. The latter possibility can be important for the atmosphere research in the mm and submm-wave regions.

8.3 Intensity Measurements with the RAD Spectrometer

RAD spectrometer has a linear amplitude scale, a high sensitivity and does not have a base line. It makes this spectrometer suitable not only for lineshape measurements, but also for the absolute intensity measurements, especially of weak lines.

8.4 Line Shape Measurements with a Fast Digital Scan

Fast digital scan of a phase-locked BWO (section 7) improved considerably the accuracy of the shape measurements of a Fabry-Perot cavity mode and increase the sensitivity of resonator spectroscopy of broad lines. The technique of the fast digital scan can be applied directly to the lineshape measurements with video-type spectrometers. This would reduce considerably the time of the measurements and can improve their quality.

8.5 Sensitivity of Lamb-dip Measurements

The scheme of Lamb-dip measurements with the Cologne THz spectrometer is not yet optimized. Its optimization could improve a signal-to-noise ratio of the saturation resonances and allow them to be observed for relatively weak lines. It is possible that the line position accuracy of Lamb-dip measurements can also be improved.

9. Acknowledgements

The authors are grateful to Edward Cohen and Brenda Winnewisser for reading the manuscript and for their valuable comments, to Svetlana Tretyakova for help with a preparation of the camera-ready typescript, to John Bevan and Andrey F. Krupnov for constructive assistance and to organizers of NATO Advanced Research workshop "Spectroscopy from Space" for publishing this review paper. Studies described in this paper were supported in part by the Deutsche Forschungsgemeinschaft (DFG) via grant SFB 301, the Air Force Office of Scientific Research, the Army Research Office and NASA, the Russian Fund for Fundamental Studies via grant 00-02-16604, joint DFG-RFBR grant No. 00-03-04001 and by State Program "Fundamental Metrology". The authors express their deep gratitude to all of this sources of support. S.P.B. would also like to thank Gisbert Winnewisser and Frank C. De Lucia for the opportunity to work in their laboratories.

10. References

1. Krupnov, A.F. (1996) Present State of Submillimeter Wave Spectroscopy at the Nizhnii Novgorod Laboratory, *Spectrochimica Acta, Part A*, **52**, 967–993.
2. Winnewisser, G. (1995) Spectroscopy in the Terahertz region, *Vibrational Spectroscopy*, **8**, 241–253.
3. Petkie, D.T.,Goyette,T.M., Bettens, R.P., Belov, S.P., Albert, S., Helminger, P., De Lucia, F.C. (1996) A fast scan submillimeter spectroscopic technique, *Rev. Sci. Instrum.*, **68** (4), 1675–1682.
4. Bogey, M., Civis, C., Delcroix, B., Demuynck, C., Krupnov, A.F., Quiguer, J.,Tretyakov, M.Yu., Walters, A. (1997) Microwave Spectrum up to 900 GHz of SO Created in Highly Excited States by Electric Discharge and UV–Laser Photolysis, *J. Mol. Spectrosc.*, **182**, 85–97.
5. Morino, I., Fabian, M., Takeo, H., Yamada, K.M.T. (1997) High-J Rotational Transitions of NNO Measured with the NAIR Terahertz Spectrometer, *J. Mol. Spectrosc.*, **185**, 142–146. Amano,T.,Maeda, A. (2000) Double modulation submillimeter-wave spectroscopy of HOC^+ in the ν_2 exited vibrational state, *J.Mol.Spectrosc.*, **203**, 140–144.

6. Belov S.P., Lewen, L., Klaus, Th.,Winnewisser, G. (1995) The $^{r}Q_4$ Branch of HSSH at 1.25 THz, *J. Mol. Spectrosc.*, **174**, 606–612.

7. Gendrisch, R., Pak, I., Lewen, F., Surin, L, Roth, D.A., Winnewisser, G. (1999) Submillimeter Detection of the van der Waals Stretching Vibration of the Ar-CO Complex, *J. Mol., Spectrosc.*, **196**, 139–145.

8. Tretyakov, M.Yu., Krupnov, A.F., Volokhov, S.A. (1995) Extension of the Range of Microwave Spectroscopy Up To 1.5 THz, *JETP Letters* (Rus), **61**(1), 75–77 (79–82 Engl. transl.).

9. Lewen, F., Belov, S.P., Maiwald, F., Klaus, Th., Winnewisser G. (1995) A quasi-optical multiplier for terahertz spectroscopy, Zeitschrift fur Naturforschung Section A-A Journal of Physical Sciences, **50**: (12) 1182–1186.

10. Lewen, L., Michael, E., Gendriesch, R.,Stutzki, J., Winnewisser, G. (1996) Terahertz Laser Sideband Spectroscopy with Backward Wave Oscillators, *J. Mol. Spectrosc.*, **183**, 207–209.

11. Winnewisser, G., Belov, S.P., Klaus, Th., Schieder, R. (1997) Sub-Doppler Measurements of the Rotational Transitions of Carbon Monoxide, *J. Mol. Spectrosc.*, **184**, 468–472.

12. Belov, S.P., Winnewisser, G, Herbst, E. (1995) The High-Resolution Rotational-Torsional Spectrum of Methanol from 0.55 to 1.25 THz, *J. Mol. Spectrosc.*, **174**, 253–269.

13. Moruzzi, G., Winnewisser, B. P., Winnewisser, M., Mukhopadhyay, I., Strumia, F. (1995) Microwave, Infrared, and Laser Transitions of Methanol: Atlas of Assigned Lines from 0 to 1258 cm^{-1}, CRC Press, Orlando.

14. Klaus, Th., Belov, S.P., Winnewisser, G. (1998) Precise Measurements of the Pure Rotational Submillimeter-Wave Spectrum of HCl and DCl in Their V=0,1 States, *J. Mol. Spectrosc.*, **187**, 109–117.

15. Klapper, G., Lewen, F., Gendriesch, R., Belov, S.P., Winnewisser, G. (2000) Sub-Doppler Measurements of the Rotational Spectrum of $^{13}C^{16}O$, *J. Mol. Spectrosc.*, **201**, 124–127.

16. Albert, S., Petkie, D.T., Bettens, R.P.A., Belov, S.P., De Lucia, F.C. (1998) FASSST: A new gas-phase analytical tool, *Analytical Chemistry*, **70**: (21) 719A–727A.

17. Priem, D., Rohart, F., Colmont, J.-M., Wlodarczak, G., Bouanich, J.-P. (2000) Lineshape study of the $J = 3 - 2$ rotational transition of CO perturbed by N_2 and O_2, *J. Mol. Structure*, 517–518, pp. 435–454.

18. A.F. Krupnov, A.V. Burenin, in *Molecular Spectroscopy: Modern Research*, K.N. Rao, Ed, Academic Press, N.Y, pp. 93 - 126 (1976).

19. A.F. Krupnov, in *Modern Aspects of Microwave Spectroscopy*, G.W. Chantry, Ed, Academic Press, L, pp. 217–256 (1979).

20. Belov, S.P., Kazakov, V.P., Markov, V.N., Melnikov, A.A., Skvortsov, V.A., Tretyakov, M.Yu. (1982) The Study of Microwave Pressure Lineshifts, *J. Mol. Spectrosc.*, **194**, 264–262.

21. Belov, S.P., Krupnov, A.F., Markov, V.N., Melnikov, A.A., Skvortsov, V.A., Tretyakov, M.Yu. (1983) Study of Microwave Pressure Line Shifts: Dynamic and Isotopic Dependencies. *J. Mol. Spectrosc.*, **101**, 258–270.

22. Belov, S.P., Krupnov, A.F., Markov, V.N., Tretyakov, M.Yu. (1984) Study of shift and broadening of lower inversion-rotation transition of $^{14}NH_3$ and $^{15}NH_3$ molecules, *Optics and spectroscopy* (in Russian), 56, No.5, 828–832.

23. Belov, S.P., Tretyakov, M.Yu., Suenram, R.D. (1992) Improved Laboratory Rest Frequency Measurements and Pressure Shift and Broadening Parameters for the *J*=2-1 and *J*=3-2 Rotational Transitions of Carbon Monoxide, *The Astrophysical Journal*, 393, N2, 848–851.

24. Nissen, N., Doose, J., Guarnieri, A., Maeder, H., Markov, V.N., Golubyatnikov, G.Yu., Leonov, I.I. Shanin, V.N. (1999) Foreign Gas Broadening Studies of the *J'-J*=1-0 rotational Line of CO by Frequency and Time Domain Technique, *Zs.Naturforschung* A, 54a, 218–224.

25. Markov, V.N. (1994) Temperature Dependence of Self - Induced Pressure Broadening and Shift of the 643–550 Line of the Water Molecule, *J. Mol. Spectrosc.*, **164**, 233–238.

26. Markov, V.N., private communication.

27. Markov, V.N., Golubyatnikov, G.Yu., Savin, V.A., Sergeev, D.A. (2000) Private communication.

28. Krupnov, A.F., Tretyakov, M.Yu., Parshin, V.V., Shanin, V.N., Kirillov, M.I. (1999) Precision Resonator Microwave Spectroscopy in Millimeter and Submillimeter Range, *Int. J. of IR and MM Waves*, Vol. 20, No. 10, 1731–1737.

29. Krupnov, A.F., Tretyakov, M.Yu., Parshin, V.V., Shanin, V.N., Myasnikova, S.E. (2000), Modern Millimeterwave Resonator Spectroscopy of Broad Lines, *J. Molec. Spectrosc.*, **202**, 107–115.

30. Liebe, H.J. (1985) An updated model for millimeter wave propagation in moist air, *Radio Science*, Vol.20. No. 5, 1069–1089. Liebe, H.J., Rosenkranz, P.W., Hufford, G.A. (1992) Atmospheric 60-GHz Oxygen spectrum: new laboratory measurements and line parameters, *J. Quant. Spectr. Radiat. Transfer*, **48**, 629–643. Rosenkranz, P. W. (1998) *Radio Science*, **33**, 919–928 & correction, (1999) *Radio Science*, **34**, 1025.

31. Tretyakov, M.Yu., Parshin, V.V., Shanin, V.N., Myasnikova, S.E., Krupnov, A.F. (In preparation) Laboratory Measurements of 118.75 GHz Oxygen Line in Real Atmosphere, *J. Molec. Spectrosc.*

32. Belov, S.P., Wood, K., Bevan, J. (2000) An adaptation of the QMC bolometer to needs of submm-wave spectroscopy (private communication).

ASSIGNMENT AND ANALYSIS OF COMPLEX ROTATIONAL SPECTRA

Z. KISIEL
Institute of Physics, Polish Academy of Sciences
Al. Lotników 32/46, 02-668 Warszawa, Poland,

Abstract. Rotational spectra of even simple molecules can pose serious interpretational difficulties, especially in the high-J conditions of spectroscopy in the environmentally relevant mm-wave region. In recent years several useful techniques for dealing with such spectra in the laboratory have become apparent. These include the choice of the most suitable experimental tools, more detailed understanding of the characteristic features of high-J rotational spectra of asymmetric top rotors, more confident prediction of centrifugal distortion and nuclear quadrupole splitting terms in the Hamiltonian, and the use of more powerful computer programs.

1. Introduction

Laboratory rotational spectroscopy is essential for understanding gas-phase environmental spectra, and it is also the source of key data for analytical applications in environmental spectroscopy. It is perhaps surprising that rotational spectra of even rather simple and rigid molecules of environmental importance can often pose severe interpretational difficulties. These may arise from lack of characteristic spectral features due to low molecular symmetry, overlaps of transitions allowed by several dipole moment components and of transitions in low lying vibrational states. In the mm-wave region of the spectrum, which is currently of particular atmospheric and astrophysical relevance there is the possibility of overlaps of transitions from plethora of transition rules and a broad range of quantum numbers. The presence of quadrupolar nuclei in the molecule may also introduce non-negligible nuclear quadrupole structure. In the room-temperature rotational spectrum of even some halogenomethanes, such as $CBrClF_2$ and CH_2I_2, such issues

J. Demaison et al. (eds.), Spectroscopy from Space, 91–106.

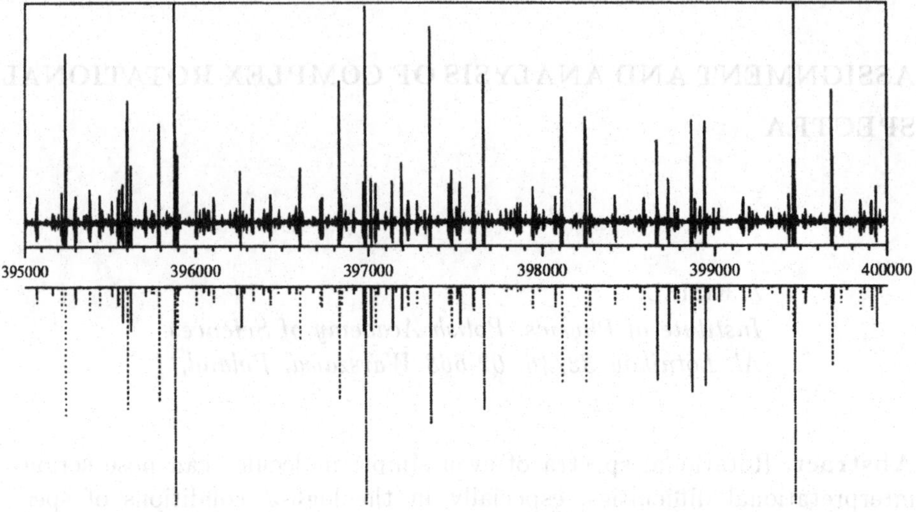

Figure 1. The observed (top) and predicted (bottom) room temperature spectrum of the HCFC-22 freon, CHF_2Cl in natural abundance. The experimental spectrum is a 5 GHz portion out of a single 10 GHz recording. The prediction encompasses the common isotopic species (dotted lines) and the ^{37}Cl isotopomer (continuous) lines and includes transitions in the ground states, and the low lying vibrational states $v_5 = 1$, $v_6 = 1$, and $v_9 = 1$.

precluded assignment with the popular Stark modulation spectrometers working in the 8-40 GHz region.

Over the recent years several molecules of this type have been successfully studied in the author's laboratory in Warsaw, and also in cooperation with several other research groups. These include the well known halon $CBrClF_2$ [1, 2], the freon HCFC-22 (CHF_2Cl) [3], trichloroethylene $Cl_2C{=}CHCl$ [4, 5] and 1,1,1-trichloroethane Cl_3CCH_3 [6, 7, 8]. Both serendipity and design led to the development of useful guidelines and techniques for facilitating such studies. The culmination of judicious application of these techniques has been the successful analysis of the rotational spectrum of methylene iodide, first of the normal isotopomer CH_2I_2[9], then of its deuterated isotopomer, CD_2I_2 [10], and finally of the extensive vibrational satellite structure [11].

2. The choice of a suitable spectrometer

Laboratory rotational spectroscopist has a powerful array of tools at his disposal. Currently some of the most useful appear to be broadband millimetre-wave rotational spectroscopy of a static sample, free-jet millimetre wave spectroscopy, and cavity Fourier transform spectroscopy of supersonic expansion. There are also many operational Stark modulation spectrometers for the region around 8-40 GHz, as well as several waveguide Fourier transform spectrometers operating at similar frequencies. In cases of complex spectra it is best to apply these techniques in an appropriate sequence to take advantage of a certain 'separability' of the spectroscopic problem, so that various parts of the spectroscopic Hamiltonian can be determined nearly independently.

Spectroscopy of the millimetre wave region is of most immediate relevance to environmental studies and the most powerful tools for this frequency region are based on broad-band backward wave oscillator (BWO) sources, and in particular on the excellent devices available from the Russian company Istok. Following the pioneering work of Krupnov and coworkers [12] and changes in the political climate spectrometers based on these sources have gained worldwide circulation. Operation well into the submillimeter region is possible and particularly active instrumental development is currently being pursued by G.Winnewisser and colleagues in Cologne [13]. A typical example of the capabilities taken from the spectrometer in Warsaw is shown in Fig.1 [3]. High resolution recordings over many GHz are possible, which is very desirable when dealing with complex spectra, since they may be either incompletely or altogether unassigned at the time of recording. Large regions of the rotational spectrum can be measured with relatively small demand on spectrometer time. If signal to noise ratio in such recordings is sufficiently high then appropriate software tools can be applied off-line in a process of analysis, which can reveal new features in the spectra for a long time after the recording. As an example the set of spectra originally recorded for the study of the ground states of $CBrClF_2$ isotopomers [1], was reprocessed in a second stage of analysis to assign transitions in three lowest excited vibrational states [2].

If the room temperature mm-wave spectrum is too rich owing to rotational transitions in low-lying vibrational states then the tool of choice is spectroscopy of a sample cooled on isothermal expansion through a nozzle. The free-jet, Stark-modulation technique pioneered by Brown and coworkers in Monash is attractive since the degree of rotational cooling (to *ca* 10-20 K) is still compatible with observation of spectra in at least the lower part of the mm-wave region. This method enabled the first analysis of the rotational spectrum of of $CBrClF_2$ [14] through elimination of transitions in vibrationally excited states. In combination with a broad-band BWO-based frequency synthesizer this technique becomes extremely useful and efficient as demonstrated in the spectrum in Fig.2 [10] obtained with the free-jet spectrometer developed by Favero and Caminati in Bologna. The difficulty in assigning the rotational spectrum of the structurally simple methylene iodide molecule was due to the combination of extensive nuclear

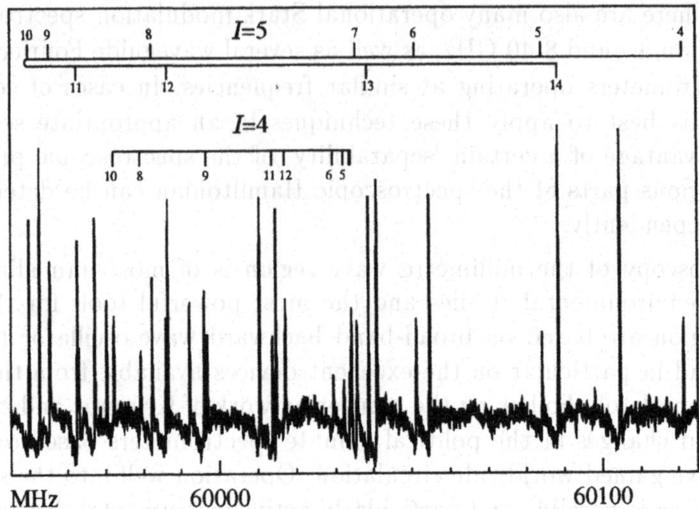

Figure 2. The $10_{2,9} \leftarrow 9_{1,8}$ rotational transition of CD_2I_2 at 60 GHz recorded with a free jet spectrometer at a rotational temperature of *ca* 10K. All of the expected 36 hyperfine components are visible, some of which are unresolved doublets. The spins are coupled using $I = I(\mathrm{I}_1) + I(\mathrm{I}_2)$, $F = I + J$ and the F quantum number assignment for the two highest values of the quantum number I is indicated.

quadrupole hyperfine patterns from the two iodine nuclei with rich structure of transitions in excited vibrational states of the ∠ICI bending mode, the frequency of which is only 121 cm^{-1}. Broad-band scans made with the free jet spectrometer eliminated the excited states, and were possible in a region in which hyperfine structure of rotational transitions was sufficiently compact so that complete, isolated quadrupole patterns of individual rotational transitions could be recorded, as in Fig. 2. This made assignment of the rotational quantum numbers reasonably straightforward. Comparison of broad band recordings of several transitions then made possible the identification of the effect of nuclear spin statistical weights on the intensities of the hyperfine components and assignment of the nuclear spin quantum numbers [9].

Finally the highest resolution and coverage of transitions with lowest values of J can be obtained if the sample is studied in full supersonic expansion, so that sub-Doppler spectra can be recorded. Since rotational cooling is typically to *ca* 1K the maximum in the rotational absorption envelope is usually well below 20 GHz. The microwave measurements can therefore be made with high efficiency with the cavity Fourier Transform (FTMW) method, as pioneered by Balle and Flygare [15]. The only drawback is that scanning is still not the strong point of such spectrometers.

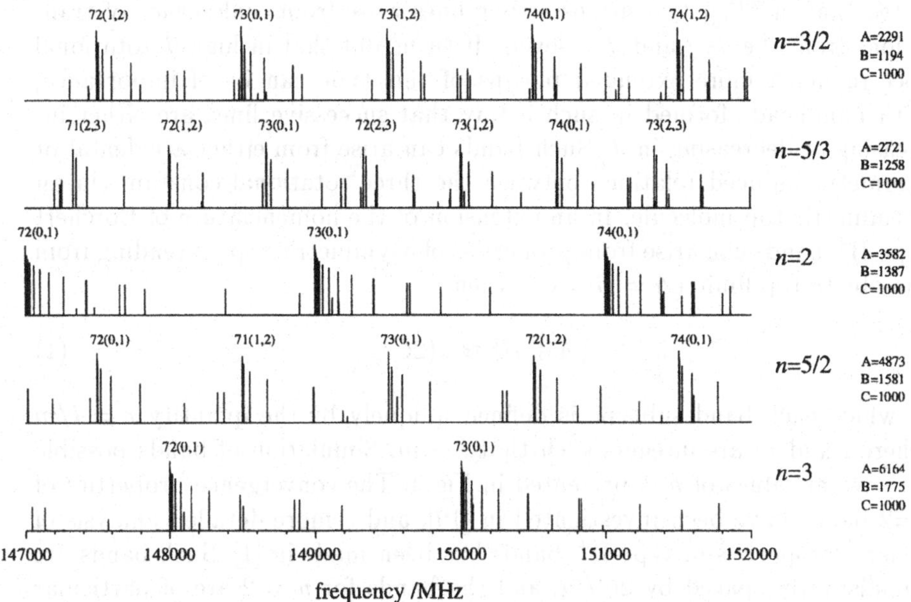

Figure 3. The oblate type-II$^+$ R-type bands, which arise for several values of the oblate limit condition $n = l/m = (A + B)/2C$. All plots are for a rotor with $\kappa = -0.7$ and C=1000 MHz. The quantum numbers $J(K^a_{-1}, K^b_{-1})$ of the two degenerate lower energy levels for the leading lines in each band are indicated.

In connection with studies of complex spectra of halogenated molecules of environmental relevance they are, however, the most powerful tool available for accurate determination of the nuclear quadrupole splitting part of the rotational Hamiltonian. Such spectrometers can of course also be used for isolation of the ground vibrational state in difficult cases, and recent advances suggest that their application for measurement of electric dipole moments in molecules with quadrupolar nuclei should be on the increase [16].

3. Band spectra

The condensation of rotational spectra into bands is a well known phenomenon, which has seen much use in the assignment process for asymmetric rotor molecules. The best established band types are type-I R-type bands, arising from pileups of $J+1 \leftarrow J\ {}^aR_{0,1}$ transitions for the higher values of K_{-1}, and Q-type bands arising from pileups of transitions for a fixed value of K_{-1} or K_{+1} and varying J. That various other types of equally spaced line pileups are also of importance has become apparent from Low Resolution Microwave (LRMW) studies [17]. Some time ago Borchert [18] noted that in ${}^aR_{0,1}$ transitions strong lines arise from coalescence of transitions from the $J_{0,J}$ and $J_{1,J}$ levels. It turns out that in high-J rotational spectra much more extended pileups of this type can be of importance, with bandheads formed in such a way that successive lines are either increasing or decreasing in J. Such bands can arise from either accidental or symmetry induced relations between the three rotational constants in an asymmetric top molecule. In an extension of the nomenclature of Borchert type-II$^+$ bands can arise from properties of asymmetric tops extending from the oblate top limit ($A = B > C$) when

$$(A + B) \approx n(2C). \tag{1}$$

in which each band subtype is defined uniquely by the quantity $n = l/m$ where l and m are integers such that $l > m$. Simulation of bands possible for several values of n is presented in Fig.3. The convergence properties of n=2 bands have been investigated in [19], and a more detailed analysis of general properties of type-II$^+$ bands has been made in [1]. Such bands are equidistantly spaced by $2C/m$, and the bands for $n = 2$ are of particular relevance since their convergence condition $(A + B) \approx 4C$ is equivalent to the planarity condition, $I_c = I_a + I_b$, expressed in rotational constants of the oblate symmetric top limit. Extended bands of this type have now been observed for several molecules, and an example for trichloroethylene can be seen in Fig.4. Even though this molecule is quite prolate ($\kappa = -0.7$) and the type-II$^+$ band in question will collapse completely only at the oblate top limit of $\kappa = +1$, the degree of convergence visible in Fig.4 is already considerable. Since these bands were the most easily identifiable feature in the mm-wave spectrum of trichloroethylene, an understanding of their properties allowed the assignment of this rather rich spectrum to be extended well beyond that of the ground state. In the end it was possible to assign all three singly substituted ^{37}Cl isotopomers, all three doubly substituted ^{37}Cl isotopomers and four low lying vibrational states in the rotational spectrum of normal trichloroethylene [4]. Since other band variants are not tied to particular molecular symmetry and are of accidental nature they are expected to be less frequent, although $n = 3/2$ type-II$^+$ bands have already been observed for $CBrClF_2$ [1].

The symmetry counterpart of type-II$^+$ bands are type-II$^-$ bands, which are possible from extension of properties extending from the prolate limit ($A > B = C$). The band formation condition is

$$2A \approx n(B + C). \tag{2}$$

where, as before, $n = l/m$. Type II$^-$ bands occur at a spacing of $2A/l$. An example of bands of this type, for $n = 3$, has also been identified in the

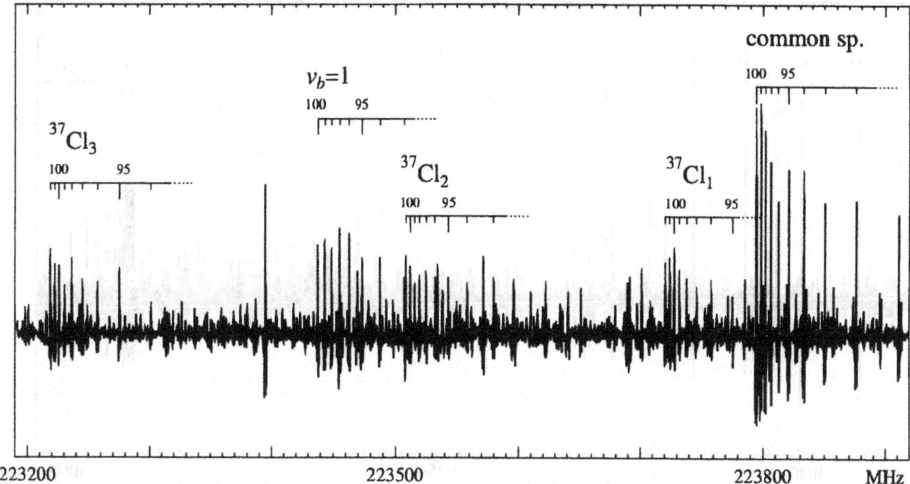

Figure 4. Type-II$^+$ $n = 2$ bands in the room temperature spectrum of trichloroethylene, visible here for the ground state, all three singly substituted ^{37}Cl species and a vibrational satellite. This band-type originates from rotor properties at the oblate rotor limit and the indicated values of J are seen to decrease away from the bandhead.

rotational spectrum of trichloroethylene, see Fig.5. A detailed discussion of properties of type-II$^-$ bands can be found in [4]. The observation of two different R-band types for trichloroethylene proved to be a bonus in determining spectroscopic constants. Transitions in the two types of bands exhibited complementary sensitivity to spectroscopic constants so that measurement strategy based on following lines along each band type was sufficient for determining decorrelated values for all constants in the sextic level rotational Hamiltonian.

A convenient summary of the properties of the four known families of R-type bands has been given in Table 9 of [4]. Actual relationships for maximum band convergence deviate somewhat from the simple form of Eq.1-2, but it was established in [4, 1] that sufficient diagnostic accuracy is obtained by replacing n with a quantity n_{eff} given by

$$\text{type}-\text{II}^- : \qquad n_{\text{eff}} = 1 + \frac{(n-1)(3+\kappa)}{2\sqrt{2(\kappa+1)}}, \ \kappa > -1 \tag{3}$$

$$\text{type}-\text{II}^+ : \qquad n_{\text{eff}} = n\left(\frac{2-b_p^2}{2-nb_p^2}\right) \tag{4}$$

where $b_p = (\kappa+1)/(\kappa-1)$.

For the most established type-I$^-$ bands there are examples in broadband mm-wave recordings for acrylonitrile [20], and chloroacetonitrile [21]. The much less common type-I$^+$ band can also be rather important in some

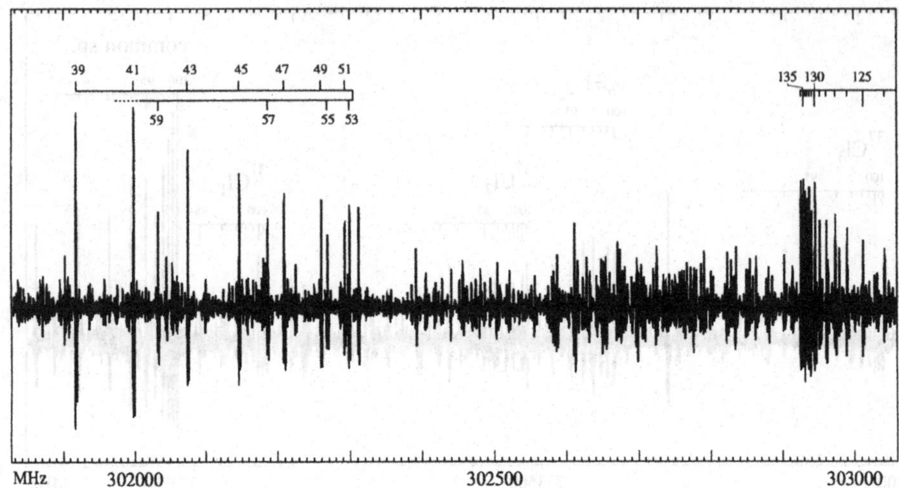

Figure 5. Type II$^-$ $n = 3$ R-type band for the ground state of trichloroethylene (left) which originates from rotor properties at the prolate limit and is characterised by J increasing away from the bandhead. A much more compact type-II$^+$ $n = 2$ band is also visible to the right.

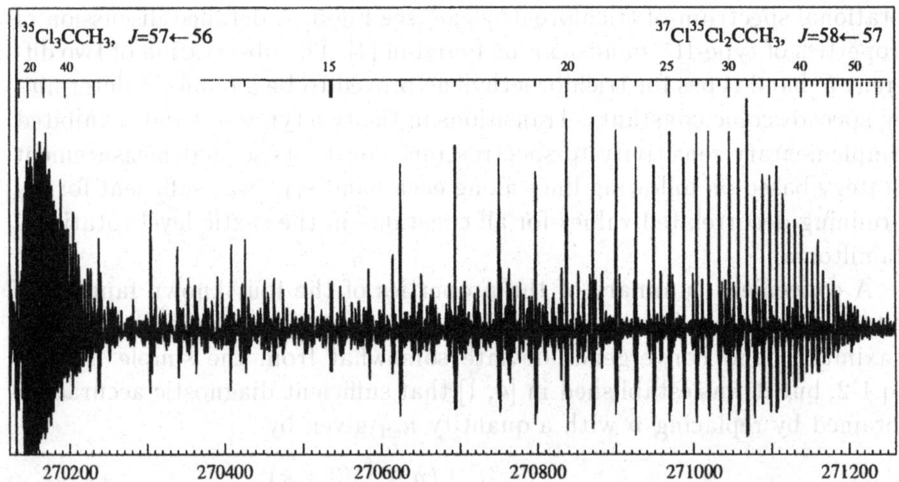

Figure 6. Type I$^+$ R-type band for the ground state of $^{37}Cl^{35}Cl_2CCH_3$ (right) compared with the symmetric top spectrum of the common isotopomer, left. The values of K_{+1} and K are marked for the two bands, respectively.

situations, for example for isotopomers of symmetric tops with a CCl_3 segment, as illustrated by the mm-wave spectrum of 1,1,1-trichloroethane in Fig.6 [7]. Note that in this case the possibility of three indistinguishable single ^{37}Cl substitutions results in comparable intensities for the common and for the naturally substituted isotopomer. In addition, in both Fig.4 and Fig.6, the isotopic bands are not a reflection of the common isotopomer band in the same window, but of a band one or two such bands to the right.

4. Complex nuclear quadrupole coupling

Plural nuclear quadrupole coupling is relatively common in trace molecules of atmospheric importance. It is normally most convenient to commence assignment of their rotational spectra with high-J, R-type transitions in the mm-wave region. The hyperfine structure in such transitions is either completely or partially collapsed so it is easy to determine the parameters in the rotational part of the molecular Hamiltonian. It is then possible to investigate the nuclear quadrupole structure itself. For ground state transitions the tool of choice is FTMW spectroscopy of the supersonic expansion which, as stated above, gives sub-Doppler access to the lowest-J transitions, which are most affected by such splitting. In addition the high accuracy with which frequencies can be measured allows off-diagonal components in the quadrupole tensor to be determined routinely. An example is provided by studies of the spectrum of trichloroethylene, which proceeded in such *top-down* manner. The rotational part of the Hamiltonian was determined first from high-J spectra, and then the nuclear quadrupole splitting itself was analysed by starting from the simplest transition $1_{1,1} \leftarrow 0_{0,0}$ [5]. Since this molecule contains three non-equivalent chlorine nuclei the splitting structure is rather complex. This is illustrated in Fig.7 which shows how the pattern of three hyperfine components for the $1_{1,1} \leftarrow 0_{0,0}$ transition in a molecule with a single chlorine nucleus develops on addition of further one and two chlorine nuclei. The complexity of the pattern of the $3_{1,3} \leftarrow 2_{0,2}$ transition in trichloroethylene is even greater, although the cavity FTMW spectrometer allows the individual hyperfine components to be readily resolved. Even though the analysis was far from trivial it was facilitated by accurate prediction of nuclear quadrupole coupling constants for the three nuclei on the basis of constants available from the mm-wave study, the molecular geometry determined therein, and quadrupole tensor elements from 1,1-dichloroethylene.

Another case of complex hyperfine coupling for three chlorine nuclei in 1,1,1-trichloroethane has also been successfully investigated in Warsaw [8], for both the symmetric top species with three identical chlorine nuclei, and the singly ^{37}Cl substituted species in which two of the three chlorine nuclei are identical. Several double nucleus cases have also been studied including $CBrClF_2$ [2], CH_2I_2 [9, 10], and CH_2Cl_2 [22]. For all of these molecules the complete nuclear quadrupole tensor in inertial axes was determined which allowed, in turn, diagonalisation to the principal quadrupole axes.

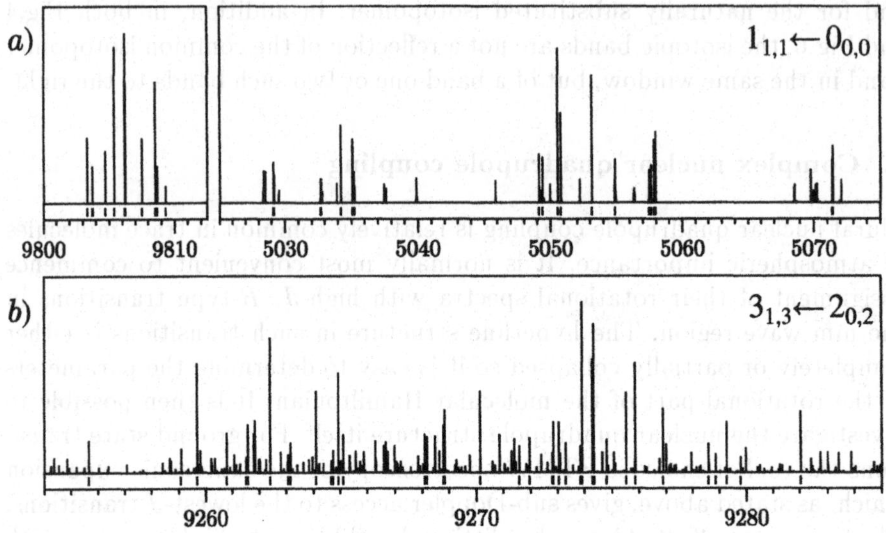

Figure 7. (a) The nuclear quadrupole splitting patterns for the $1_{1,1} \leftarrow 0_{0,0}$ rotational transition in $^{35}Cl_2C{=}CH_2$ (left) and $^{35}Cl_2C{=}CH^{35}Cl$ (right). (b) The central portion of the $3_{1,3} \leftarrow 2_{0,2}$ transition of $^{35}Cl_2C{=}CH^{35}Cl$. The splitting patterns are calculated from the final spectroscopic constants and the small ticks underneath each diagram denote the measured components. The frequency axes are in MHz.

If there is accidental near coincidence between suitable rotational levels then the strong perturbations in the hyperfine patterns allow considerable enhancement in the precision of measurement of the splitting constants. In this way the principal nuclear quadrupole tensor for ^{14}N, which is normally rather elusive in an asymmetric top molecule, could be determined precisely for 2-chloro-acrylonitrile [23].

Synthesis of such results allowed useful conclusions to be drawn concerning the relative orientation of the principal axes of the nuclear quadrupole coupling tensor and the principal inertial axes [5]. Diagonalisation of the inertial quadrupole tensor allows the angle between these two sets of axes to be determined accurately, often to better than 0.01°. In the first approximation it was anticipated that, at least for a quadrupolar nucleus terminal to a chemical bond, the z axis of the principal quadrupole tensor will be aligned with the bond. Accurate experimental data revealed deviations between bond direction and the z axes at the level of *ca* 1-2°. These deviations have been confirmed to be experimentally reliable and it was found that their magnitude and direction are easily reproduced in *ab initio* calculations [5]. The reason for the small deviations turns out to be distortion of the field gradient by electron density further away from the quadrupolar nucleus, and its understanding allows the use of quadrupolar angles as valuable data in determination of molecular geometry. In addition transfer of quadrupole tensors between molecules for the purposes of prediction can be made with increased confidence.

5. Centrifugal distortion constants

In high-J rotational spectra the contributions to transition frequencies from even small values of quartic centrifugal distortion constants can be appreciable. For example for chlorobenzene Δ_J = 60 Hz contributes 250 MHz at J=100 [19]. Estimates of centrifugal distortion constants are therefore most useful since their values may have a bearing even on the J assignment. In the absence of highly anharmonic motions, it is now possible to make routine, reliable predictions of quartic constants from the *ab initio* harmonic force field. This has been demonstrated quite some time ago for the rather rigid ring molecules furan, pyrrole, pyridine [24]. Equally encouraging results have been obtained for the environmentally relevant molecules $CBrClF_2$, trichloroethylene and 1,1,1-trichloroethane discussed presently, as summarised in Table 1.

TABLE 1. The observed and calculated quartic centrifugal distortion constants (kHz) for several selected molecules

	Cl_2C=CHCl, [4]		$^{35}Cl_2{}^{37}ClCCH_3$, [7]	
	obs.	calc.	obs.	calc.
Δ_J	0.09309(3)	0.0962	0.26797(4)	0.2777
Δ_{JK}	1.2662(2)	1.266		0.0782
Δ_K	1.6268(8)	1.548	0.07021(9)	0.0722
δ_J	0.02098(2)	0.0222	0.06634(5)	0.0669
δ_K	0.6734(2)	0.680	0.05013(11)	0.0511
	$C^{79}Br^{35}ClF_2$, [1]		CD_2I_2, [10]	
	obs.	calc.	obs.	calc.
D_J	0.19872(3)	0.2029	0.067109(8)	0.0622
D_{JK}	0.0700(2)	0.0711	-4.3281(3)	-4.28
D_K	1.0988(2)	1.237	166.72(3)	176.9
d_1	-0.03939(3)	-0.0402	-0.003355(4)	-0.00306
d_2	-0.003034(8)	-0.00314	-0.0000430(5)	-0.0000324

In our laboratory we have routinely used theoretically predicted quartic constants in *top-down* assignment commenced from high-J transitions and it was particularly fortunate to find that the quartics for CH_2I_2 and CD_2I_2 were reliably calculated with even the rudimentary basis sets which are available for iodine, see Table 1 and [9, 10]. The reason for this success is that appreciable contributions to quartic constants arise from only the several lowest frequency vibrational modes, so that only a small part of the total force field is of importance.

To further aid the analysis of rotational spectra it is also possible to use *ab initio* harmonic force fields to calculate reliable values of Coriolis coupling constants and inertia defects. Calculated Coriolis constants can serve as a check of the analysis of spectra in doubly degenerate excited states of symmetric top molecules, or of fitting of Coriolis perturbations between vibrationally excited states of molecules, as in $CHClF_2$ [25]. Although moments of inertia can contain appreciable anharmonic contributions, these are known to cancel on evaluation of inertia defects, so that calculation of harmonic contributions to moments of inertia can yield reliable inertia defects. This feature has been used, for example, to assign with confidence excited vibrational states in the mm-wave rotational spectrum of pyrimidine [26].

Encouraging progress has been made in the prediction of sextic centrifugal distortion constants and of the vibration-rotation α constants, but this is yet far from routine. That reasonably satisfactory results can be obtained has been demonstrated for several molecules, including $CHClF_2$ [27], but efforts in the direction of bringing such calculations closer to spectroscopist are still required.

6. Computer programs

The use of efficient software tools is always desirable, and becomes mandatory in spectroscopic problems involving many transitions, vibrational states, high values of quantum numbers etc. Appropriate computer programs are necessary to process spectra, fit data to determine spectroscopic constants, and to compare known data or predictions with spectra. Traditionally many computer codes have been circulated informally within the rotational spectroscopic community, but the internet age is bringing considerable improvement in access to such information.

The fitting side of the spectroscopic problem has benefited considerably from the availability [28] of the powerful package SPFIT/SPCAT written by H.M.Pickett [29]. These programs were originally written for the purpose of cataloguing rotational spectra of many different molecules. This requirement enforced a very general way of setting up the molecular Hamiltonian and efficient factorisation of its energy matrix. Easy scalability then allowed a multitude of research grade applications. An example from this discussion is that all studies of complex hyperfine splitting discussed above have been carried out with that package. The importance of the package is such that various accessory programs, worked examples and help files are now

ASCP 9:54:23 30/8/2000

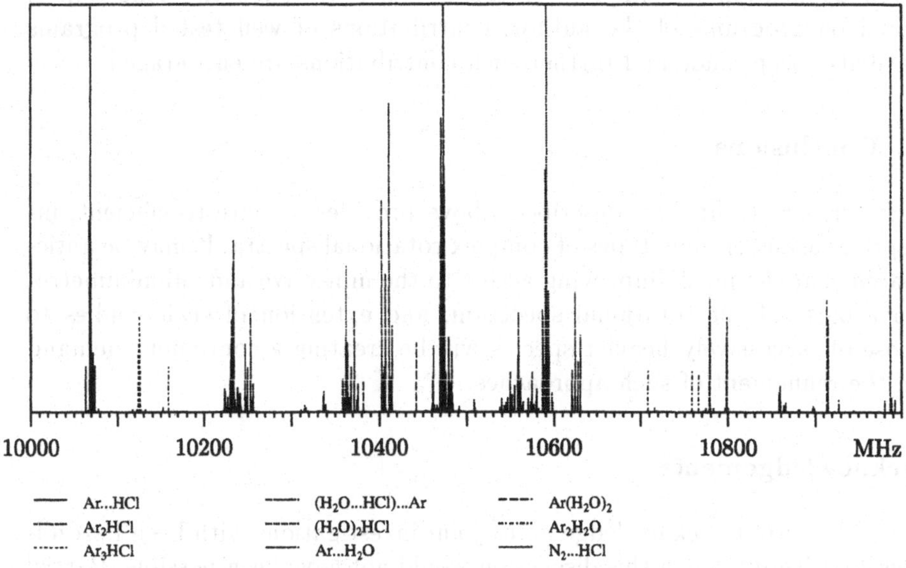

Figure 8. Illustration of common impurities observed during FTMW work involving HCl, as an example of the use of the graphics viewing program ASCP to inspect a catalogue of known spectra. The relative intensities of the various molecular species have been adjusted in the control file for the program to obtain comparable displayed intensities.

available [30, 31].

When it is necessary to deal with complex spectra flexible graphical display of spectroscopic predictions becomes essential. This is possible with the program ASCP [30], which can display predictions from SPCAT and, in fact any other predictive programs, once conversion to a standard input format is made. The program allows merged display of lines from many data files, and has been found invaluable in investigations ranging from complex hyperfine structure to overlaps of transitions from many vibrational states and isotopic species. A summary of some previous applications is given in [11]. The program has been used to produce the bottom part of Fig.1, reproducing the spectrum of $CHClF_2$, which is devoid of characteristic features. An example of the use of ASCP to display a catalogue of spectra is also shown in Fig.8.

ASCP is one of the programs available from the PROSPE (Programs for ROtational SPEctroscopy) web site, which has recently been established by this author. The aim of the site has been to make available well tested, clearly documented and satisfactorily commented computer programs in this field. The programs deal with various aspects of the rotational problem, including structure fitting and display, dipole moment determination, vibrational calculations. The programs range from state of the art for a given application, to simple, yet useful tools. Although the collection is based on programs of the author, contributions of well tested programs have also been made and further such contributions are encouraged.

7. Conclusions

The array of techniques described above provides a route to efficient, in-depth analysis of some types of complex rotational spectra. It may be anticipated that the much improving access to the mm-wave and submillimetre-wave parts of the rotational spectrum, and extension of such studies to those of successively heavier species will be creating a continuous demand on the refinement of such approaches.

Acknowledgements

It is a pleasure to acknowledge many joint investigations with Lech Pszczółkowski, without which this discussion would not have been possible. Partial support of this work by the Institute of Physics and the grant KBN-3T09A-126-17 is acknowledged.

References

1. Kisiel, Z., Białkowska-Jaworska, E., and Pszczółkowski, L. (1996) The mm-wave rotational spectrum of $CBrClF_2$ (halon BCF): observation of a new R-type band and generalization of conditions for oblate-type band formation, *J.Mol.Spectrosc.* **177**, 240-250.
2. Kisiel, Z., Białkowska-Jaworska, E., and Pszczółkowski, L. (1997) The rotational spectrum of $CBrClF_2$ (halon BCF); II. The lowest excited vibrational states and nuclear quadrupole coupling tensors, *J.Mol.Spectrosc.* **185**, 71-78.
3. Kisiel, Z., Alonso, J.L., Blanco, S., Cazzoli, G., Colmont, J.M., Cotti, G., Graner, G., Lopez, J.C., Merke, I., and Pszczółkowski, L. (1997) Spectroscopic constants for HCFC-22 from rotational and high-resolution vibration-rotation spectra: $CHF_2{}^{37}Cl$ and ${}^{13}CHF_2{}^{35}Cl$ isotopomers, *J.Mol.Spectrosc.* **184**, 150-155.
4. Kisiel, Z. and Pszczółkowski, L. (1996) Assignment and analysis of the mm-wave rotational spectrum of trichloroethylene: observation of a new, extended ${}^{b}R$-band and an overview of high-J, R-type bands, *J.Mol.Spectrosc.* **178**, 125-137.
5. Kisiel, Z., Białkowska-Jaworska, E., and Pszczółkowski, L. (1998) Nuclear quadrupole coupling in $Cl_2C{=}CHCl$ and $Cl_2C{=}CH_2$; Evidence for systematic differences in orientations between internuclear and field gradient axes for terminal quadrupolar nuclei, *J.Chem.Phys.* **109**, 10263-10272.
6. Cazzoli, G., Cotti, G., Dore, L., and Kisiel, Z. (1995) The high frequency rotational spectrum of 1,1,1-trichloroethane and the observation of $K{=}3$ splitting, *J. Mol. Spectrosc.* **174**, 425-432.
7. Kisiel, Z. and Pszczółkowski, L. (1997) Millimeter wave rotational spectra of the ${}^{37}Cl$ species of 1,1,1-trichloroethane, *J. Mol. Spectrosc.* **181**, 48-55.

8. Dore, L. and Kisiel, Z. (1998) Nuclear quadrupole coupling in 1,1,1-trichloroethane: inertial and principal tensors for ^{35}Cl and ^{37}Cl, *J. Mol. Spectrosc.* **189**, 228-234.
9. Kisiel, Z., Pszczółkowski, L., Caminati, W., and Favero, P.G. (1996) First assignment of the rotational spectrum of a molecule containing two iodine nuclei: spectroscopic constants and structure of CH_2I_2, *J. Chem. Phys.* **105**, 1778-1785.
10. Kisiel, Z., Pszczółkowski, L., Favero, L.B., and Caminati, W. (1998) Rotational spectreum of CD_2I_2: An isotopomer of the first molecule containing two iodine nuclei investigated by microwave spectroscopy, *J. Mol. Spectrosc.* **189**, 283-290.
11. Kisiel, Z., Białkowska-Jaworska, E., and Pszczółkowski, L. (2000) The ∠ICI bending satellites in the millimeter-wave rotational spectra of CH_2I_2 and CD_2I_2, *J. Mol. Spectrosc.* **199**, 5-12.
12. Krupnov, A.F. and Burenin, A.V. (1976) New Methods in Submillimeter Microwave Spectroscopy, in K.N. Rao (ed.), *Molecular Spectroscopy: Modern Research*, Academic Press, New York, Vol.2, pp.93-126.
13. Belov, S.P, Liedtke, M., Klaus, Th., Schieder, R., Saleck, A.H., Behrend, J., Yamada, K.M.T., Winnewisser, G., and Krupnov, A.F. (1994) Precision measurement of the $^{r}Q_2$ branch at 700 GHz and the $^{r}Q_3$ branch at 980 GHz of HSSH, *J. Mol. Spectrosc.* **166**, 489-494.
14. Bettens, R.P.A and Brown, R.D. (1992) The microwave spectrum and the molecular structure of bromochlorodifluoromethane, *J. Mol. Spectrosc.* **155**, 55-76; for the experimental method see Brown, R.D., *et.al.* (1988) A Stark-modulated supersonic nozzle spectrometer for millimetre-wave spectroscopy of larger molecules of low volatility, *J. Mol. Struct.* **190**, 185-193.
15. Balle, T.J. and Flygare, W.H. (1981) Fabry-Perot cavity pulsed Fourier transform microwave spectrometer with a pulsed nozzle particle source, *Rev. Sci. Instrum.* **52**, 33-45.
16. Kisiel, Z., Kosarzewski, J., Pietrewicz, B.A., and Pszczółkowski, L. (2000) Electric dipole moments of the cyclic trimers $(H_2O)_2HCl$ and $(H_2O)_2HBr$ from Stark effects in their rotational spectra, *Chem. Phys. Lett.* **325**, 523-530.
17. Gillies, C.W. (1982) Low resolution microwave spectroscopy, *Applied Spectroscopy Reviews* **18**, 1-58.
18. Borchert, S.J. (1975) Low-resolution microwave spectroscopy: a new band type, *J.Mol.Spectrosc.* **57**, 312-315.
19. Kisiel, Z. (1990) The millimeter wave spectrum of chlorobenzene: analysis of centrifugal distortion and of conditions for oblate-type bandhead formation, *J. Mol. Spectrosc.* **144**, 381-387.
20. Cazzoli, G. and Kisiel, Z. (1988) The rotational spectrum of acrylonitrile in excited states of the low frequency CCN bending vibrational modes, *J. Mol. Spectrosc.* **130**, 303-315.
21. Kisiel, Z. and Pszczółkowski, L. (1994) The millimeter-wave rotational spectrum of 2-chloroacrylonitrile, *J. Mol. Spectrosc.* **166**, 32-40.
22. Kisiel, Z., Kosarzewski, J., and Pszczółkowski, L. (1997) Nuclear quadrupole coupling tensor of CH_2Cl_2: Comparison of quadrupolar and structural angles in methylene halides, *Acta Phys. Pol. A* **92**, 507-516.
23. Kisiel, Z. and Pszczółkowski, L. (1997) Nuclear quadrupole coupling in 2-chloroacrylonitrile: inertial and principal quadrupole tensor components for Cl and N, *J. Mol. Spectrosc.* **184**, 215-220.
24. Wlodarczak, G., Martinache, L., Demaison, J., and van Eijck, B.P. (1988) The millimeter-wave spectra of furan, pyrrole and pyridine: experimental and theoretical determination of the quartic centrifugal distortion constants, *J. Mol. Spectrosc.* **127**, 200-208.
25. Kisiel, Z., Pszczółkowski, L., Cazzoli, G., and Cotti, G. (1995) The millimeter-wave rotational spectrum and Coriolis interaction in the two lowest excited vibrational states of $CHClF_2$, *J. Mol. Spectrosc.* **173**, 477-487.
26. Kisiel, Z., Pszczółkowski, L., Lopez, J.C., Alonso, J.L., Maris, A., and Caminati, W.

(1999) Investigation of the rotational spectrum of pyrimidine from 3 to 337 GHz: Molecular structure, nuclear quadrupole coupling, and vibrational satellites, *J. Mol. Spectrosc.* **195**, 332-339.
27. Palmieri, P., Tarroni, R., Hühn, M.M., Handy, N.C., and Willets, A. (1995) An improved anharmonic force field of $CHClF_2$, *Chem. Phys.* **190**, 327-344.
28. Pickett, H.M., http://spec.jpl.nasa.gov/ftp/pub/calpgm.
29. Pickett, H.M. (1991) The fitting and prediction of vibration-rotation spectra with spin interactions, *J. Mol. Spectrosc.* **148**, 371-377.
30. Kisiel, Z., *PROSPE - Programs for Rotational SPEctroscopy,* http://info.ifpan.edu.pl/~kisiel/prospe.htm.
31. http://www.ph1.uni-koeln.de/vorhersagen/pickett/index.html

DETERMINATION OF MOLECULAR PARAMETERS FROM EXPERIMENTAL SPECTRA

Possible Problems and Solutions

J. DEMAISON[†] and K. SARKA[*]
[†] *Laboratoire de Physique de Lasers, Atomes and Molécules, UMR CNRS 8523, Bâtiment P5, Université de Lille I, F-59655 Villeneuve d'Ascq, France;*
[*] *Department of Physical Chemistry, Faculty of Pharmacy, Comenius University, 832 32 Bratislava, Slovakia*

1. Introduction

When fitting the rotation-vibrational spectra we try to determine such a set of parameters $\boldsymbol{x}$, $[x_i, i = 1, \cdots m)]$ that would give the best agreement between the observed and calculated frequencies of transitions $\boldsymbol{y}$, $[y_j, j = 1, \cdots n]$ between the rotation-vibrational energy levels . In most cases the relation between the two sets is nonlinear

$$y_j = F_j(x_1, \cdots x_m), \tag{1}$$

and we have to apply the iteration method, starting with some initial estimate of the parameters $\boldsymbol{x}^0$, which upon substituting in Eq. (1) provides a set of calculated values $\boldsymbol{y}^{\mathrm{calc}}$. Assuming that our initial estimate is good we can use the Taylor expansion

$$\boldsymbol{y}^{\mathrm{exp}} - \boldsymbol{y}^{\mathrm{calc}} = \boldsymbol{J}\Delta\boldsymbol{x}, \tag{2}$$

to linearize the relation between $\boldsymbol{y}$ and $\boldsymbol{x}$. In Eq. (2) the rectangular matrix $\boldsymbol{J}$ has dimension $(n \times m)$, its elements are the derivatives

$$J_{ji} = \left(\frac{\partial F_j}{\partial x_i}\right)_{x=x^0}, \tag{3}$$

and

$$\Delta\boldsymbol{x} = \boldsymbol{x} - \boldsymbol{x}^0. \tag{4}$$

The matrix $\boldsymbol{J}$ is usually called Jacobian or design matrix.

In the studied cases the nonlinearity follows from the fact that the energy levels E_k are the eigenvalues of the Hamiltonian matrix $\mathbf{H}$

$$\tilde{V}_k \mathbf{H} V_k = E_k, \tag{5}$$

J. Demaison et al. (eds.), Spectroscopy from Space, 107–129.

where the elements of the Hamiltonian matrix are the linear combinations of the parameters x_i

$$H_{pr} = \sum_{i=1,\cdots m} Z_i^{pr} x_i, \tag{6}$$

and the coefficients Z_i^{pr} are the appropriate functions of the rotational and vibrational quantum numbers J, k, v, l following from the theory of rotation-vibration energy levels. The derivatives of the energy levels E_k with respect to the parameters x_i are

$$\frac{\partial E_k}{\partial x_i} = \sum_{p \leq r} V_{pk} V_{rk} (2 - \delta_{pr}) Z_i^{pr}, \tag{7}$$

and their appropriate differences provide the elements J_{ji} defined in Eq. (3). In some cases,usually for isolated vibrationally nondegenerate states of the molecules with higher symmetry, the matrix H is to a high degree of approximation diagonal and Eqs.(2,7) are replaced by simple linear equations

$$\boldsymbol{y} = \boldsymbol{J}\boldsymbol{x}, \tag{8}$$

$$\frac{\partial E_k}{\partial x_i} = Z_i^{kk}. \tag{9}$$

Whether the problem is defined by Eq. (2) or Eq. (8) we have to solve in both cases the system of linear equations for $\Delta \boldsymbol{x}$ or $\boldsymbol{x}$ where the number of equations n (number of experimental data) usually significantly exceeds the number of parameters m, therefore all equations cannot be satisfied simultaneously and an extra criterion is required to define what is the "best average" simultaneous solution of all equations. Most often this criterion is chosen to be a minimum of the sum of weighted squares of residuals S

$$Min\{\mathrm{S} = \sum_{\mathrm{j}=1\cdots \mathrm{n}} \mathrm{w_j}(\mathrm{y_j^{exp}} - \mathrm{y_j^{calc}})^2\}, \tag{10}$$

where the weights w_j are used to distinguish between the accuracy and "importance" of the data. Well known solution (Ref. [1]) satisfying this criterion and Eq. (8) is given by the equation

$$(\tilde{\boldsymbol{J}}\boldsymbol{W}\boldsymbol{J})\boldsymbol{x} = \tilde{\boldsymbol{J}}\boldsymbol{W}\boldsymbol{y}. \tag{11}$$

Numerous standard programs are available for solution of the system of linear equations (11), e.g. EXCEL software package also contains such program. The matrix $\boldsymbol{W}$ is an $(n \times n)$ diagonal matrix of weights w_i.

If the problem is nonlinear the solution of Eq. (2) satisfying criterion [10] is obtained if $\Delta \boldsymbol{x}$ replaces $\boldsymbol{x}$ and $\Delta \boldsymbol{y} = \boldsymbol{y}^{\text{exp}} - \boldsymbol{y}^{\text{calc}}$ replaces $\boldsymbol{y}$ in Eq.(11). In this case the solution $\Delta \boldsymbol{x}$ is added to the original estimate of parameters

$$\boldsymbol{x} = \boldsymbol{x}^0 + \Delta \boldsymbol{x} \tag{12}$$

and in a new iteration the vector $\boldsymbol{x}$ is used as a new, improved estimate. The iterations continue, and if the process converges well, usually only a few iterations are required to obtain a final converged set of parameters $\boldsymbol{x}$. Again, a number of programs exist for this iteration calculation, developed in various laboratories, some of them more sophisticated or more general than others, e.g. SIMFIT [2], SYMTOP [3], CALPGM [4].

A significant portion of the above text is applicable to any problem where experimental data are fitted with a model with several parameters. There are however following specific features that, although not always unique, distinguish fitting the rotation-vibrational spectra from other problems

- *Extremely high accuracy.* The infrared frequencies are routinely determined with eight valid digits and even significantly higher accuracy is achieved with various modern techniques in microwave, millimeterwave and submillimiterwave spectroscopy.
- *Highly sophisticated models.* Theory of rotation-vibrational spectra benefited from the fortunate fact that zero-order approximation rigid rotor+harmonic oscillator is a good one and at the same time the wavefunctions and the matrix elements of the Hamiltonian can be expressed analytically. This made it possible to develop high-order perturbation formulae that in turn made it possible to match the extremely high accuracy of the experimental data [5,6].
- *Parameters with a clear physical meaning.* It follows from the preceding item that the parameters to be fitted such as rotation and distortion constants B, D^J, Coriolis constants ζ, or doubling constants q_t, to name just a few examples, are not just the numbers that come out of the computer, but if the things are done properly they have a clear physical meaning and can be expressed in terms of the basic molecular properties such as atomic masses, internuclear distances, interbond angles, and potential surface constants [5,6,7]. This in turn offers a possibility to use them for determination of highly precise molecular structures. At the same time it offers an opportunity to check their correctness by independent quantum chemical calculations [8,9]. With the tremendeous development in both hardware and software computing techniques this feature is becoming more and more important.

On the other hand some of the mentioned specific features create also specific problems. We have presented above rather idyllic picture of the fitting process where everything ran smoothly. While this is often the case, when using more and more sophisticated models with increasing number of parameters we encounter frequently also difficulties that may origin from different causes, but share one common feature, singularity or near-singularity of the normal equations matrix $\tilde{\boldsymbol{J}}\boldsymbol{W}\boldsymbol{J}$. Formally, this matrix has to be inverted in Eq. (11), to provide a solution for $\boldsymbol{x}$ or $\Delta\boldsymbol{x}$. However, only a non-singular matrix can be inverted and if a near-singular matrix is inverted, the matrix elements of the inverted matrix are too large and may cause divergence of the iteration process. These problems can be quantified in a following way. One of the methods used for solving the system of linear equations (2) with a least-squares criterion (10) is a *singular value decomposition* (SVD) method. The singular-value decomposition of a rectangular $(n \times m)$ matrix $\boldsymbol{J}$ can be always written as [10,11]

$$\boldsymbol{J} = \boldsymbol{U}\boldsymbol{D}\tilde{\boldsymbol{V}} \tag{13}$$

where $\boldsymbol{U}$ is an $(n \times m)$ matrix, $\boldsymbol{D}$ is an $(m \times m)$ diagonal matrix whose elements are the singular values μ_i of $\boldsymbol{J}$ and $\boldsymbol{V}$ is an $(m \times m)$ orthonormal matrix. The condition indexes of $\boldsymbol{J}$ are defined by:

$$\eta_k = \frac{\mu_{\max}}{\mu_k} \quad k = 1, \cdots m \tag{14}$$

The highest condition index is the condition number $\kappa(\boldsymbol{J})$.

If in the SVD process we use a scaled matrix $\boldsymbol{J}^*$, which is obtained if the columns of the Jacobian matrix $\boldsymbol{J}$ are scaled to have unit length (each term of the vector column J_i is divided by the norm $||J_i||$) then the condition number κ of the scaled matrix is an error magnification factor and it is used to determine whether a matrix is ill-conditioned or not. *If the data are known to d significant figures and if the condition number of J is* 10^r, *then a small change in the data in its least significant digit can affect the solution in the* $(d-r)$*th place.* Thus the condition number is a quantitative measure of singularity of a problem.

In this paper we shall focus on tracing the origins of the problems that cause singularity or near-singularity of the normal equations matrix, and consequently problems and failures in a fitting process and we shall try to provide solutions or at least a better insight into the problems. The causes of the problems may be broadly divided into two classes.

- *The problems where the origin is in the model used.* We shall see that these may be indeterminacies among certain parameters which are inherent property of the Hamiltonian, or near-singularities if high degree polynomials are used, or the problems that follow from a slow convergence of the perturbation expansions of the Hamiltonian
- *The origin of the problem is with the data.* This may be caused by using an overambitious set of parameters where the information on certain parameters simply is not present in data used, whether because the contribution of some parameters is too small to be detected or because of unfavourable selection rules for the transitions, or because determination depends too strongly on only a limited number of data. We shall discuss first these types of problems.

2. Problems Caused by the Data

When one least-squares fits a spectrum to determine molecular parameters, several problems may be encountered. When the measurements and most of the assignments are correct, the most frequent ones are :

i) some parameters are not determined, i.e. they are smaller (in absolute value) than their corresponding standard deviation,

ii) the system of equations is singular and the parameters cannot be determined,

iii) the parameters seem to be well determined (i.e. they are larger in absolute value than about 3 standard deviations), but some of them are not accurate, i.e. they are rather far from their true value.

The first case is rather obvious: the available data do not allow us to determine these parameters (for instance when one tries to obtain sextic centrifugal distortion constants from low J transitions). The solution is either to measure higher J transitions or to fix these parameters to some predefined value (often to zero, when there is no a priori knowledge of the value of these parameters). When possible, it is better not to fix any parameter but to use the method of mixed regression or predicate observations where an estimated value of these parameters is used as supplementary data together with an estimated standard deviation [12].

The second case is more complex : a) it may happen that the measured transitions do not depend on the parameter one tries to determine, e.g. the axial A rotational constant of a symmetric top molecule when one has only $\Delta K = 0$ transitions. This case is still easy to solve because it is enough to fix this parameter. It has no influence on the other parameters. b) some parameters may be fully correlated leading to a singular system. This happens e.g. when the Hamiltonian is not fully reduced. This will be discussed in details in the next section.

The third case is the most vicious one because it can be only detected by a careful analysis and it will be discussed in details. Obviously, the problem may be more easily pointed out when there are several, independent determinations of the same parameter. It is easy to find many examples in the literature and we will present only briefly two typical ones. The centrifugal distortion constants of methyl fluoride, CH_3F, have been determined many times and Ref. [8] makes a critical comparison of the different determinations. It is obvious that the oldest determinations (up to 1986 included) are not compatible with the most recent one (1993) which is also the most accurate one. The second example is the centrifugal distortion of formaldehyde which has been recently reanalyzed. We will limit our discussion to the sextic constant Φ_J which is quite typical, but similar conclusions may also be reached for the other sextic (and higher) constants. For H_2CO, the last determination gives $\Phi_J = 0.0949(11)$ Hz whereas the previous one is $\Phi_J = 0.0314(70)$ Hz [13] and for HDCO the last determination gives $\Phi_J = 0.1671(16)$ Hz whereas the previous one is $\Phi_J = -31(1)$ Hz [14]. Here again, the different values are fully incompatible and what is striking is that the uncertainty of the parameter is always too small.

2.1 INFLUENTIAL OBSERVATIONS

It may happen that one parameter is determined by only one (or very few) datum which is then called influential observation. In this case, changing the value of the datum only changes the value of the corresponding parameter without affecting the overall quality of the fit. If this datum is an outlier (misassignment), the parameter is completely wrong. This problem often occurs when one tries to determine higher order constants. To detect it, it is in principle enough to vary each datum after the other and to examine the effect on the quality of the fit. This method, although conceptually simple, is rather cumbersome although it might be easy to implement using the jackknife method [15]. However, it is relatively easy to identify this problem with the help of the diagonal terms of the "hat" matrix H. To explain the procedure, we use in the following the notations of the tutorial by Albritton et al. [1]. If we want to determine m parameters x_i $(i = 1 \ldots m)$ from n experimental data y_j $(j = 1 \ldots n)$, the starting equation of the linear least-squares method is

$$\boldsymbol{y} = \boldsymbol{J}\boldsymbol{x} + \boldsymbol{\varepsilon}, \tag{15}$$

where $\boldsymbol{J}$ is the design matrix (or Jacobian with the notation $\boldsymbol{J} = \boldsymbol{W}^{1/2}\boldsymbol{J}'$) and $\boldsymbol{\varepsilon}$ the vector of errors. The hat matrix $\boldsymbol{H}$ is defined by :

$$\boldsymbol{H} = \boldsymbol{J}(\tilde{\boldsymbol{J}}\boldsymbol{J})^{-1}\tilde{\boldsymbol{J}}. \tag{16}$$

As $\boldsymbol{H}$ is symmetric and idempotent, it is easy to show that the following equalities hold for its diagonal terms :

$$0 \leq h_{jj} \leq 1. \tag{17}$$

$\boldsymbol{H}$ transforms the observed vector y into its least-squares estimate:

$$\hat{y}_j = h_{j1}y_1 + \cdots + h_{jj}y_j + \cdots + h_{jn}y_n. \tag{18}$$

If $h_{jj} = 1$, $\hat{y}_j = y_j$ and the corresponding residual is zero even if y_j is an outlier. h_{jj} is called leverage of case j. A simple example will illuminate the usefulness of the hat matrix. Assume that we try to determine two unkown parameters α and β in measuring several y_j which obey the following relation:

$$y_j = \alpha f_j + \beta k_j + \varepsilon_j. \tag{19}$$

The f_j and k_j, which are the coefficients of the $\boldsymbol{J}$ matrix, are known but it may happen that the following conditions hold :

$$\begin{cases} f_j \neq 0 \text{ for all } j \neq n & f_n = 0 \\ k_j = 0 \text{ for all j } \neq n & k_n \neq 0 \end{cases} \tag{20}$$

This corresponds for instance to the case where we measure many $\Delta J \neq 0$, $\Delta K = 0$ transitions and one $\Delta J = 0$, $\Delta K \neq 0$. The parameter β is obviously determined by only y_n and its estimate is

$$\beta = \frac{y_n}{k_n} \tag{21}$$

but, more important, its standard deviation is given by

$$s(\beta) = \frac{s}{k_n} \tag{22}$$

where s is the standard deviation of the fit

$$s^2 = \frac{1}{n-m} \sum w_j (y_j - \hat{y}_j)^2. \tag{23}$$

The standard deviation s may be quite small even if y_n is an outlier. And thus, one may have the impression that the fit is good and β well determined. But, in this case, it is easy to verify that $h_{nn} = 1$. This indicates that y_n is an influential observation that we cannot trust. The only solution is to add further measurements which lower the value of h_{nn}. It may also happen that $f_m \gg f_j$ for all $j \neq m$. In this case

$$h_{mm} = \frac{f_m^2}{\sum f_j^2} = \frac{f_m^2}{f_m^2 + \varepsilon} \simeq 1. \tag{24}$$

This often occurs when one tries to determine higher order terms. This schematic example, although extremely simple, is not far from the reality. It is important to note that neither the parameter nor its standard deviation are reliable. h_{jj} theoretically can range from $1/n$ to 1. Most statistics textbooks suggest that high-leverage be identified when $h_{jj} > 2m/n$. This cutoff value is obviously too small for most spectroscopic applications. Actually, it is enough to carefully check the data for which $h_{jj} > 0.5$ and to eliminate data for which $h_{jj} > 0.9$ (this can be satisfactorily solved by reducing the weight of these data). h_{ji} indicates that there is a problem with the jth measurement but it does not indicate which parameter is affected. There is another diagnostic which may give this information :

$$\mathrm{DFBETAS_i(j)} = \frac{\mathrm{x_i - x_i(j)}}{\mathrm{s(x_i)}} \tag{25}$$

where $x_i(j)$ is the estimate of the ith parameter when the jth measurement is omitted. A simple plot of DFBETAS$_i(j)$ against j allows us to find the measurement which plays a much more important role than the others for the ith parameter. A striking example of the usefulness of the hat matrix is given by the analysis of the rotational spectrum of the anti form of butyronitrile, $CH_3CH_2CH_2CN$ [16]. A fit of 136 transitions is at first sight rather satisfactory, the standard deviation of the fit being : $s = 91$ kHz. But, for two transitions h_{ii} is extremely high, see TABLE 1. These transitions are influential points and they have indeed a tremendous influence on the value of $\Delta_K = -219.5(27)$ kHz. If we add to the fit high-J microwave transitions which were measured later [18], the standard deviation of the fit raises to 2149 kHz and it appears that the $23_{3,20} \longleftarrow 22_{2,21}$ transition is definitely an outlier. After elimination of two dubious transitions, the fit becomes extremely good with a standard deviation of only 24 kHz and the value of Δ_K is extremely different : +240.616(34) kHz.

TABLE 1. Centrifugal distortion analysis of the *anti* form of butyronitrile.

	136 lines			150 lines		
transition	$\Delta\nu_j$ [a]	t_j [b]	h_{jj}	$\Delta\nu_j$ [a]	t_j [b]	h_{jj}
$23_{3,20} \longleftarrow 22_{2,21}$	-0.005	+1.69	0.99	20.603	-11.05	0.23
$43_{2,42} \longleftarrow 42_{1,41}$	+0.056	-3.20	0.92	2.226	-1.40	0.41

[a]) Residual : $\Delta\nu_j = y_j - \hat{y}_j$, [b]) Standardized residual : $t_j = \Delta\nu_j / s(\Delta\nu_j)$

When the problem of influential observation concerns only one parameter, it is still easy to handle but the reality is sometimes more complicated. It often happens that some parameters are highly correlated. In this case, the diagnostic h_{jj} is not enough to detect the problem.

2.2 COLLINEARITY OR ILL CONDITIONING

Indeed, it often happens that some parameters cannot be estimated with precision and that they are very sensitive to small perturbations in the data. This is due to collinearity which increases the variances of the estimated parameters and is responsible for important round-off errors. It would be useful to have a diagnostics that determines whether a collinearity exists and identifies the parameters affected. The correlation matrix is often employed for that purpose. But the absence of high correlations does not imply the absence of collinearity. Therefore many different procedures have been proposed by statisticians and used by spectroscopists. Belsley [19] has critically reviewed these procedures and has concluded that "none is fully successful in diagnosing the presence of collinearity and variable involvement or in assessing collinearity's potential harm". To palliate the weaknesses of the existing diagnostics, he has introduced the condition indexes (Eqs.(13,14)) and has shown that they can be easily used to determine the strength and number of near dependencies. The number of near-dependencies is equal to the number of high scaled condition indexes of the scaled matrix $\boldsymbol{J}$.

To determine which parameters are involved in the collinearities, Belsley defines the variance-decomposition proportions. The variance-covariance matrix Cov of the least-squares estimator $\hat{\boldsymbol{x}}$ of $\boldsymbol{x}$ is

$$\boldsymbol{Cov}(\hat{\boldsymbol{x}}) = \boldsymbol{s}^2(\tilde{\boldsymbol{J}}\boldsymbol{J})^{-1} = \boldsymbol{s}^2\boldsymbol{V}\boldsymbol{D}^{-2}\tilde{\boldsymbol{V}}, \tag{26}$$

$$Var(\hat{x}_k) = s^2 \sum_j \frac{V_{ki}^2}{\mu_i^2}. \tag{27}$$

Let $\varphi_{ki} = V_{ki}^2/\mu_i^2$ and $\varphi_k = \sum_{i=1}^m \varphi_{ki}$, $k = 1 \cdots m$. The variance-decomposition proportions are

$$\pi_{ik} = \frac{\varphi_{ki}}{\varphi_k}. \tag{28}$$

Belsley proposes the following rule of thumb: estimates are degraded when two or more variances have at least half of their magnitude associated with a scaled condition index of 30 or more.

A textbook example is given by the determination of the average structure (r_z) of the linear triatomic molecule HBO [20]. The r_z structure is derived from the average moments of inertia which are obtained by correcting the ground state moments of inertia for the harmonic part of the vibrational contribution. This correction is calculated from the harmonic force field. To determine the two bond lengths in HBO, it is necessary to know at least the moments of inertia of two isotopic species. The problem is that the r_z structure is not isotopic independent.

This is particularly true when an X-H bond is present. Usually, the isotopic dependence of the average bond lengths is estimated using the approximate formula of Kuchitsu [21]

$$\delta r_z = \frac{3}{2} a \delta(u^2) - \delta K, \tag{29}$$

where u^2 is the mean square amplitude for the bond concerned, and K the mean square perpendicular amplitude correction, both obtained from the harmonic force field, while a is the Morse anharmonicity parameter which is often assumed to be equal to that of the corresponding diatomic molecule. But it is also tempting to try to directly determine δr_z from the experimental data. As Kuchitsu's formula is only an approximate one, one might think that this second method is better. This should be particularly true for HBO because we have at our disposal the moments of inertia of six different isotopic species. The results of the two methods are given in TABLE 2. The two structures are incompatible and the isotopic dependence does not even have the same sign.

To determine which structure is the best, one may extrapolate the r_z structure to obtain the r_e structure (which was also accurately determined from the equilibrium moments of inertia). It appears that the structure obtained with δr_z fixed is significantly more accurate. To understand what happens when one fits r_z(B-H), r_z(B-O), and δr_z together, one calculates the condition number which is rather large : $\kappa = 96$ and the corresponding variance-decomposition proportions which are given at the last line of TABLE 2. The explanation is now obvious : although δr_z seems to be determined from the relative small value of its standard deviation, actually it is not because it is fully correlated with the two bond lengths and it explains why the results of the fit are not reliable.

TABLE 2. Structure of HBO (in Å).

	r(B-H)	r(B-O)	$\delta r_z = r$(B-D)$-r$(B-H)
r_z	1.1771(18)	1.20277(36)	-0.00470(75)[a]
r_z	1.1653(5)	1.2053(2)	0.0022[b]
$r_e(z)$[c]	1.1674(20)	1.2018(20)	
r_e	1.16667(41)	1.20068(10)	
ρ[d]	1.000	0.999	0.993

[a]) fitted together with the two bond lengths. [b]) fixed at the value obtained from Eq. (29). [c]) extrapolated from the r_z structure of previous line (δr_z fixed). [d]) Variance-decomposition proportions of the global least-squares fit (first line of the table).

Another striking example is the determination of the sextic centrifugal distortion constant H_J in SiH_3Br which is discussed in Ref. [6].

2.3 WEIGHTING

When measurements are of different precision, it is usual to fit them with the weighted least squares method, the weights being taken equal to the inverse of the square of the experimental precision, but the usefulness of the weighted least squares method is not limited to measurements of different precision. It is indeed a powerful tool to alleviate the problems we have just discussed because

- it allows us to take into account the imperfections of the model. For instance, it frequently happens that it is not possible to significantly determine one (or several) parameter, it is then usual to fix this parameter, but, in some cases, it slightly deteriorates the quality of the fit because some data are slightly sensitive to this parameter. A satisfactory solution is to lower the weight of these data.
- when some data are influential (high value of h_{ii}), statistics textbooks advise to drop these data which is a pity because it is often a waste of information. When the assignment of these data is certain, a better solution is to lower the weight of these data until they are no more influential.

In many cases, the weights can be automatically adjusted using the iteratively reweighted least squares method (IRLS) as described in Ref. [23]. The principle of this method is that data with large residuals receive lower weights. For an homogeneous data set (i.e. data of the same origin, if necessary, the data may be subdivided in several data sets), the median of absolute deviations (MAD) is first calculated

$$\mathrm{MAD} = \mathrm{median}\left|\mathrm{e_i} - \mathrm{median}(\mathrm{e_i})\right|, \tag{30}$$

where e_i are the residuals of a first ordinary least-squares fit. Then, the standard deviation is estimated

$$s = 1.4826 \cdot \mathrm{MAD}, \tag{31}$$

and the scaled residuals u_i are calculated by dividing the residuals e_i by s

$$u_i = \frac{e_i}{s}. \tag{32}$$

Then the new weights are estimated using either the Huber function or the biweight function (or the Huber function for the first iterations and the biweight function for the last cycles):

	Huber function(c = 1.345)	biweight function(c = 4.685)
if $\lvert u_i \rvert \le c$	$w_i = 1$	$w_i = \left[1 - (u_i/c)^2\right]^2$
if $\lvert u_i \rvert > c$	$w_i = \frac{c}{\lvert u_i \rvert}$	$w_i = 0$

These weights are then modified on the basis of leverage. Let c^H be the tuning constant (for instance $c^H = 0.9$ or the 90th sample percentile of h_{ii}). We define the leverage based weights w_i^H:

$$\begin{cases} w_i^H = 1 \text{ if } h_{ii} \le c^H \\ w_i^H = \left(\frac{c^H}{h_{ii}}\right)^2 \text{ if } h_{ii} > c^H \end{cases}$$

We proceed with IRLS estimation as usual, but instead of applying biweight weights w_i, at each step we weigh by $w_i \cdot w_i^H$. This procedure is obviously iterative. It stops when the maximum change in weights from one iteration to the next becomes negligible. In some cases, it is safer to adjust the weights manually. Eventually, it is recommended to use a normality plot [24] to check that the weights are correctly chosen: the expected values of the standardized residuals are calculated using the normal law and the ordered values of the experimental standardized residuals are plotted as a function of the calculated ones. This plot should be approximately a straight line if the weighted errors are normally distributed.

3. The Problems Originating from the Model

3.1 COLLINEARITIES

Probably the most severe problem that may be encountered is linked to collinearity, which means that the columns of the matrix $\boldsymbol{J}$ are not linearly independent and the following equation is satisfied or nearly satisfied for all values of j

$$c_1 \frac{\partial y_j}{\partial x_1} + c_2 \frac{\partial y_j}{\partial x_2} + c_3 \frac{\partial y_j}{\partial x_3} + \cdots c_m \frac{\partial y_j}{\partial x_m} = 0, \tag{33}$$

where c_i are some constants. It can be shown easily that if Eq. (33) is satisfied for all values of j, the matrix $\tilde{\boldsymbol{J}}\boldsymbol{W}\boldsymbol{J}$ is singular. In simple terms it means that although we may have many more equations (n) than parameters (m), the structure of the equations is such that not all the parameters can be determined independently. Eq. (33) describes the problem in strict mathematical terms but perhaps a better insight into the heart of the problem may be obtained from a simple hypothetical example.

Let us suppose that it follows from the theory that the frequencies y_J of the observed transitions depend on the angular momentum quantum number J as follows

$$y_J = \mathbf{Q}J + \mathbf{P}J^2 + \mathbf{R}J(J+1) \tag{34}$$

and the physical meaning of all three molecular parameters $\mathbf{Q}, \mathbf{P}, \mathbf{R} = f(r, \alpha, f)$ is clearly established by equations expressing them in the terms of more basic molecular parameters such as atomic masses, interatomic distances r, angles α and potential constants f so that these three parameters can be calculated independently by quantum chemical methods and they have definite values independent of our fitting process. Let us assume that the true values of the parameters are

$$\mathbf{Q} = 1000, \ \ \mathbf{P} = -1, \ \ \mathbf{R} = 3, \tag{35}$$

and the reader is encouraged to substitute these values to generate from Eq. (34) a set of pseudo-experimental values y_J for various values of J. Let us assume further that theoretical *ab initio* calculations provided the values

$$\mathbf{Q}_c = 995, \ \ \mathbf{P}_c = -0.9, \ \ \mathbf{R}_c = 3.2, \tag{36}$$

and we would like to compare these values with those determined from fitting Eq. (34) for many values of J. We shall find out that the fitting procedure fails and error message such as "division by zero" appears. What happened? We can see easily that

$$J + J^2 - J(J+1) = 0, \tag{37}$$

Eq. (33) is satisfied with

$$c_1 = 1, \;\; c_2 = 1, \;\; c_3 = -1$$

and the matrix of normal equations is singular. Looking at the problem from another angle we can say that following equation is clearly satisfied for arbitrary value of s

$$s[J + J^2 - J(J+1)] = 0, \tag{38}$$

and this equation can be added to Eq. (34) giving

$$y_J = (\mathbf{Q} + s)J + (\mathbf{P} + s)J^2 + (\mathbf{R} - s)J(J+1) \tag{39}$$

Because the value of s is arbitrary we can have infinitely many sets of parameters $\mathbf{Q} + s$, $\mathbf{P} + s$, $\mathbf{R} - s$, one for every value of s and all sets will provide exactly the same fit to the values y_J. We could almost say that there is no way for a poor computer how to decide which of these sets is a correct one and therefore the computer fights back with an error message.

In situations like this it is quite common to take a course of action which could be almost given the name "a golden rule of fitting", namely "when in trouble, constrain one parameter". Suppose that we constrain, say, parameter $\mathbf{R}$ to zero and try to fit the equation

$$y_J = \mathbf{Q}J + \mathbf{P}J^2. \tag{40}$$

This time we encounter no problems and if we have used generated pseudo experimental values y_J the following values should be obtained from the fit

$$\mathbf{Q}_{fit} = 1003, \;\; \mathbf{P}_{fit} = 2. \tag{41}$$

It is easy to see why the "golden rule" worked. If you constrain one parameter x_k, the correponding term on the left-hand side of Eq. (33) $c_k \partial y_j / \partial x_k$ disappears and the remaining terms do not add up to zero, hence the collinearity has been removed. This is true clearly only if $c_k \neq 0$, thus we have to constrain a parameter really involved in a collinearity.

Although the fit was successful, the determined parameter values are not very satisfactory. While the parameter $\mathbf{Q}_{fit}$ has a value that is not far from its true or computed values the second parameter $\mathbf{P}_{fit}$ has a value that is widely at variance with true and computed values, differing from them not only by about 100% in absolute value but also in sign. Normally, we do not know the true values of the parameters and have at our disposal for comparison at best only the fitted and *ab initio* values. Because in general the estimated standard deviations of the fitted parameters are usually much lower than uncertainties of the same parameters from theoretical calculations one could suspect that the fault is with theoretical values even though it exceeds significantly estimated uncertainty. Better insight into the problem is provided by alternative approach where we discuss Eq. (39) rather then Eq. (33).

Another way of looking at success of the constraint $\mathbf{R}_{fit} = 0$ is that it assigns to arbitrary parameter s in Eq. (39) one definite value out of infinitely many posible values. In this case

$$\mathbf{R}_{fit} = 0 \Longrightarrow s = \mathbf{R} \tag{42}$$

and it follows then from Eq. (39) that

$$\mathbf{Q}_{fit} = \mathbf{Q} + \mathbf{R}, \ \mathbf{P}_{fit} = \mathbf{P} + \mathbf{R}. \tag{43}$$

The last equations express the parameters $\mathbf{Q}_{fit}$, $\mathbf{P}_{fit}$ obtained from the fit with the constraint $\mathbf{R} = 0$ as combinations of the true molecular parameters. In this way the physical meaning of the fitted parameters is reestablished and they can be compared with the calculated parameters. The theoretical values $\mathbf{Q}_{fit} = 998.2$, $\mathbf{P}_{fit} = 2.1$ are then in reasonably close agreement with those determined from the fit. The point just made is of extreme importance *when the calculated values of the molecular parameters are compared with the values determined from fitting experimental data.* Excellent examples can be found in several recent papers [8,9].

Geometrical description of the collinearities provides an alternative, and very illustrative picture of the problem. The column vectors of the design matrix have often been called the "fit vectors" which span the "fit space" (dimension $= m$), a subspace of the "observation or measurement space" (dimension $= n > m$). This permits a vivid picture of the collinearity as being the true geometrical collinearity of two fit vectors which necessariy reduces the dimension of the fit space. The fit space collapses from dimension m to $m - 1$. That means, one of the m fit vectors is surplus and any vector in fit space (in particular, the solution vector which comes nearest to the vector of the observations) can be constructed by an infinitely large number of different linear combinations of the m original fit vectors. Since the m coefficients of the solution vector's linear combination are the parameters originally desired, they cannot be unambiguously determined. However, constraining the coefficient of one of the two parallel vectors to an arbitrary value (e.g., zero), the remaining $m - 1$ coefficients of the solution vector become all fixed and determinable (but they depend on the arbitray value given to one partner of the two parallel fit vectors). The picture can be easily expanded to one or more bunches of parallel vectors. The authors are indebted to Prof. Rudolph [25] for this geometrical picture of the collinearities.

The key results of this subsection can be summarized as follows

- The collinearities cause the singularity of the normal equations matrix.
- The collinearity can be removed by constraining one of the parameters involved in it.
- Such constraint changes the meaning of the parameters retained in a fit.

- When the values of the fitted parameters are compared with the theoretical ones or used in structure and potential determination one must give them true physical meaning.

3.2 COLLINEARITIES CAUSED BY INDETERMINACIES. REDUCTIONS

In the preceding subsection we have introduced the notion of collinearity and showed some important implications of this term. It may seem however that it did not help us in our search for the causes of singularity of the matrix $\tilde{J}WJ$, because it only transformed the original question to the question "what are the causes of collinearity "?

In the simple example we have presented, the core of the problem could be apparent even to an untrained eye and it might generate an impression that the collinearity is a trivial problem. Actually, in rotation-vibration spectroscopy collinearity represents certainly a non-trivial problem and we shall see that the indeterminacy problems are caused by the fact that some of the terms in the Hamiltonian are not independent and may be inherently bound together by some relations independent of the data. It can be shown that such collinearities follow from the fundamental principles of quantum mechanics.

The problem of indeterminacies among the spectroscopic constants in fitting the experimental data has been known for a long time. The problem has been explained by Watson [5,26] who has shown that the observed indeterminacies were caused by existence of a *block-diagonal unitary transformation*

$$\tilde{H} = e^{iS} H^{\text{eff}} e^{-iS} \tag{44}$$

Such transformation changes the values of parameters – spectroscopic constants x_i $(i = 1 \ldots m)$ – in the effective Hamiltonian H^{eff} for the vibrational state v in question without changing its eigenvalues.

The operator S must be block-diagonal, Hermitian, invariant with respect to the molecular symmetry group operations and change sign under the time reversal operation. These restrictive requirements usually limit the number of terms in the operator S to just a few, say r terms

$$iS = \sum_{i=1}^{r} s_i (P_i - P_i^{\dagger}), \tag{45}$$

where $P_i(\hat{\mathcal{L}}^+, \hat{\mathcal{L}}^-, \hat{J}_\alpha)$ are generally the products of the vibrational operators $\hat{\mathcal{L}}^+$, $\hat{\mathcal{L}}^-$, and the rotational operators $\hat{J}_\alpha$, $\hat{P}_k^{\dagger}$ is the Hermitian conjugate of $\hat{P}_k$, and the coefficients of these terms s_i $(i = 1 \ldots r)$ are real and must satisfy order-of-magnitude limitations [27,28] . However, within these limitations the coefficients are free to have any value without changing the eigenvalues.

The actual evaluation of Eq. (44) is carried out by expanding the exponentials and evaluating commutators and the transformation results in the transformed values $\tilde{x}_i$ of the parameters

$$\tilde{x}_i = x_i + f_i(x_1, \ldots x_m;\ s_1, \ldots s_r). \tag{46}$$

The indeterminacy problem is caused by free parameters s_k [27,28,29] and comparing Eqs. (39,46) shows that the free (arbitrary) parameters s_k play in this more complex problem the same role as does the arbitrary parameter s in the simple example and the problem can be solved in the same way by constraining r molecular parameters $\tilde{x}_i$ to the predetermined values, r being the number of the free parameters s_i.

The direct link can be established between existence of a unitary transformation (44) which uses the operator S in Eq. (45) with the free parameters s_i, and the collinearity equation (33).

Energies E_i which are eigenvalues of the Hamiltonian H^{eff}, depend on the transformed parameters $\tilde{x}_i$, but are invariant with respect to the unitary transformation [44]

$$\frac{\partial E_i}{\partial s_k} = \sum_{l=1}^{m} \frac{\partial E_i}{\partial \tilde{x}_l} \frac{\partial \tilde{x}_l}{\partial s_k} = 0 \tag{47}$$

for all i and for all the parameters s_k in Eq. (45). In the fitting of spectroscopic data, the experimental data can be energies, frequencies or wavenumbers of transitions. Because the latter two correspond to differences between energies, Eq. (47) can be generalized to

$$\sum_{l=1}^{m} \frac{\partial y_j}{\partial \tilde{x}_l} \frac{\partial \tilde{x}_l}{\partial s_k} = 0 \tag{48}$$

for all y_j $(j = 1, \ldots n)$, where y_j is the calculated value of of the jth experimental quantity, be it energy, frequency or a wavenumber. When Eq. (48) is compared with Eq. (33) it becomes apparent that the derivatives $(\partial \tilde{x}_l / \partial s_k)$ in Eq. (48) can be identified with the constants c_l in Eq. (33) and the transformed parameters $\tilde{x}_i$ are the parameters x_i in Eq. (33).

We can formulate then following important statement. Every parameter s_i in the operator S causes one independent collinearity described by Eq. (48) causing in turn the singularity of the matrix $\tilde{\boldsymbol{J}}\boldsymbol{W}\boldsymbol{J}$ and indeterminacy among the parameters involved in Eq. (48).

The unitary transformation described by Eqs. (44-46) is usually called a *reduction* because the operator S may be chosen to eliminate as many terms as possible from the transformed Hamiltonian, which then becomes a reduced Hamiltonian appropriate to the fitting of the spectra.

The reduction process starts with the symmetry analysis for the vibrational state or polyad of states in question which provides the number of terms and allowed structure of the transformation operator S for a required level of approximation . Such analysis can be accomplished in a fashion similar to that used for construction of the phenomenological Hamiltonians. The number of independent terms in this operator equals the number of free parameters and this in turn equals the number of necessary constraints. The operator S is applied in Eq. (44) and the expressions are derived for the transformed parameters in Eq. (46) in terms of the original ones and free parameters s_i of the operator S. These expressions are used in making decisions on which of the parameters are to be constrained and for describing the effect of the constraints on the physical meaning of the remaining parameters.

The reduction method provides the answers to the very practical questions related to spectroscopic data fitting, namely: how many constraints are to be applied and which parameters should be constrained, if collinearities are to be avoided? If such constraints are applied, what is the physical meaning of the remaining parameters? For these reasons the reduction theory has been applied to the effective Hamiltonians for various systems of states. The original papers on the reduction of the rotational Hamiltonian of asymmetric rotors [26,29,30] were followed by a review paper summarizing the results for rotational Hamiltonians of semirigid molecules of all symmetries [27], and *Watson's A-reduced Hamiltonian* and *Watson's S-reduced Hamiltonian* have become part of the textbooks on rotation spectroscopy [31,32]. The special case of the quasi-spherical symmetric top has been treated in Refs. [33,34].

The reduction method has been later extended also to the Hamiltonians for the rotation-vibration states of asymmetric top molecules [35,36], spherical top molecules [37,38,39,40] symmetric top molecules [40,41,42,43], linear molecules [44,45], and to molecules with large-amplitude motions [46,47].

The numerous applications of the reductions to the various states demonstrated repeatedly that the reductions are an appropriate solution for the indeterminacy problems and therefore they should be applied wherever necessary. However, reductions are "state specific" in the sense that each type of vibrational state, or polyad of vibrational states requires specific reduction. It can be anticipated that computer assisted algebraic method [40], or numerical reductions will probably play important role in reductions of complex polyads.

3.3 COLLINEARITIES CAUSED BY HIGH DEGREE POLYNOMIALS IN J. WHEN SHOULD ONE APPLY THE ORTHOGONAL POLYNOMIALS?

In fitting the highly accurate data on the transitions with high J values according to Eq. (8) often polynomials of high degree in quantum number J have to be employed. However, it is a well known fact that if the experimental data y_j are fitted to a polynomial of a degree m in the independent variable z_j

$$y_j(z_j) = \sum_{i=0}^{m} x_i z_j^i, \quad (j = 1, \cdots n), \tag{49}$$

the higher is m, the more "ill–conditioned" becomes the system of normal equations for determination of the coefficients x_i [48]. The term "ill–conditioned" implies that solution is unstable with respect to small changes in experimental data and that the matrix of normal equations is nearly singular. The coefficients x_i should minimize the weighted sum of residuals S

The *design or jacobian matrix* [10] of the fitting problem J is $(n, m+1)$ matrix

$$J_{ji} = \frac{\partial y_j(z_j)}{\partial x_i} = \frac{z_j^i}{\sigma_j}, \tag{50}$$

where σ_j are measurement errors. The vector $\boldsymbol{y}$ of length n is defined by $y_j = y_j^{\exp}/\sigma_j$ and the components of the vector $\boldsymbol{x}$ of length $m+1$ are the parameters to be fitted $x_0, \ldots, x_m$. If the *design matrix* $\boldsymbol{J}$ is decomposed according to Eq.(13), the ratio of the largest and smallest element of the matrix $\boldsymbol{D}$ is called a *conditional number* of the *design matrix* and it is a quantitative expression of the "conditioning" of a matrix. We shall denote it by symbol κ. According to Ref. [48] a matrix is *ill-conditioned* if its condition number is too large, that is, if its reciprocal approaches the computer's floating point precision.

The orthogonal polynomials $p_k(z)$, where k denotes the polynomial degree, satisfy for $(k \neq r)$ the orthogonality relation

$$\sum_{j=1}^{n} w_j p_k(z_j) p_r(z_j) = 0, \tag{51}$$

and fitting the experimental data with the expression

$$y_j(z_j) = \sum_{i=0}^{m} c_i p_i(z_j) \quad (j = 1, \cdots n), \tag{52}$$

has been recommended in Ref.[48] as a superior alternative to Eq. (49). The orthogonal polynomials are generated easily by recursive procedure described in detail on pp.256-258 in Ref.[48]. Because we shall need the orthogonal polynomials with odd powers of z only, we must modify slightly the standard method of their construction and the recursive procedure for their generation is

$$\begin{aligned} p_1(z) &= z, \; p_0(z) = 0, \\ p_{k+1}(z) &= (z^2 - \alpha_{k+1}) p_k(z) - \beta_k p_{k-1}(z), \; k = 1, 2, \ldots \end{aligned} \tag{53}$$

where the coefficients β_k, α_{k+1} are generated for $k = 1, 2, \ldots$ by a sequence

$$\begin{aligned} \beta_1 &= 0, \\ \gamma_k &= \sum_{j=1}^{n} w_j [p_k(z_j)]^2, \\ \beta_{k+1} &= \gamma_{k+1}/\gamma_k, \\ \alpha_{k+1} &= \sum_{j=1}^{n} w_j z_j^2 [p_k(z_j)]^2/\gamma_k \end{aligned} \tag{54}$$

and the fitted parameters c_i in Eq. (52] are determined easily by

$$c_k = \sum_{j=1}^{n} w_j y_j p_k(z_j)/\gamma_k \tag{55}$$

with understanding that p_k for $k > 0$ is now actually a polynomial of a degree $2k - 1$.

Thus the first two polynomials are,

$$p_1(z_j) = z_j, \tag{56}$$
$$p_2(z_j) = z_j^3 - \alpha_2 z_j, \tag{57}$$
$$\alpha_2 = \sum_j w_j z_j^4 / \sum_j w_j z_j^2, \tag{58}$$

The frequencies of rotation-vibration transitions in linear molecules between two vibrationally nondegenerate states can be expressed as polynomials in rotational quantum number J. In particular purely rotational transitions in such states can be described by a polynomial having only odd powers of $J = J_{upper}$

$$\begin{aligned}\nu(J \leftarrow J-1) &= 2BJ - 4DJ^3 + 2H(3J^5 + J^3) + 8L(J^7 + J^5) \\ &\quad + 2M(5J^9 + 10J^7 + J^5) + 4N(3J^{11} + 10J^9 + 3J^7)\ldots\end{aligned} \tag{59}$$

For light molecules like HF and higher J values higher order terms with powers of J up to J^{11} have to be employed to describe adequately the frequencies and thus this molecule may be used as a simple and convenient testing ground for comparing direct fitting of B, D, H,... constants with the fit using the orthogonal polynomials as an intermediate step. The actual application of the orthogonal polynomials is done by identifying $z_j \leftarrow J$ for transition j and for example Eqs. (56-58) become

$$p_1(J) = J, \tag{60}$$
$$p_2(J) = J^3 - \alpha_2 J, \tag{61}$$
$$\alpha_2 = \sum_J w_J J^4 / \sum_J w_J J^2, \tag{62}$$

where the sum in the equation (62) runs over J for all transitions (59) included in the fit. The recursive formula used for generation of the orthogonal polynomials is ideally suited for easy computer programming. We have made four fits of $n = 30$ ground state rotational lines of HF for $J = 1-7, 13-35$ [49]. No attempt has been made to improve the values of rotational constants which were determined accurately from a larger data set in Ref. [49]. Our interest was to see the difference between the two fits, one obtained from the direct fit of Eq. (59) and the other one using the orthogonal polynomials according to Eq. (52).

In one pair of fits we have fitted five rotational constants B, D, H, L, M and coefficients $c_1, \ldots c_5$ of the orthogonal polynomials and the results are presented in TABLE 3.

In the other pair of fits presented in TABLE 4 we have attempted to fit in addition also the constant N and the coefficient c_6. The application of the orthogonal polynomials in all cases brought down by several orders of magnitudes the conditional number of the design matrix $\boldsymbol{J}$. For the fits with five coefficients the conditional numbers were $\kappa_{dir} = 5.9 \times 10^{13}$ and and $\kappa_{ortho} = 1.0 \times 10^{10}$ for direct fit of Eq. (59] and fit using the orthogonal polynomials respectively, and the same numbers for the fits with 6 parameters are $\kappa_{dir} = 9.5 \times 10^{16}$ and

$\kappa_{ortho} = 3.1 \times 10^{13}$. If the design matrix is scaled by dividing each element of the column by the norm of the column then the conditional numbers of these scaled design matrices are $\kappa_{dir} = 740$ and 3990 respectively for the fits with 5 and 6 rotational constants while $\kappa_{ortho} = 1$ always for the orthogonal polynomials.

TABLE 3. The Fits up to Decadic Terms ($m = 5^a$).

Digits	Standard[b]		Orthogonal[c]	
	$10^5.\sigma_{fit}^d$	$10^{15}.M_0^e$	$10^5.\sigma_{fit}^d$	$10^{14}.c_5^f$
6	1886(4844)[g]	0.7(1.1)[g]	52	1.322
8	33.9(12.4)[g]	1.0	4.6	1.051
10	4.223(4.154)[g]	0.997	4.061	0.9993
12	4.239	0.998	4.062427	0.999847
14	4.066	0.9972	4.062404	0.999832
16	4.169	1.0069	⋮	⋮
18	4.0627	1.0002	⋮	⋮
20	4.062404	0.999832	⋮	⋮
⋮	⋮	⋮	⋮	⋮
40	4.062404	0.999832	4.062404	0.999832

[a] Five parameters were included in the fit.
[b] The parameters were fitted according to Eq. (59)
[c] The orthogonal polynomials were used in the fit according to Eq. (52)
[d] The standard deviation of a fit.
[e] The decadic parameter M_0 of the ground state.
[f] The coefficient of $p_5(J)$ in Eq.(52). ($c_5 = 10M_0$)
[g] The round-off errors can produce different results

MapleV algebraic software provides a suitable environment for calculations where dependence of the results on the precision expressed by number of decadic digits, the program uses during calculations, is of interest, because the number of digits can be varied by a software command. The results are presented for various number of digits in TABLE 3 and TABLE 4 for the fits with five and six parameters respectively. The standard deviation of a fit σ_{fit} has been calculated as

$$\sigma_{fit} = \left(\frac{\sum_J w_J (\nu_J - \nu_J^{calc})^2}{n - m} \right)^{1/2} \tag{63}$$

and this value together with the value of the highest fitted parameter are shown to demonstrate how the fits approach with increasing number of digits used in the calculation the exact values which have been obtained with 40 decadic digits precision. For low number of digits sometimes the round-off errors are such that the results are not reproducible and two numbers for some entries document it. For 16 decadic digits in Table 2 in direct approach the design matrix is singular and the system blows up. We can see that the number of decadic digits required for exact result is in both cases by six digits lower for the orthogonal polynomials (14 versus 20 for five parameters and 20 versus 26 for six parameters). The answer to the question whether to apply the orthogonal polynomials or not depends on your hardware. If you are using hardware carrying high enough number of decimal digits in the calculation then the orthogonal polynomials are not necessary. If that is not the case the orthogonal polynomials provide a simple, easily programmable way how to improve the precision.

TABLE 4. The Fits up to Dodecadic Terms ($m = 6^{a}$).

Digits	Standard[b]		Orthogonal[c]	
	$10^5.\sigma^d_{fit}$	$10^{20}.N^e_0$	$10^5.\sigma^d_{fit}$	$10^{18}.c^f_6$
6	2486(8645)[g]	-1090(-1260)[g]	7281	-5164
8	488(484)[g]	-604(-602)[g]	45.0	-318.0
10	1928	1470	44.9	-317.8
12	1459	-915	1518	1077
14	1807	233	4.172	-0.3688
16	–[h]	–[h]	4.322	0.6856
18	34.6	0.1611	4.14296	-0.1090
20	4.1464	-0.7472	4.1429379	-0.11623
22	4.14293	-0.9673	⋮	⋮
24	4.1429379	-0.968592	⋮	⋮
26	4.1429379	-0.968584	⋮	⋮
⋮	⋮	⋮	⋮	⋮
40	4.1429379	-0.968584	4.1429379	-0.11623

[a] Six parameters were included in the fit.
[b] The parameters were fitted according to Eq. (59)
[c] The orthogonal polynomials were used in the fit according to Eq. (52]
[d] The standard deviation of a fit.
[e] The dodecadic parameter N_0 of the ground state.
[f] The coefficient of $p_6(J)$ in Eq. (52].$c_6 = 12N_0$
[g] The round-off errors can produce different results
[h] The matrix of normal equations was singular

4.References

1. Albritton, D.L., Schmelltekopf, A.L., and Zare, R.N. (1976) in K. Narahari Rao (ed.), *Molecular Spectroscopy: Modern Research* Vol. II, p. 1, Academic Press, New York.
2. Pracna, P., Sarka, K., Demaison J., Cosléou J., Herlemont, F., Khelkhal, M., Fichoux, D., Papoušek, D., Paplewski M., and Bürger H. (1997), *J. Mol. Spectrosc.* **184**, 93-105.

3. Halonen, L., Kauppinen, J., and Caldow, G.L. (1987), *J. Chem. Phys.***81**, 2257-2269. (1984) ; **86**, 5888-5889.

4. Pickett,H. M. (1991) The Fitting and Prediction of Vibration-Rotation Spectra with Spin Interactions, *J. Mol. Spectrosc.***148**, 371-377.

5. Aliev, M.R., and Watson, J.K.G. (1985) in K. Narahari Rao (ed.),*Molecular Spectroscopy: Modern Research*, Vol.III, Academic Press, San Diego, pp. 1-67.

6. Sarka, K. and Demaison J. (2000) Perturbation theory, Effective Hamiltonians and force constants, in P. Jensen and P.R. Bunker (eds.), *Computational Molecular Spectroscopy*, John Wiley & Sons, Chichester, pp. 255-303.

7. Sarka,K., and Demaison,J. (1997), *J. Mol. Spectrosc.***185**, 194-96.

8. Schneider,W., and Thiel, W. (1992) *Chem. Phys.*, **159**, 49-66.

9. Demaison, J., Margulès, L., Breidung, J., Thiel, W., and Bürger, H. (1999)*Mol. Phys.* **97**, 1053-1067.

10. Press, W.H., Flannery, B.P., Teukolsky, S.A., and Vetterling, W.V. (1988) *Numerical Recipes, The Art of Scientific Computing*, Cambridge: Cambridge University Press.

11. Brodersen, S. (1990), *J. Mol. Spectrosc.* **142**, 122-138.

12. Demaison, J., Le Guennec, M., Wlodarczak, G., and Ban Eijck, B.P. (1993), *J. Mol. Spectrosc.* **159**, 357-362.

13. Bocquet, R., Demaison, J., Poteau, L., Liedtke, M., Belov, S., Yamada, K.M.T., Winnewisser, G., Gerke, C., Gripp, J., and Köhler, Th. (1996), *J. Mol. Spectrosc.* **177**, 154-159.

14. Bocquet, R., Demaison, J., Cosléou, J., Friedrich, A., Margulès, L., Macholl, S., Mäder, H., Beaky, M.M., and Winnewisser, G. (1999), *J. Mol. Spectrosc.* **195**, 345-355.

15. Kinsella, A. (1986), *Amer. J. Phys.* **54**, 464-464.

16. Demaison, J., Burie, J., Boucher, D., and Wlodarczak, G. (1991), *J. Mol. Spectrosc.* **146**, 455-464.

17. Wlodarczak, G., Martinache, L., Demaison, J., Marstokk, K.-M., and Møllendal, H. (1988), *J. Mol. Spectrosc.* **127**, 178-185.

18. Vormann, K., and Dreizler, H. (1988), *Z. Naturforsch.* **43a**, 338-344.

19. Belsley, D.A. (1991)*Conditioning Diagnostics*, Wiley, New York.

20. Kawashima, Y., Endo, Y., and Hirota, E. (1989), *J. Mol. Spectrosc.* **133**, 116-127.

21. Kuchitsu,K. (1968), *J. Chem. Phys.* **49**, 4456-4466.

22. Kuchitsu, K., Fukuyama, T., and Morino, Y. (1969), *J. Mol. Struct.***4**, 41-50.

23. Branham, R.L. (1990) *Scientific data Analysis*,Springer, New York.

24. Sen, A., and Srivastava, M.(1990) *Regression Analysis*, Springer, New York.

25. Rudolph, H. D., (2001) Private communication.

26. Watson, J.K.G., (1967), *J. Chem. Phys.***46**, 1935-1949.

27. Watson, J. K. G. (1977) in Durig, J. (ed.), *Vibrational Spectra and Structure* Vol.6, Elsevier, Amsterdam, pp. 1-89.

28. Sarka, K., Papoušek, D., Demaison, J., Mäder, H., and Harder, H. (1997), in Papoušek, D. (ed.) *Advances in Physical Chemistry, Vibration-Rotational Spectroscopy and Molecular Dynamics*, Vol.9, World Scientific Publishing, Singapore, pp. 116-239.

29. Watson, J. K. G., (1966), *J. Chem. Phys.***45**, 1360-1361.

30. Watson, J. K. G., (1968), *J. Chem. Phys.***48**, 181-185.

31. Papoušek, D., and Aliev, M. R. (1982) *Molecular Vibrational-Rotational Spectra*, Elsevier, Amsterdam, New York.

32. Bunker, P. R., and Jensen, P., (1998) *Molecular Symmetry and Spectroscopy*, NRC Research Press, Ottawa .

33. Sarka, K., (1989), *J. Mol. Spectrosc.***133**, 461-466.

34. Sarka, K., (1989), *J. Mol. Spectrosc.***134**, 354-361.

35. Perevalov, V. I., and Tyuterev, V. G., (1981), *Opt. Spectrosc.***51**, 354-357.

36. Perevalov, V. I., and Tyuterev, V. G., (1982), *J. Mol. Spectrosc.* **96**, 56-76.

37. Perevalov, V. I., and Tyuterev, V. G., and Zhilinskii, B. I., (1984), *J. Mol. Spectrosc.* **103**, 147-159.

38. Perevalov, V. I., and Tyuterev, V. G., (1984), *Chem. Phys. Lett.***104**, 455-461.

39. Tyuterev, V. G., Champion, J. P., Pierre, G., and Perevalov, V. I., (1984), *J. Mol. Spectrosc.* **105**, 113-138.

40. Nikitin, A., Champion, J. P., and Tyuterev, V. G., (1997), *J. Mol. Spectrosc.* **182**, 72-84.

41. Lobodenko, E. I., Sulakshina, O. N., Perevalov, V. I., and Tyuterev, V. G., (1987), *J. Mol. Spectrosc.* **126**, 159-170.

42. Bürger, H., Cosléou, J., Demaison, J., Gerke, C., Harder, H., Mäder, H., Paplewski, M., Papoušek, D., Sarka, K., and Watson , J. K. G., (1997), *J. Mol. Spectrosc.* **182**, 34-49.

43. Watson, J. K. G., Gerke, G., Harder, H., and Sarka, K., (1998), *J. Mol. Spectrosc.* **187**, 131-141.

44. Teffo, J.-L., Sulakshina, O. N., and Perevalov, V. I., (1992), *J. Mol. Spectrosc.* **156**, 48-64.

45. Teffo, J.-L., Perevalov, V. I., and Lyulin, O. M., (1994), *J. Mol. Spectrosc.* **168**, 390-403.

46. Sarka, K., and Schrötter, H. W., (1996), *J. Mol. Spectrosc.* **179**, 195-204.

47. Tang, J., and Takagi, K., (1993), *J. Mol. Spectrosc.* **168**, 487-498.

48. Ralston, A., and Rabinowitz, P. (1978) *A First Course in Numerical Analysis*, 2nd ed. New York: Mc Graw Hill.

49. Ram, R.S., Morbi, Z., Guo, B., Zhang, K.Q., Bernath, P. F., Vander Auwera, J., Johns, J.W.C., and Davis, S.P. (1996), *Astrophys. J. Suppl. Series***103**, 247-254.

50. LeBlanc, R.B., White, J.B., and Bernath, P.F. (1994), *J. Mol. Spectrosc.* **164**, 574-576.

METHANOL SPECTROSCOPY FROM MICROWAVE TO INFRARED — FUNDAMENTALS AND APPLICATIONS

LI-HONG XU

Department of Physical Sciences
University of New Brunswick
Saint John, N.B., Canada E2L 4L5 (xuli@unb.ca)

1. Introduction

The importance of methanol microwave and millimeter wave spectroscopy to space science and astrophysics can be traced back to several decades ago when methanol was first discovered in interstellar clouds and star forming regions [1]. The rich variety of torsion-rotational transitions falling in the frequency bands accessible to most radio telescopes leads to a dense interstellar spectrum and demands an accurate knowledge of the methanol energy levels so that the interstellar "methanol weeds" can be removed. The infrared (IR) spectroscopy of methanol has also acquired renewed importance in wide areas of application in recent years, such as the recent observations of the 10 μm feature in forest fires [2] and the 3 μm features in several comets and the icy mantles of interstellar dust grains [3-6]. Cometary spectra have been seen at different resolving powers in different telescopes with the observed methanol column density varying significantly from one comet to another. For these applications, increased demands for high quality methanol spectroscopic information are emerging not only simply for line positions but also for line intensities, for which no information is presently available in the HITRAN database [7].

At the University of New Brunswick in Canada, we have been engaged in a broad range of experimental and theoretical studies of the spectroscopy of methanol aimed at supplying benchmark laboratory data to user communities and to the HITRAN database. Our investigations can be grouped into five principal categories discussed in detail below in Sections 2 to 6.

The notation used in the following discussion is briefly summarized here. Energy levels are labeled with the quantum numbers (v_i, v_t, τ, K, J) $^{vt}\Gamma$ where v_i (i = 1 to 11) denotes the 11 small-amplitude vibrations in contrast to the large-amplitude torsion v_{12} or v_t. (n is often used as an alternative notation to v_t.) K is the projection of the overall rotational angular momentum J along the *a*-axis, which points approximately along the symmetry axis of the methyl top. τ is Dennison's torsional index [8], with τ = 1, 2 or 3, in terms of

J. Demaison et al. (eds.), Spectroscopy from Space, 131–146.

which the torsional energies vary smoothly with K along three identical but differently phased τ-curves. The τ notation is related to the torsional symmetry species ($\sigma = 0, 1, -1$ for ${}^{t}\Gamma$ = A, E_1 and E_2) by the rule $(\tau+K)\text{mod}3 = 1 - \sigma = 0, 1, 2$ for E_1, A and E_2, respectively. Note that the $+KE_1$ levels are degenerate with the $-KE_2$ levels, hence it is important to realize that only the A and E species constitute distinguishable torsional families and that E_1 and E_2 species belong to the same E torsional family. In the customary molecular symmetry group notation [9], ${}^{vt}\Gamma$ represents the overall symmetry of a vibration-torsion state, for which the symmetry of the small-amplitude vibration (${}^{v}\Gamma$) is combined with that of the large-amplitude torsion (${}^{t}\Gamma$). The latter is either ${}^{t}A_1$ or ${}^{t}E$ in the G_6 permutation-inversion (PI) group (isomorphic to C_{3v}). The symmetry species for the 11 small-amplitude vibrations are ${}^{v}\Gamma = 5A_1+3E$ in G_6 (C_{3V}: ${}^{v}\Gamma = 8A_1+3A_2$ or Cs: ${}^{v}\Gamma = 8A'+3A''$). These above notations are used interchangeably in the literature.

2. Large-Amplitude Torsion and One-Dimensional Torsion-Rotation Hamiltonian – Interstellar Methanol

Methanol is a slightly asymmetric molecule with a plane of symmetry traditionally chosen as the *a-b* plane where the *a*-axis is the C_3 axis of the methyl rotor and the *c*-axis is perpendicular to the plane as shown in Fig. 1. The three-fold methyl rotor ($-CH_3$) undergoes a large-amplitude hindered internal rotation against its framework (HO-C) which is frequently referred to as torsion. The potential energy varies with period $2\pi/3$ in the torsional angle γ as depicted in Fig. 2. The torsional potential barrier height (V_3) is about 373.6 cm^{-1} (1.068 kcal/mol) for CH_3OH, representing a medium barrier case. The ground torsional level ($v_t = 0$ or $v_{12} = 0$, called the ground state of the molecule where $v_i = 0$ for $i = 1,\ldots 11$) lies at the zero-point torsional energy (ZPE) of 127 cm^{-1}, with an A-E torsional splitting of 9.12 cm^{-1}, well below the top of the barrier, while the A and E sublevels of the first excited torsional state ($v_t = 1$ or $v_{12} = 1$) straddle the top of the barrier. Because the *a* and *b* axes lying in the plane of symmetry are not principal axes, the kinetic energy initially contains non-zero P_aP_γ and P_bP_γ coupling terms when written in the 4-dimensional angular momentum space (P_a, P_b, P_c, P_γ), where P_a, P_b, P_c are the components of the rotational angular momentum and P_γ is the torsional angular momentum. Traditionally, to solve the torsion-rotational problem, one (i) eliminates the P_bP_γ term by rotating the *a-b* plane about the *c*-axis; and then may (ii) further separate torsion and rotation by eliminating the P_aP_γ term in the Hamiltonian by additional rotation about the new *a*-axis after step (i). The case employing step (i) alone is called the rho axis method (RAM) in which a separation between the rotational wavefunction $|J,K\rangle \propto (e^{iK\chi})$ and torsional wavefunction $|m\rangle \propto (e^{im\gamma})$ is retained. For case (ii), however, called the internal axis method (IAM), the additional rotation about the *a*-axis destroys the periodic boundary condition for the torsion so that the overall wavefunction now requires a "make-up" phase factor ($e^{i\rho K\gamma}$) to retain its mandatory periodicity. Here, χ is the rotational angle and ρ is the ratio of the axial moment of inertia of the methyl top (I_γ) to that of the whole molecule (I_a). The variable $(1 - \rho)$ acts as a scaling factor for the τ-curves, which oscillate as a function of $(1 - \rho)K$ with period 3 [10].

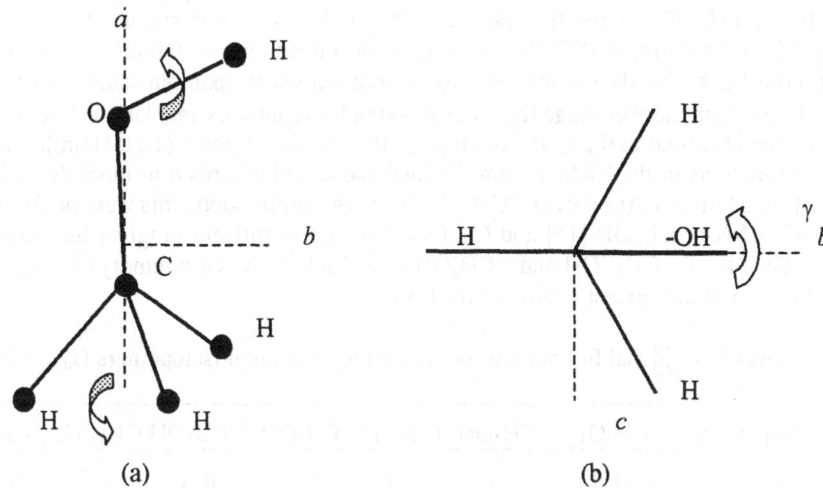

Fig. 1 (a) Methanol molecule and (b) Norman projection with *a, b, c* axes and torsional angle γ.

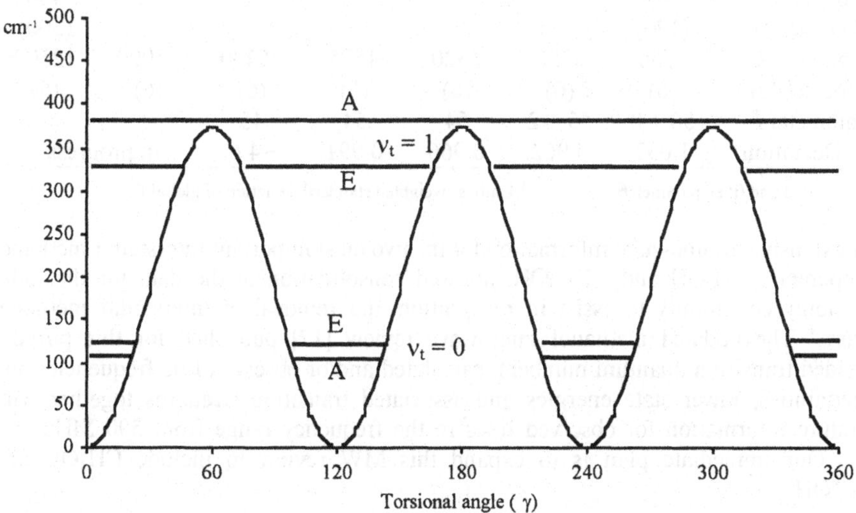

Fig. 2 Torsional potential as a function of torsional angle γ. Horizontal lines indicate K=0 torsional energy levels with respect to the bottom of the barrier.

In our analysis so far of the pure torsional motion in the vibrational ground state of CH_3OH, we have succeeded in globally modeling all reported spectroscopic observations for the first two torsional levels (v_t = 0 and 1) up to J_{max} = 20 to within experimental accuracy. Our global fit includes A and E torsional species simultaneously, and employs 56 adjustable and 8 fixed parameters [11]. The data set contains a large body of about 7000 carefully scrutinized literature measurements of microwave (MW — mostly assigned a 50 kHz measurement uncertainty), tunable far-infrared (TUFIR — mostly assigned 100 or 200 kHz uncertainties) and Fourier-transform far-infrared (FTFIR — assigned an uncertainty of 0.0002 cm^{-1} = 6 MHz) spectral lines. The data cover not only a large frequency span but more importantly a wide quantum number range ($K \leq 12$) allowing for reliable extrapolation. The computer program [12] used in the global fit employs the one-dimensional (1-D) Hamiltonian and wavefunctions in the RAM system outlined above and described in more detail in Ref. [13]. In addition to the parent $^{12}CH_3{}^{16}OH$ species, similar global fits were performed for $^{13}CH_3OH$ [14], CD_3OH [15] and CH_3OD [16], and significant progress has been made on global fits for $CH_3{}^{18}OH$ and $^{13}CD_3OH$ also. Table 1 gives a summary of the global fit data sets and the current quality of the fitting.

TABLE 1. Global fit data sets and results for methanol isotopomers (J_{max} = 20)

Isotopomers	CH_3OH	$^{13}CH_3OH$	CD_3OH	CH_3OD	$^{13}CD_3OH$	$CH_3{}^{18}OH$	CH_3OH
v_t	0 1	0 1	0 1	0 1	0 1	0 1	0 1 2
# of MW	909	725	472	447			1520
(Unc. kHz)	(50)	(50)	(100)	(100)			(50)
# of TuFIR	197						450
(Unc. kHz)	(100)						(300)
# of FTFIR	5550	6283	5320	4575	~5300	~5000	12500
(Unc. MHz)	(6)	(6)	(6)	(6)	(6)	(6)	(6)
Parameters[a]	56 + 8	55 + 2	54	54	45		
Std. Deviation[b]	1.032	0.962	0.966	0.991	~4.6	in progress	

[a] Fitted and fixed parameters [b] Unitless weighted standard deviation of global fit.

Our extensive ground-state information for the two most important interstellar methanol isotopomers, CH_3OH and $^{13}CH_3OH$, allowed consolidation of the data for the radio astronomy community to assist in recognition and removal of interstellar methanol "grass." The updated methanol microwave review [17] published for this purpose provided transition quantum numbers, calculated and/or observed line frequencies and uncertainties, lower state energies and associated transition strengths together with literature information for observed lines in the frequency range from 500 MHz to 1 THz. Our immediate plan is to expand this MW review to include CH_3OD and $CH_3{}^{18}OH$.

In future, to test further the existing 1-D model and to prepare for modeling of the small-amplitude vibrations, we need to extend our global fitting to higher torsional states, i.e. v_t = 2, 3 and up, which are located above the torsional barrier. Global fitting of the v_t = 2 state is of particular interest on theoretical grounds, as it lies in the transition region between the strongly bound torsional case and the almost completely free rotor limit so will serve as a stringent test of the 1-D model. Thanks to the large body of methanol literature information available for these highly excited torsional levels [18], transition sets associated with the high torsional states are ready to be included in our global fit data. We have so far achieved a unitless weighted standard deviation approaching unity in preliminary fitting of an augmented data set including more than 7,000 FTFIR transitions terminating on $v_t = 2$ levels. We are currently working on the MW data for the associated states and for $v_t = 3$ levels in order to complete the modeling of the torsional bath to our satisfaction.

To conclude discussion of the purely torsional motion and to sum up the present situation, we believe it is fair to say that the existing 1-D Hamiltonian and computer program are capable of modeling the torsional energy ladder to an accuracy sufficient for most applications. What seemed the major stumbling block in the torsional modeling turned out in most cases to be misassignments or, more subtle and difficult to establish, misreported measurements involving discrepancies as large as ±3 MHz [11]. The need to critically evaluate all reported measurements and to carefully "clean" the data sets prior to fitting resulted in a significant slowing of the global modeling process. Nevertheless, we are now confident that, with some effort, the analysis and understanding of the torsional bath will be under control in the not-too-distant future.

3. Infrared Fundamentals and Observed vs. Calculated Vibration-Torsion-K-Rotational Energy Structure

The locations of the fundamental IR absorption bands of the 11 small-amplitude vibrations of CH_3OH between 10 and 2.6 μm are indicated in the low resolution spectrum of Fig. 3. The bands are clustered in two regions. The lower frequency group

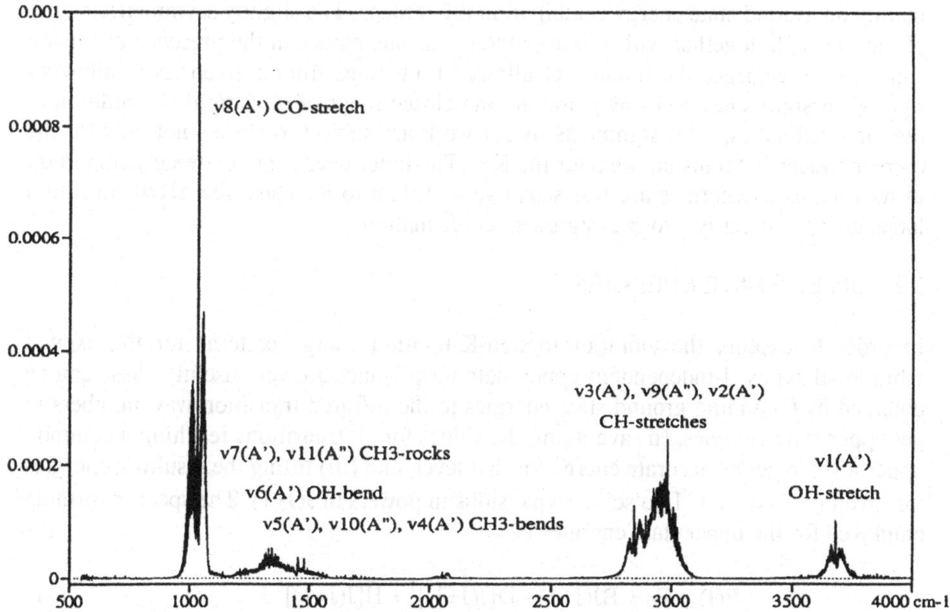

Fig. 3. Low resolution IR spectrum of CH_3OH (recorded by Dr. Pam Chu at NIST), showing absorption (in arbitrary unit) regions associated with the various vibrational fundamentals.

begins with the strongest ν_8 (A_1) C-O stretch mode centered at 1033 cm^{-1}, and then extends upwards to about 1600 cm^{-1} in a broad band of weak and relatively featureless absorption. It is in this region that 6 fundamental modes are predicted, namely the ν_7 (A_1) in-plane and ν_{11} (A_2) out-of-plane CH_3 rocks, the ν_6 (A_1) OH bend, and the ν_5 (A_1), ν_{10} (A_2), and ν_4 (A_1) CH_3 bends. In the higher frequency cluster in the 3 μm region, the fundamental motions are the stretches involving the light H-atoms, i.e. the ν_3 (A_1), ν_9 (A_2), and ν_2 (A_1) trio of C-H stretches, and the ν_1 (A_1) O-H stretch. These are especially important to the understanding of the molecular internal dynamics and the transfer channels for intra-molecular vibration energy redistribution (IVR). In our high resolution spectra, the individual line profiles are generally fully resolved for the lower frequency region, but gradually become more and more overlapped in moving to the high frequency range. Here, numerous underlying weak features arise from excited torsional bands and/or overtones and combination bands from the low-lying vibrations. Currently, detailed spectroscopic analyses have either been carried out or are actively in progress for each of the spectral regions above.

3.1. SUBBAND ASSIGNMENT

The first step in the spectral analysis of a band is to group lines into J-sequences forming R-, P- and/or Q-subbranches. Related trios of subbranches are then matched and linked into subbands in which the $(v_i', v_t', K') \leftarrow (v_i'', v_t'', K'')\, ^{vt}\Gamma$ transitions share all the same transition quantum numbers but J. Here, the ' and "denote the upper and lower states. The subbranch linking and consequent subband assignments are based mainly on ground state energy combination differences. The slightly asymmetric nature of the molecule together with mixing among various modes in the presence of torsion significantly enlarges the number of allowed transitions from a given level, allowing rigorous assignment checks by going around closed loops of transitions. In addition to the torsional states, all assignments in our work are subject to checks not only for the correct J identifications but also for the K's. The latter checks are extremely important in most cases as energies are less sensitive to J than to K. Thus, the $\Delta K \neq 0$ transition loops give particularly strong assignment confirmation.

3.2. UPPER STATE ENERGIES

In order to explore the vibration-torsion-K-rotation energy patterns for the excited vibrational states, J-independent upper state term values are very useful. These can be obtained by (i) adding ground state energies to the infrared transition wavenumbers to get upper state energies, (ii) averaging the values for all transitions reaching a common upper level to get an accurate energy for that level, and (iii) fitting the resulting energies for given $(v_i', v_t', K')\, ^{vt}\Gamma$ to series expansions in powers of J(J+1). The specific formula employed for the upper state energies is:

$$E(J) = E_0 + BJ(J+1) - D[J(J+1)]^2 + H[J(J+1)]^3 + \ldots \quad (1)$$

where E_0 is the J-independent term value, and B, D and H are constants. The former can be written as

$$E_0 = E_{vib} + E_{tor} + (A - B)K^2 \quad (2)$$

where E_{vib} is purely vibrational energy, E_{tor} is the upper state torsional energy, and (A-B)K^2 is the K-rotational energy. To highlight the torsional problem, the K^2 term can initially be calculated using the ground state rotational constants and then subtracted off. The remainder, $E_0-(A-B)K^2=E_{vib}+ E_{tor}$ is often called the K-reduced vibration-torsion energy, and can readily be plotted as a function of K for each τ as shown in Fig. 4.

Fig. 4 collects all observed K-reduced vibrational substate energies to date as derived from assignments for the methanol IR regions up to 1650 cm^{-1} covering most of the fundamentals as well as torsional hot bands. In this frequency region, only the two CH_3-bending modes ν_5 and ν_{10} remain to be identified. However, an extensive network of torsion-mediated interactions couples and mixes the various modes, perturbing the expected regular oscillatory τ-curve patterns, so that the precise torsional and/or vibrational identities of a number of the upper substates are still not definitely established. The coupling is particularly complex in the higher frequency 3 μm region, the K-reduced energy pattern for which will be discussed in Sec. 5, where the overtones and combination states of the CH_3-bending modes lie very close to and interact strongly with the CH-stretching fundamentals [19]. So far, the $2\nu_4$ overtone has been firmly identified and the $\nu_4 + \nu_{10}$ combination tentatively, but this still leaves a further 4 overtone and combination bands to be located. This enhances the need to identify the two missing bending fundamentals in order to characterize their behaviour and torsional energy structure, and the search for these bands is actively in progress.

3.3. COMPARISON OF VIBRATION-TORSION-K-ROTATION ENERGIES — OBSERVED VS. CALCULATED

The vibration-torsion-K-rotational energy patterns which are calculated using the 1-D Hamiltonian discussed in section 2 above, assuming no interaction between the vibrations and torsion, are plotted in Fig. 5. In the absence of any coupling, the torsional patterns in Fig. 5 are simply the ground-state torsional energies E_{tor} superimposed onto the vibrational energies E_{vib}. Clearly, this is not the situation in reality. Comparing Figs. 4 and 5, we see that the observed curves resemble the calculated ones in some cases, but deviate significantly in others. The discrepancies in the observed curves range from isolated points being pushed away from their predicted positions to more extreme cases where a set of curves at first glance seems regular but on closer examination is found to have a pattern completely reversed from the expected one (such as the out-of-plane CH_3-rocking mode at 1290 cm^{-1}).

If we concentrate only on the K = 0 torsional E and A energy ordering for the ground torsional substates ($\nu_t = 0$) of the small-amplitude vibrations, we find that the vibrations assigned at high resolution fall into two categories: those with normal E and A ordering similar to the $\nu_t = 0$ ground state (i.e. degenerate E levels lying above the A level), and others possessing inverted E and A ordering (i.e. the degenerate E levels now located below the A level). This inverted torsional pattern was very surprising to us in the beginning, but has now been successfully explained both by a local mode approach for

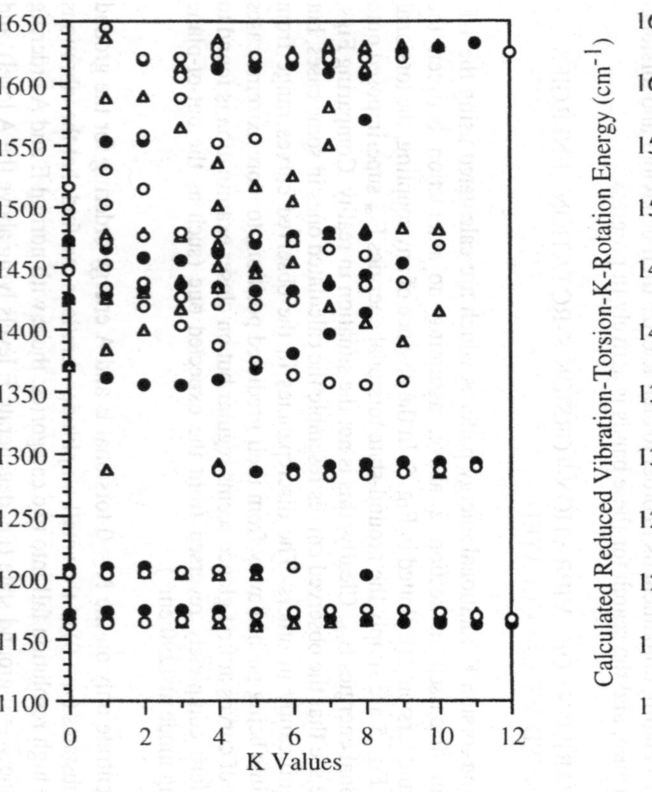

Fig. 4 Observed methanol reduced vibration -torsion-K-rotation energy up to 1650 cm^{-1}.

Fig. 5 Calculated methanol reduced vibration -torsion-K-rotation energy up to 1650 cm^{-1}.

the 3 CH-stretches [20], and from a purely group theoretical point of view covering more general cases [21].

At the present time, extensive assignments are available for various IR fundamental regions. Thus, the main challenge now facing us is how to model the observed vibrational state energy patterns. While new ideas [20,21] are emerging for possible explanations of some of the observations, the implementation process will need some time. In the meantime, one can still proceed along two alternative directions in order to meet the urgent needs of the various communities: (i) For those states having normal torsional patterns, we can tentatively ignore the local perturbations and treat the torsion in the excited vibrational states in the same way as described in section 2. Only slight modifications are needed in the computer program in order to model the IR line positions for a hypothetical non-perturbed energy structure, hence intensities can also be predicted as discussed below in the next section; (ii) For the states with inverted torsional patterns such as those observed in the 3 μm region, we are exploring a multi-state treatment using an extended local mode formalism which is discussed further in section 5.

4. Isolated Band Modeling — Positions and Intensities

4.1. LINE POSITIONS

Among the assigned IR bands, "regular" torsional oscillatory patterns similar to that of the ground state are found for the ν_8 (A_1) CO-stretch, ν_3 (A_1) symmetric CH-stretch and ν_1 (A_1) OH-stretch modes, with only occasional points perturbed away from their expected positions. Analysis of these vibrational modes, therefore, could be attempted with an isolated band model employing the following steps: (i) points associated with local perturbations were excluded from our fit data sets; (ii) the ground-state parameters were fixed to the values from Ref. [11]; (iii) the vibrational state energy, E_{vib}, was set to the band center value determined from low resolution spectra [22] to remove the serious correlation problem with the excited state barrier known to be a general difficulty [23]; and (iv) the low order parameters for the excited vibrational state were adjusted, with the remaining high order terms fixed to the ground state values, by minimizing the overall obs. – calc. residuals. The results were reasonably encouraging, as we achieved global fits with unitless rms deviations of about 94x and 190x for ν_8 and ν_3, respectively. In the case of ν_8, our global fit data set includes the ν_8 fundamental (i.e. $(\nu_8, \nu_t) = (1, 0) \leftarrow (0, 0)$) as well as the $\nu_8+\nu_t-\nu_t$ torsional combination band (i.e. $(\nu_8, \nu_t) = (1, 1) \leftarrow (0, 1)$). In addition, we covered quantum numbers up to $K_{max} = 12$ and $J_{max} = 20$ with altogether about 1600 fitted lines. The best fit was achieved with 28 adjusted parameters. In the case of ν_3, our global fit data set includes the main ν_3 band (i.e. $(\nu_3, \nu_t) = (1, 0) \leftarrow (0, 0)$) up to $K_{max} = 6$ and $J_{max} = 15$ with 510 fitted lines altogether. The best fit was achieved with 15 adjusted parameters.

4.2. LINE INTENSITIES — 10 μm FOR APPLICATION TO FOREST FIRES

Our focus has now shifted significantly towards line intensity measurement and modeling, for application to forest fires in the 10 μm region and comets in the 3 μm region. In collaboration with Linda Brown at JPL, measurements of individual line intensities are in progress for both the 10 μm and 3 μm regions. Our ultimate aim is to produce a line list for user communities, similar to our methanol MW review aimed at the radio astronomy community, which will include line positions, transition quantum numbers, line intensities and lower state energies. As a warm-up exercise, we started first with the 10 μm CO-stretching region for application to forest fires. The methanol 10 μm feature together with O_3 and NH_3 have been observed in several California forest fires by NASA overflights [2]. For individual line intensity measurements, methanol spectra at different optical depths were recorded on the Kitt Peak Observatory Fourier-transform instrument by Linda Brown at JPL. Although the absolute line intensities still remain to be established by appropriate averaging over several spectra at different optical densities, some of the initial 10 μm intensity measurements have now been fitted to the isolated band model described in Sec.4.1, employing a two-parameter μ_a dipole moment function. A preliminary fit to about 300 measured intensities gave a mean residual of 8x the estimated measurement uncertainty for unperturbed lines, but at the same time exposed certain limitations of the non-interacting band approach. A segment of simulated spectrum is plotted in Fig. 6 where the bottom window shows the simulated spectrum (dashed line) superimposed on the observed one (solid line) and the top window plots the relative % errors.

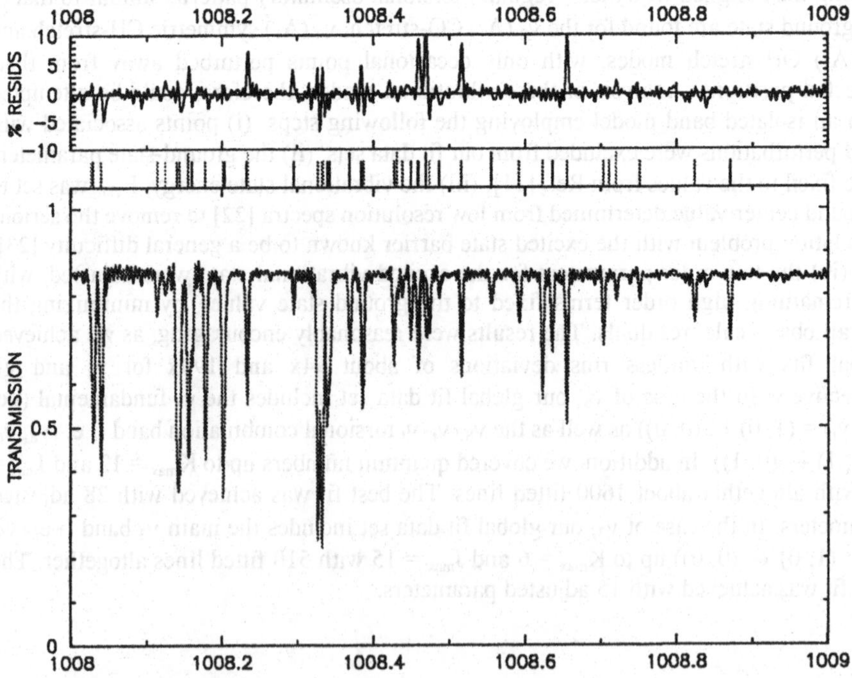

Fig. 6 Comparison of observed and computed CH_3OH CO-stretching spectra using Kitt Peak data (recorded by Dr. Linda Brown at JPL).

4.3. VIBRATIONAL BAND STRENGTHS AND BAND TYPES

A knowledge of the band intensity for each IR fundamental is of interest on its own, and even more so for the transition moment components in separate directions as they give information on the types of band one should expect. To investigate these, we carried out a detailed *ab initio* intensity study for the *A*-, *B*- and *C*- principal directions in order to gain information on the transition types for each IR mode. The calculations were performed using the Gaussian 98 (G98) *ab initio* package, with the focus here on the dipole derivatives $(\partial\mu_g/\partial\mathbf{x})$, where g = *A, B, C*, with respect to the atom displacement **x**, and the derivatives $(\partial\mathbf{x}/\partial Q_i)$ of those displacements with respect to the normal coordinates, Q_i (i=1, …12). The G98 calculation treats the torsion (ν_{12}) as one of the 12 independent vibrations and assumes the harmonic approximation. Once all *ab initio* quantities, $(\partial\mu_g/\partial\mathbf{x})$ and $(\partial\mathbf{x}/\partial Q_i)$, are transferred to the same coordinate system, the square of the transition moment $|(\partial\mu_g/\partial\mathbf{x})\cdot(\partial\mathbf{x}/\partial Q_i)|^2$ can be evaluated for each mode, giving information on the relative intensity contributions from the components in each direction. Table 2 presents our *ab initio* intensity calculations in the principal axis system, where I(*A*), I(*B*) and I(*C*) represent the contributions from the dipole components in the three separate directions and the sum, I_{tot} = I(*A*)+I(*B*)+I(*C*), is the overall band strength in units of km/mol. In Table 2, we also list the observed IR frequencies and relative intensities from Ref. [22] for comparison. It is interesting to note that the *ab initio* transition dipole ratios for several vibrations are in good agreement with the observations.

TABLE 2. Methanol IR band frequencies and intensities from *ab initio* calculations and comparison with literature experimental values.

Mode	Description[a]	Symm.	Ab Initio Results Frequency (cm^{-1})	Intensity (km/Mol) Total	A^b	B^b	C^b	Experiment (Ref.22) Frequency[c] (cm^{-1})	Intensity obs.	Intensity calc.
ν_{12}	CH_3 tor.	A_2	294.1	111.0			111.0	271.5		2.3
ν_8	C-O str.	A_1	1073.2	113.5	112.4	1.1		1033.5	2.4	2.0
ν_7	CH_3 rock	A_1	1102.5	8.4	8.2	0.2		1074.5	<0.05	0.08
ν_{11}	CH_3 rock	A_2	1197.6	0.4			0.4	1145±4	<0.05	0.01
ν_6	OH bend	A_1	1377.9	21.3	20.1	1.2		1332.0 1339.5	0.31	0.38
ν_5	s. CH_3 bend	A_1	1504.9	3.3	3.2	0.1		1454.5	0.21	0.18
ν_{10}	as. CH_3 bend	A_2	1532.4	3.4			3.4	1465±3		0.04
ν_4	as. CH_3 bend	A_1	1542.5	5.5	0.4	5.1		1477.2 1479.5		0.02
ν_3	s. C-H str.	A_1	3070.9	54.4	49.8	4.6		2844.2	0.77	1.1
ν_9	as. C-H str.	A_2	3140.5	40.4			40.4	2970±4		2.0
ν_2	as. C-H str.	A_1	3200.7	18.8	0.5	18.3		2999.0		0.85
ν_1	O-H str.	A_1	3906.6	42.5	17.5	25.0		3681.5	0.62	0.62

[a] Abbreviations are: as. - asymmetric, s. - symmetric, str. - stretch. CH_3-bend modes are often referred to as CH_3-deformations in the literature.

[b] Principal axis *A*, *B* and *C* directions.

[c] Gas phase experimental band origins from Ref. 22.

5. Stretch-Bend Interaction Model for the Methanol 3 μm Spectra for Application to Cometary Spectra

The 3 μm region is the location not only for the 3 CH-stretching fundamentals but also for the 6 CH_3-bending overtones and combinations, giving numerous possibilities for coupling among the states and significant perturbations to the energy patterns. In order to account for these perturbations properly, we need to treat the interacting states simultaneously. An initial approach to this problem is underway, with the 3 fundamental CH-stretches ((v_3,v_9,v_2) = (1,0,0), (01,0) and (0,0,1)) now coupled to the 6 CH_3-bending overtones and combinations ((v_5,v_{10},v_4) = (2,0,0), (0,2,0), (0,0,2), (1,1,0), (1,0,1), and (0,1,1)). We chose to attack this region in view of its important application to cometary spectra, with the aim of reproducing the observed vibration-torsion-K-rotation structure of both energies and intensities using an extended local mode approach. The initial Hamiltonian matrix is set up according to the polyad number, p = $2(v_3+v_9+v_2)+(v_5+v_{10}+v_4)$, which classifies the states into groups of similar energy. Symmetry-allowed Fermi and Coriolis interaction terms are then introduced to couple the states within a polyad. More specifically, p = 0 means the 1x1 torsional problem, p = 1 refers to the 3x3 submatrix for the CH_3-bending fundamentals, and p = 2 corresponds to the 9x9 situation for the 3 μm region in which the 3 CH-stretch fundamental modes (3x3) are now allowed to interact with the 6 CH_3-bending overtone and combination states (6x6) through Fermi and Coriolis coupling. In both the p = 1 and p = 2 cases, *a*-type Coriolis coupling off-diagonal terms are included in the 3x3 submatrices. While specific symmetry properties must be carefully taken into consideration for full implementation of the $\{v_{\text{CH-stretch}}=1/v_{\text{bending}}=2\}$ interactions, initial results are encouraging. Fig. 7 shows the comparison of experimental reduced energies for the 3 CH-stretches of CH_3OH with early results from trial calculations based on a 3x3 model including a single *a*-type Coriolis interaction term with coupling parameter ζ=1.0 cm^{-1}. The 9x9 calculation including off-diagonal terms of both Fermi and Coriolis form is in progress.

In the extended local mode model above, the variation of the dipole moment as a function of torsional angle is needed to calculate the torsion-K-rotational structure of the vibrational intensities. This information has now been obtained using *ab initio* Gaussian 98 calculations. Initially, we will incorporate it into a completely dark state model for the CH_3-bending overtones and combinations interacting with the fundamental CH-stretches. Under the dark state assumption, any intensity seen for the overtones and combination bands (observed so far for $2v_4$ and v_4+v_{10}) must arise solely through intensity borrowing from the bright CH-stretch fundamentals via wavefunction mixing. We hope to use this model as a guide to search for still missing CH_3-bending overtone and combination bands in the region.

6. *Ab Initio* Assisted Molecular Dynamics

The complicated vibration-torsion-K-rotation structure observed in the excited state region, illustrated in Figs. 4, 5 and 7, indicates large torsional effects on the small-

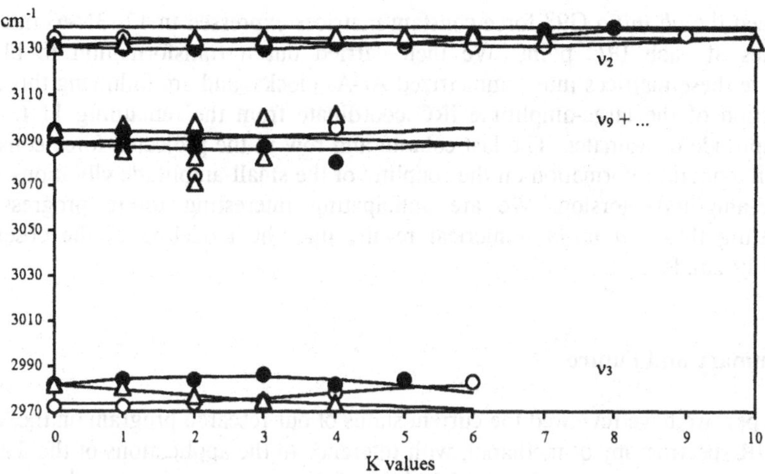

Fig. 7 Reduced vibration-torsion energies in the CH-stretch fundamental region for CH_3OH. Points represent experimental observations; the curves are calculated by a 3x3 local mode approach including an *a*-type Coriolis term with coefficient $\zeta = 1.0\ cm^{-1}$.

amplitude vibrational motions. Thus, any proper vibrational treatment must take torsional dependence into consideration. In practice, however, comprehensive information on the torsional coupling is difficult to obtain experimentally, hence we have been exploring an *ab-initio*-assisted approach as an alternative route to obtain the variation of structural and force field parameters with torsional angle γ. Again, the Gaussian 98 package is being used for this purpose. In the first phase of this study, to test the accuracy that the *ab initio* technique could deliver, we calculated the structural flexing for CH_3OH at the top and the bottom of the torsional potential barrier, and interpreted the results in terms of a set of vibration-torsion-rotation interaction constants with a $(1\text{-}\cos 3\gamma)$ dependence [24]. Good agreement was found between the *ab initio* values of these parameters and our global fit results, leading us to believe that the current *ab initio* calculation is capable of delivering accurate information for high resolution spectroscopy. In the second phase [25], torsion-dependent projected vibrational frequencies for the CH_3-bends and the CH-stretches were calculated, treating them as sets of three equivalent local oscillators in a local mode environment, and were used to obtain higher order parameters appearing in the local mode Hamiltonian designed initially for the three CH-stretches [20]. We are currently in the third phase of this project [26], in which the variations of the structure and force field have been mapped from top to bottom of the torsional potential as a function of the *ab initio* intrinsic reaction coordinate (IRC). The detailed relation between the IRC and the torsional angle γ used in the spectroscopic Hamiltonian has been obtained, and contains a non-negligible *sine* contribution. These results are needed in order to prepare for the implementation and application to methanol of a small-amplitude vibrational formalism [27] for internal rotor molecules in which resolution of the small-amplitude vibrational problem involves expansion of all quantities as functions of the torsional angle. First, we obtained the *ab initio* G98 force constant matrices expressed in 12 (3N-6) internal coordinates at each IRC point. We then carried out a transformation to block-diagonalize these matrices into symmetrized A_1-A_2 blocks, and are following this next by projection of the large-amplitude IRC coordinate from the remaining 11 (3N-7) small-amplitude coordinates. The last column and row in the projected force constant matrix will contain information on the coupling of the small-amplitude vibrations with the large-amplitude torsion. We are anticipating interesting future progress on incorporating these *ab initio* numerical results into the modeling of the observed methanol IR bands.

7. Summary and Future

In this paper, we have reviewed the current status of our research program on the MW, FIR and IR spectroscopy of methanol, with reference to the applications of the data in astronomy and space science. The initial phase of the work, on critical evaluation and global modelling of literature MW and FIR data for the vibrational ground state, is in an advanced stage with publication of a MW spectral atlas for CH_3OH and $^{13}CH_3OH$ for the radio astronomical community and successful global fits of the $v_t = 0$ and 1 torsional states for several isotopic species. Inclusion of higher torsional states is in progress to enable confident treatment of the full torsion-rotation manifold. This should provide benchmark laboratory-based data sufficient to support all future astrophysical MW and FIR programs as well as laying the groundwork for future IR modelling.

In the IR region, our high-resolution studies are achieving identification of substantial fractions of the observed spectral lines in the 10 μm and 3 μm regions and have revealed new insights into the detailed structure of the vibrational bands and the complex network of interactions coupling different vibrations. Here, we are still largely in data-gathering mode and full analyses and modelling of the excited vibrational energies, while very encouraging, are only preliminary. Nevertheless, our IR spectral atlases will provide an important foundation for interpretation of the rapidly increasing body of astronomical and cometary methanol observations.

In parallel with spectroscopic assignments, we have been pursuing *ab initio* calculations to explore the coupling of the large-amplitude torsion with the other vibrations and its effects on the vibrational band structures and intensities. Calculations of the relative transition moment components are in reasonable agreement with observation. Results of initial modelling of individual line intensity measurements are promising, with potential application to simulation of band profiles at cometary and interstellar conditions.

A major remaining frontier for methanol IR spectroscopy is the CH_3-bending modes, both for the fundamentals at 6 μm and the overtones and combinations at 3 μm where strong interactions occur with the CH-stretches. Our group is actively attacking the spectral assignments, and has achieved significant progress. In the 3 μm region, we are exploring schemes to treat the 3 CH-stretches and the 6 overtone and combination CH_3-bending states simultaneously, with appropriate Fermi and Coriolis interaction terms.

In conclusion, we can say that spectral identification and modelling are actively being pursued for methanol in all spectral regions, and that the understanding of the level structure is advancing rapidly to meet the challenges posed by new astronomical and cometary observations. We plan in the near future to complement our microwave databases with extensive catalogues of IR line wavenumbers and intensities which will serve as a solid foundation for interpretation of the exciting new results from space.

8. Acknowledgements

This research was financially supported by the Natural Sciences and Engineering Research Council of Canada. The author is grateful to Drs. M. Abbouti Temsamani, L.R. Brown, J.T Hougen, I. Kleiner, R.M. Lees, G.M. Moruzzi, and D.S. Perry for their interest and valuable discussions concerning the project, and thanks Drs. Abbouti, Brown, Lees and Perry for provision of data prior to publication.

9. References

1. Ball, J.A., Gottlieb, C.A., Lilley, A.E., and Radford, H.E. (1970) Detection of methyl alcohol in Sagittarius, *Astrophys. J. (Letters)* **162**, L203.
2. Worden, H., Beer, R., and Rinsland, C.P. (1997) Airborne infrared spectroscopy of 1994 western wildfires, *Journal of Geophysical Research* **102**, 1287-1299.
3. Mumma, M.J, et al. (2000) A survey of organic volatile species in comet C/1999 H1 (Lee) using NIRSPEC at the Keck Observatory, *Astrophys. J.*, in press.
4. Reuter, D.C. (1992) The contribution of methanol to the 3.4 μm emission feature in comets, *Astrophys. J.* **386**, 330-335.
5. Gibb, E.L., Whittet, D.C.B., Schutte, W.A., Boogert, A.C.A., Chiar, J.E., Ehrenfreund, P., Gerakines, P.A., Keane, J.V., Tielens, A.G.G.M., van Dishoeck, D.F., and Kerkhof, O. (2000) An inventory of interstellar ices toward the embedded protostar W33A, *Astrophys. J.* **536**, 347-356.
6. Bockelée-Morvan, D., et al. (2000) New molecules found in comet C/1995 O1 (Hale-Bopp) – Investigating the link between cometary and interstellar material, *Astron. Astrophys.* **353**, 1101-1114.
7. Rothman, L.S., et al. (1998) The HITRAN molecular database and HAWKS (HITRAN atmospheric workstation): 1996 edition, *J. Quant. Spectrosc. Rad. Transf.* **60**, 665-710.
8. Ivash, E.V., and Dennison, D.M. (1953) The methyl alcohol molecule and its microwave spectrum, *J. Chem. Phys.* **21**, 1804-1816.
9. Bunker, P.R., and Jensen, P. (1998) *Molecular Symmetry and Spectroscopy, 2nd Ed.*, NRC Research Press, Ottawa.
10. Lees, R.M., and Baker, J.G. (1968) Torsion-vibration-rotation interactions in methanol. I. Millimeter wave spectrum, *J. Chem. Phys.* **48**, 5299-5319.
11. Xu, L.H., and Hougen, J.T. (1995) Global fit of torsional-rotational transitions in the ground and first excited torsional states of methanol, *J. Mol. Spectrosc.* **173**, 540-551.

12. Kleiner, I.J., and Godefroid, M., private communication.
13. Herbst, E., Messer, J.K., DeLucia, F.C., and Helminger, P. (1984) A new analysis and additional measurements of the millimeter and submillimeter spectrum of methanol, *J. Mol. Spectrosc.* **108**, 42-57.
14. Xu, L.H., Walsh, M.S., and Lees, R.M. (1996) Global fit of torsion-rotation transitions in the ground and first excited torsional states of ^{13}C-methanol, *J. Mol. Spectrosc.* **179**, 269-281.
15. Walsh, M.S., Xu, L.H., and Lees, R.M. (1998) Global fit of torsion-rotation transitions in the ground and first excited torsional states of CD_3OH methanol, *J. Mol. Spectrosc.* **188**, 85-93.
16. Walsh, M.S., Xu, L.H., Lees, R.M., Mukhopadhyay, I., Moruzzi, G., Winnewisser, B.P., Albert, S., Butler, R.A.H., and DeLucia, F.C. (2000) Millimeter-wave spectra and global torsion-rotation analysis for the CH_3OD isotopomer of methanol, *J. Mol. Spectrosc.* **204**, 60-71.
17. Xu, L.H., and Lovas, F.J. (1997) Microwave spectra of molecules of astrophysical interest. XXXIV. Methanol (CH_3OH and $^{13}CH_3OH$), *J. Phys. Chem. Ref. Data* **26**, 17-156.
18. Moruzzi, G., Winnewisser, B.P., Winnewisser, M., Mukhopadhyay, I., and Strumia, F (1995) *Microwave, Infrared and Laser Transitions of Methanol: Atlas of Assigned Lines from 0 to 1258 cm^{-1}*, CRC Press, Boca Raton, FL.
19. Xu, L.H., Abbouti-Temsamani, M., Wang, X., Ma, Y., Chirokolava, A., Cronin, T.J., and Perry, D.S. (2000) The 3 μm vibration-torsion-rotation energy manifold of methanol, 55th Int. Symp. on Mol. Spectrosc., Columbus, Ohio, Paper RB12.
20. Wang, X., and Perry, D.S. (1998) An internal coordinate model of coupling between the torsion and C-H vibrations in methanol, *J. Chem. Phys.* **109**, 10795-10805.
21. Hougen, J.T., submitted to *J. Mol. Spectrosc.*
22. Serrallach, A., Meyer, R., and Günthard, H.H. (1975) Methanol and deuterated species: infrared data, valence force field, rotamers and conformation, *J. Mol. Spectrosc.* **52**, 94-129.
23. Xu, L.H., and Lees, R.M. (1994) FTIR spectroscopy, torsion-vibration coupling, and new FIR laser assignments for $^{13}CD_3OH$ methanol, *J. Opt. Soc. Am. B* **11**, 155-169.
24. Xu, L.H., Lees, R.M., and Hougen, J.T. (1999) On the physical interpretation of torsion-rotation parameters in methanol and acetaldehyde: comparison of global fit and *ab initio* results, *J. Chem. Phys.* **110**, 3835-3841.
25. Xu. L.H (2000) Can a local mode picture account for vibration-torsion coupling? *Ab initio* test based on torsional variation of methyl stretching and bending frequencies in methanol, *J. Chem. Phys.* **113**, 3980-3989.
26. Xu, L.H., Lees, R.M., Mekhtiev, M.A., and Hougen, J.T., in preparation.
27. Hougen, J.T. (1997) Coordinates, Hamiltonian, and symmetry operations for the small-amplitude vibrational problem in methyl-top internal rotor molecules like CH_3CHO, *J. Mol. Spectrosc.* **181**, 287-296.

ATMOSPHERIC CHEMISTRY EXPERIMENT (ACE): AN OVERVIEW

PETER BERNATH
Department of Chemistry
University of Waterloo
Waterloo, ON
Canada, N2L 3G1
tel: (519)-888-4814
email: bernath@uwaterloo.ca
web: www.ace.uwaterloo.ca

Abstract

ACE is a Canadian satellite mission that will measure and help to understand the chemical and dynamical processes that control the distribution of ozone in the stratosphere. The ACE instruments are a Fourier transform infrared spectrometer, a UV/visible/near IR spectrograph and a two channel solar imager, all working in solar occultation mode.

1. Introduction

The principal goal of the Atmospheric Chemistry Experiment (ACE) mission is to measure and to understand the chemical and dynamical processes that control the distribution of ozone in the upper troposphere and stratosphere. Anthropogenic changes in atmospheric ozone are increasing the amount of ultraviolet radiation received on the ground and may also affect the climate. A comprehensive set of simultaneous measurements of trace gases, thin clouds, aerosols and temperature will be made by solar occultation from a small satellite in a low earth orbit.

A high resolution (0.02 cm^{-1}) infrared Fourier transform spectrometer (FTS) operating from 2 to 13 microns (750-4100 cm^{-1}) will measure the vertical distribution of trace gases and temperature. During sunrise and sunset, the FTS measures infrared absorption signals that contain information on different atmospheric layers and thus provides vertical profiles of atmospheric constituents. Aerosols and clouds will be monitored using the extinction of solar radiation at 1.02 and 0.525 microns as measured by two filtered imagers. The vertical resolution will be about 3 - 4 km from the cloud tops up to about 100 km.

A second instrument called MAESTRO (Measurement of Aerosol Extinction in the Stratosphere and Troposphere Retrieved by Occultation) has recently been added to the ACE mission with T. McElroy of the Meteorological Service of Canada (MSC) as the

J. Demaison et al. (eds.), Spectroscopy from Space, 147–160.

principal investigator. MAESTRO is a dual optical spectrograph that will cover the 285-1030 nm spectral region. It will have a vertical resolution of about 1 km and will measure primarily ozone, nitrogen dioxide and aerosol/cloud extinction.

A high inclination (74°), circular low earth orbit (650 km) will give ACE coverage of tropical, mid-latitude and polar regions. Because reference spectra of the sun will be recorded outside the earth's atmosphere, the ACE instruments will be self-calibrating.

The ACE-FTS and imagers will be built by ABB-Bomem in Quebec City and the satellite bus will be made by Bristol Aerospace in Winnipeg. MAESTRO will be designed and built in a partnership between MSC and the Ottawa-based company EMS Technologies. The satellite will be launched by a NASA using a Pegasus XL rocket in 2002 for a 2-year mission. ACE is the first mission in the Canadian Space Agency's SCISAT-1 program. SCISAT-1 is a small (150 kg) dedicated Canadian science satellite as shown in Figure 1.

Figure 1. Artistic view of SCISAT-1 in flight. The circular solar panel and the passive cooler of the FTS (on top) are apparent (from Tom Doherty).

2. Science Goals

2.1 SCIENCE SUMMARY

2.1.1 *Priority 1*

a) Measurement of regional polar O_3 budget to determine the extent of O_3 loss. This will require measurements of O_3, tracers (CH_4 and N_2O), and meteorological variables (pressure and temperature).
b) Measurement / inference of details of O_3 budget by detailed species measurements (for O_3, H_2O, NO, NO_2, N_2O_5, HNO_3, HNO_4, HCl, $ClNO_3$, ClO) and modelling.
c) Measurement of composition, size and density of aerosols and PSCs (Polar Stratospheric Clouds) in the visible, near IR and mid IR.
d) Comparison of measurements in the Arctic and Antarctic with models to provide insight into the differences, with emphasis on the chlorine budget and denitrification.

2.1.2 *Priority 2*

a) Mid-latitude O_3 budget.
b) Measurement of Arctic vortex descent.

2.1.3 *Priority 3*

a) Study of upper tropospheric chemistry.
b) Monitoring of CFCs (chlorofluorocarbons), CFC substitutes and greenhouse gases.

2.2 DETAILED SCIENCE GOALS

ACE is focussing on one important and serious aspect of the atmospheric ozone problem - the decline of stratospheric ozone at northern mid-latitudes and in the Arctic. Average ozone declines have been measured over much of Canada using ground-based Brewer spectrophotometers [1]. Since 1980 a statistically significant decrease of about 6% has been found by all five long-term Canadian stations (see Figure 2), including Toronto (44°N, 79°W). Ozone sonde measurements show that most of the decline has occurred in the lower stratosphere.

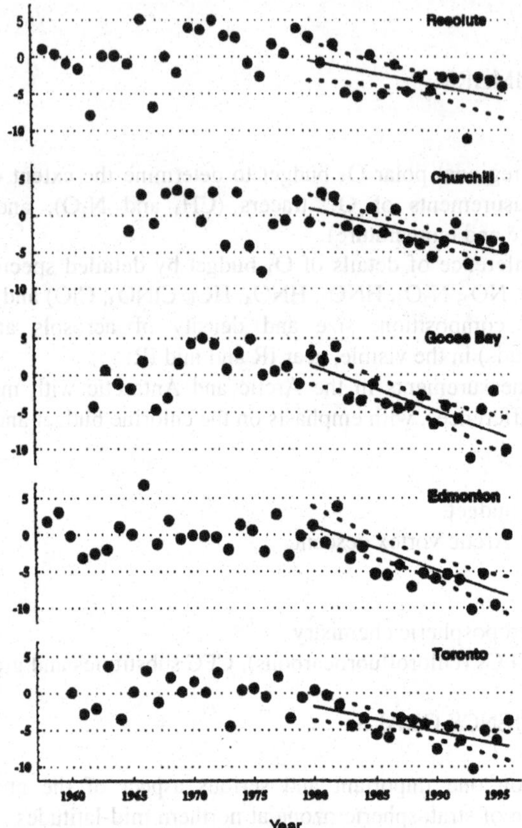

Figure 2. Trends in ozone at Canadian stations [1].

The decrease in Arctic ozone is most severe in the spring as can be seen from the ground and from satellites. The total ozone column obtained from satellite-based TOMS instruments shows this decrease (averaged over the entire polar region) in the month of March (see Figure 3). In March 1997, the ozone column was 21% less than normal and in a small region near the pole the decrease was 40% [2]. Although dynamics plays a major role in redistributing ozone in the Arctic stratosphere, it seems clear that chemical loss of ozone due to heterogeneous chemistry is playing a major role. In fact, several recent publications have indicated that much of the chemical loss in the winter and springtime Arctic is often masked by transport of ozone-rich air from above [3–5]. Specifically, Muller *et al.* [3] estimate that 120-160 DU (Dobson Units) of ozone were chemically destroyed between January and March 1996 overwhelming the dynamical increase leading to a net loss of about 50 DU. This chemical loss is greater than that which was occurring over Antarctica when the ozone hole was first observed in 1985 [6].

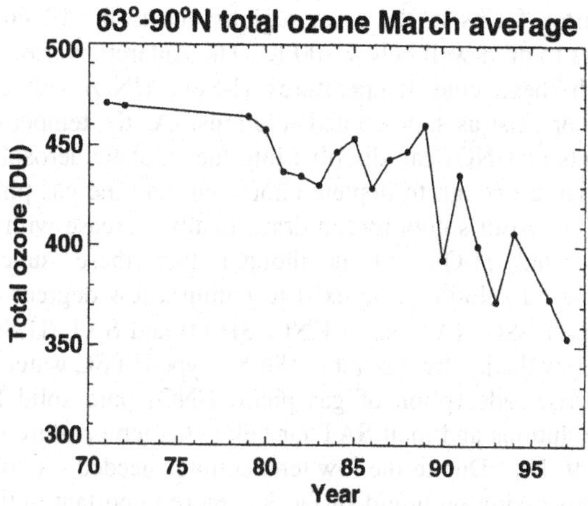

Figure 3. Polar ozone column in March [2].

The major goal of ACE will be to address the question of Arctic ozone loss, and, by means of modelling and measurements (both atmospheric and laboratory) attempt to quantify the contributions from dynamics and chemistry. Since sulphate aerosol and PSCs play a major role in ozone loss in the Arctic, distinguishing between PSCs, aerosols and cirrus and other clouds will be important. There will be an important role for laboratory measurements. The role of modelling, either forward modelling or data assimilation, will also be critical. A variety of atmospheric models ranging from simple box models for the chemistry to atmospheric global circulation models (GCMs) will be used to interpret our data.

2.3 STRATOSPHERIC CHEMISTRY AND ARCTIC OZONE DECLINE

The anthropogenic release of CFCs affects the stratospheric ozone layer through gas phase chemical reactions (such as the Cl/ClO cycle) as well as by heterogeneous chemistry on PSCs and aerosols [1]. CFCs, transported to the stratospheric, are broken down largely by UV light and release "active" chlorine species such as Cl and ClO. These species destroy odd oxygen (O + O_3) via various gas phase catalytic cycles. However, the full extent of the destruction is ameliorated by the storage of the active chlorine in the reservoir species HCl and $ClONO_2$, neither of which reacts efficiently with O_3 or O. If only the gas phase chemical reactions are considered then the predicted effect on the total ozone column is relatively small. For example, Brasseur [7] predicts with a 2D model (latitude-elevation) that the ozone column would decrease by less than 1% over the ten-year period 1980-1990 in the absence of heterogeneous chemistry.

The discovery of the Antarctic ozone hole in 1985 by Farman *et al.* [6] lead to the realisation that heterogeneous reactions were important in determining the ozone budget in polar regions. Later work [8] indicated that heterogeneous reactions on and within sulphate aerosols also play an important role in the stratosphere. Volcanic SO_2 from "random" eruptions penetrating the stratosphere also is important in maintaining the stratospheric sulphate layer budget.

Each winter the vortex forms in the polar regions due to the IR cooling that occurs causing temperatures to drop well below 200 K. The Antarctic vortex is colder than the Arctic vortex. With these cold temperatures H_2SO_4, HNO_3 and H_2O and various mixtures can freeze or exist as supercooled solutions. As the temperature drops below 195 K, large amounts of HNO_3 can dissolve into the sulphate aerosols to form ternary solutions, amounts large enough to deplete almost entirely the gas phase HNO_3 levels. The volume of these aerosol solutions can dramatically increase with this large uptake of gaseous HNO_3 and H_2O. It is thought that these super-cooled ternary $H_2SO_4/HNO_3/H_2O$ liquid solutions can exist to within a few degrees of the frost point without forming type I PSC NAT (solid $HNO_3 \cdot 3H_2O$) and SAT ($H_2SO_4 \cdot 4H_2O$). If the temperature falls below the ice frost point (~188 K), type II PSC water ice can form. As temperatures again rise, adsorption of gas phase HNO_3 onto solid SAT can help to regenerate ternary solutions and melt SAT far below the temperature (210-215 K) SAT is believed to melt [9, 10]. Due to the low temperatures needed for solid PSCs to form, it is expected that processing on liquid surfaces is more important in the Arctic than the Antarctic.

Heterogeneous reactions on these condensed phases can activate chlorine and bromine while tying up odd-nitrogen as $HNO_3(a)$ [11]. The active chlorine and to a lesser extent bromine drive the reactions that form the ozone hole. For example, on the ice crystals, inactive or reservoir forms of the halogen catalysts are freed,

$$ClONO_2 + HX\ (ice) \rightarrow HNO_3\ (ice) + XCl$$

(where X = Cl, Br). Low temperatures drive these processes. As the polar lower stratospheric temperatures drop at the end of the fall season, the aerosol reactions become important. For example, the solubility of HCl is very temperature sensitive and as the temperature drops, it begins to dissolve in the ternary solution / sulphate aerosol. These and similar reactions drive the formation of the ozone hole in the polar late winter (Northern Hemisphere) and austral springtime. These reactions can occur during the night-time and when polar sunrise occurs, species such as Cl_2, BrCl and $ClNO_2$ are readily photolysed into more labile species such as Cl. The processed air can persist for several weeks since the reformation of the reservoir species HCl and $ClNO_3$ is a relatively slow process at polar latitudes. Gravitational sedimentation of the PSCs removes stratospheric HNO_3 and H_2O (denitrification and dehydration).

One of the main features of ozone loss in polar regions is that it is not rate limited by the low abundance of atomic oxygen. One of main loss mechanisms involves self-reaction of ClO [12]

$$
\begin{array}{l}
ClO + ClO \rightarrow Cl_2O_2 \\
Cl_2O_2 + h\nu \rightarrow Cl + ClO_2 \\
ClO_2 + M \rightarrow Cl + O_2 \\
2(Cl + O_3 \rightarrow ClO + O_2) \\
\text{--------------------------} \\
\text{net: } O_3 + O_3 \rightarrow 3O_2
\end{array}
$$

As long as temperatures are less than ~210 K then thermal decomposition of Cl_2O_2

$$Cl_2O_2 + M \rightarrow ClO + ClO$$

which does not lead to ozone loss, does not play a major role.

Another important ozone loss mechanism is due to the synergistic reaction between BrO and ClO [13]:

$$BrO + ClO \rightarrow Br + Cl + O_2$$
$$BrO + ClO \rightarrow BrCl + O_2$$
$$BrCl + h\nu \rightarrow Br + Cl$$
$$Cl + O_3 \rightarrow ClO + O_2$$
$$Br + O_3 \rightarrow BrO + O_2$$

As before this loss rate is not limited by the abundance of atomic oxygen.

The inclusion of heterogeneous reactions appears to be able to account for the severe ozone loss in the Antarctic spring [1]. There have been problems accounting for the ozone decline in the Arctic [15] although recent modelling work by Lefevre *et al.* [5] using a chemical transport model produces quite good agreement with column ozone measurements. Springtime ozone depletion differs between the Arctic and Antarctic due to differences in seasonal temperature extremes largely caused by atmospheric dynamics related to topography. The northern polar vortex is somewhat larger in extent, less well defined spatially, and more unstable than the southern vortex. More of the northern vortex is exposed to sunlight during the boreal winter, increasing the complexity of both its chemistry and physics. Another effect that may be important is processing that occurs at mesoscales (< 50 km). Most model simulations to date are run at rather low resolution. High-resolution (mesoscale) model runs by Carslaw *et al.* [16] suggest that temperatures sufficiently low for PSCs to form can be induced by mountain waves even when the synoptic temperatures appear to be too high for the formation of PSCs. And as noted above the air, once processed, they will remain processed for several weeks. In this manner large volumes of stratospheric air may undergo processing.

There is no dramatic "Arctic ozone hole" in the spring because downward transport of ozone-rich air masks the strong ozone depletion [3–5]. Additionally, Arctic stratospheric temperatures are generally warmer than those in the Antarctic because the vortex is often in sunlight. The PSCs associated with strong ozone depletion in the springtime form in the stratosphere at temperatures below about 195 K (see above). Such temperatures are common in the Antarctic winter but are rarer in the Arctic. Typical Arctic winter temperatures lie just above 195 K so that there is a strong correlation between springtime Arctic ozone loss and the formation of PSCs. Salawitch has noted the correlation between Arctic ozone column and the minimum temperature measured in March at 50 mbar (about 20 km) (see Figure 4). When this temperature dropped below 195 K as it did in 1994, 1995 and 1997, strong ozone loss was detected. The spring of 1998, however, was relatively warm (198 K) and the ozone loss was much reduced (Figure 4). These observations make a strong case for the importance of heterogeneous PSC chemistry in Arctic ozone declines. In addition to gases, ACE will thus strongly focus on the chemistry and physics of PSC particles. Note that the ozone declines at mid-latitudes (Figure 2) cannot be attributed to the polar chemistry, although transport of ozone-depleted air might be a contributory factor.

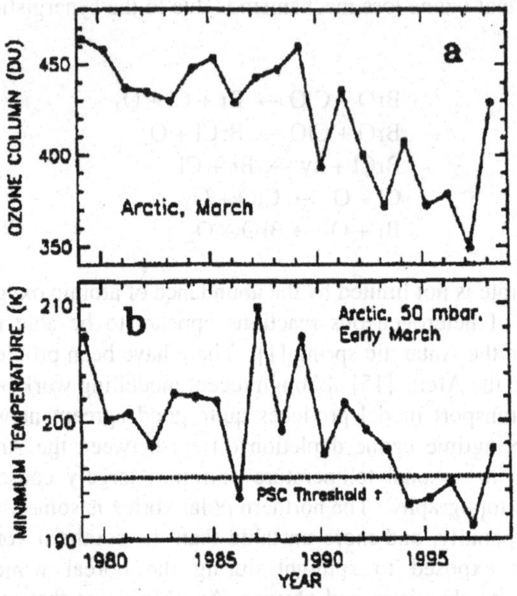

Figure 4. Correlation between minimum Arctic temperature and springtime ozone column (from Salawitch).

2.4 STRATOSPHERIC OZONE LOSS AND CLIMATE CHANGE

Since it was first realised that low temperatures (and high chlorine amounts) drove the polar ozone loss, there has been speculation regarding possible connections between the build-up of CO_2 and other IR-active gases and polar ozone loss. The increase of these greenhouse gases, while leading to increased IR heating in the troposphere leads to enhanced cooling in the stratosphere and stratosphere temperatures are expected to decrease due to cooling to space from the increased levels of CO_2. In fact, this change in the vertical temperature profile has been detected and is taken as strong evidence for a human influence on climate [17]. This increased stratospheric cooling could affect stratospheric dynamics by making the boreal vortex more robust [18] and could lead to an Arctic ozone hole if CFCs continued to increase. The recent work by Shindell *et al.* [19] using a GCM with chemistry suggests that the stratospheric cooling will lead to a prolongation of ozone loss due to cooler temperatures and PSC formation even though chlorine levels decrease.

As noted above, Shindell *et al.* [19] used a Global Climate/Middle Atmospheric Model (GCMAM) with simplified chemistry and constrained ozone transport to investigate the coupling between chemistry and climate. For the period 2010-2019 they found that the increase due to radiative cooling of the stratosphere was 1-2 K poleward of 70°N while there was an additional cooling of 8–10 K attributable to the increased stability of the Arctic vortex. This dramatic decline in stratospheric temperatures in their model caused an Arctic ozone hole in the spring with a 2/3 loss of the total ozone column. An Arctic ozone hole is thus predicted to occur in years 2010–2019, in spite of anticipated decreases of CFC concentrations because of the implementation of the Montreal protocol.

The possibility of an Arctic ozone hole is a very disturbing development with strong political and social implications. In contrast to the Antarctic ozone hole, an Arctic ozone hole would effect heavily populated parts of the Northern Hemisphere. Our experimental and theoretical understanding of Arctic chemistry is still in a primitive state. The possibility of the development of an Arctic ozone hole clearly will require more detailed modelling studies combined with a detailed complement of measurements. ACE can contribute to our understanding of physics, chemistry and dynamics of the Arctic polar stratosphere.

3. The Orbit

The selection of an appropriate orbit is critical for the ACE mission to achieve its science goals. Several orbit options were analysed before the baseline mission orbit was chosen during the Phase A period. These orbits included a 57° inclined, 65° inclined, and a 10:30 am sun-synchronous orbit. The baseline orbit was chosen to be circular with an inclination of 65° and an altitude of 650 km.

The sun-synchronous orbit was rejected due to the lack of global coverage. The 57°-inclined orbit was not acceptable due to the lack of polar coverage in the Northern Hemisphere, compared to the baseline orbit. The baseline orbit provided global coverage and coverage of the Arctic region in the spring when ozone depletion is most severe. Ultimately the 65° inclined orbit was shifted to 74° in order to give slightly more polar occultations (at the expense of those in the tropics and mid-latitudes) as well as for easier launch by the Pegasus rocket.

4. The FTS Instrument and Imagers

The main instrument on ACE is a Michelson interferometer with a custom design to meet the requirements of the ACE mission. The interferometer uses two corner cubes rotating on a center flex pivot to produce the optical path difference. A folding mirror inside the interferometer is used to increase the optical path difference. The ACE design is fully compensated for tilt and shear of both moving and stationary optics inside the interferometer. The optical position is measured with a laser diode operating at 1500 nm. The instrument is also equipped with a visible and a near-infrared imager. A pointing mirror, controlled by a suntracker servo-loop, will lock on the sun and track it while the instrument is taking measurements.

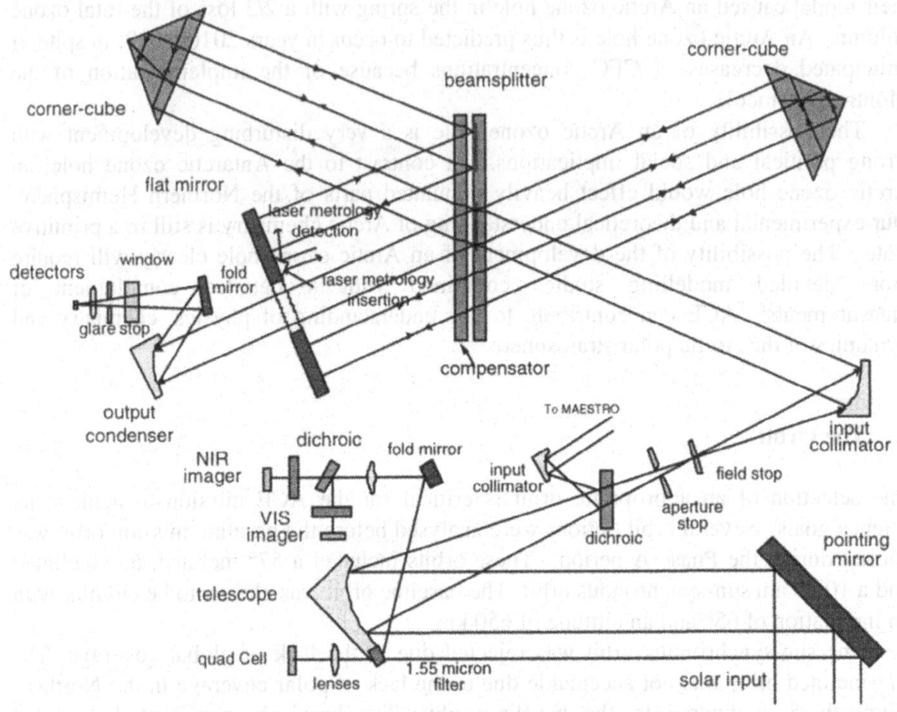

Figure 5. The optical layout of the ACE-FTS and imagers (from ABB-Bomem).

A summary of the main characteristics for the FTS are provided in Table 1.

TABLE 1: Main characteristics of the ACE FTS spectrometer

Spectral range:	750 – 4100 cm^{-1}	Noise equivalent radiance:	< 0.5% of the radiance of a blackbody at 5800 K
Spectral bandwidth:	0.02, 0.04, 0.08, 0.4 cm^{-1}	Detectors:	InSb and HgCdTe
Spectral resolution:	< 0.028, 0.056, 0.11, 0.55 cm^{-1}	Detectors cooling system:	Passive cooling to < 110 K
Sweep duration:	2, 1, 0.5, 0.1 seconds	Field of view:	1.25 mrad
Spectral stability (relative):	3×10^{-7} (rms) for 180 sec.	Mass:	35 kg
Transmittance uncertainty:	< 1% (rms)	Volume:	58440 cm^3
Dynamic range:	0 – 5800 K	Power consumption:	< 40 W

Double-sided interferograms will be Fourier transformed on the ground to obtain the desired atmospheric transmission spectra. The FTS covers the 750-4100 cm^{-1} range using two detectors (InSb and HgCdTe) in a sandwich configuration. The detectors will be cooled to less than 110 K (typically 90 K) by a passive radiator pointing towards deep space. The detector/cooler sub-assembly will be built by Ball Aerospace in Colorado.

The satellite will always be pointing towards the sun. This means that during eclipse the FTS will be pointing towards the earth and near-nadir measurements will be possible. Although the signal-to-noise ratio of a single scan will not be high, a rapid series of short scans can be recorded and co-added on the ground.

The visible/near infrared imager has two filtered channels at 0.525 and 1.02 μm, chosen to match two of the wavelengths monitored by the SAGE II satellite instrument [21]. These two wavelengths are useful for the study of clouds and aerosols because they are relatively free of absorption by atmospheric molecules. The detectors in the imagers will be a 256 x 256 active pixel sensors from Fill Factory, a Belgian company. The total field of view of the imagers will be 30 mrad, to be compared to the 9 mrad angular size of the sun. The signal-to-noise ratio of the solar image will be greater than 1000.

5. MAESTRO

MAESTRO is a small (about 5 kg) spectrograph that will cover the 285 – 1030 nm region in two overlapping pieces. The use of two spectrographs (280 – 550 nm, 500 – 1030 nm) improves the stray light performance. The spectral resolution is about 1–2 nm and the detectors are linear EG & G Reticon photodiode arrays with 1024 elements. The design is based on a simple concave grating with no moving parts. The entrance slit will be held horizontal to the horizon during sunrise and sunset by controlling the spacecraft roll with a startracker and a momentum wheel on the satellite bus. The FTS, imagers and MAESTRO will all share a single suntracker and will have approximately the same field of view. The vertical resolution of MAESTRO will, however, be about 1 km and will have a signal-to-noise ratio in excess of 1000 as compared to about 200 for the FTS. While the ACE mission will work primarily by solar occultation, MAESTRO will also be able to make some near-nadir solar backscatter measurements like the GOME instrument on the European ERS-2 satellite [22].

6. Ground Segment and Data Analysis

The ACE mission is based on the successful (but now retired) ATMOS (Atmospheric Trace Molecule Spectroscopy) instrument which flew four times on the NASA Space Shuttle [23]. ATMOS recorded some remarkable high resolution solar occultation spectra as illustrated by Figure 6. The ACE-FTS instrument has been miniaturized by nearly a factor of 10 in terms of mass, power and volume as compared to ATMOS. Note, however, that ATMOS had a resolution of 0.01 cm^{-1} (±50 cm optical path difference) as compared to ACE-FTS (0.02 cm^{-1}, ±25 cm optical path difference).

The ACE mission has been augmented by the two imagers and the MAESTRO spectrograph as compared to ATMOS. By utilizing a small science satellite in a high inclination orbit rather than the shuttle in a low inclination orbit, continuous coverage of the globe is possible. The ACE orbit improves polar coverage for an occultation instrument.

The imagers will give high signal-to-noise measurements of atmospheric extinction and will lead to SAGE-like aerosol and cloud data products [24]. In addition, the imagers will provide important pointing information low in the atmosphere where refraction and extinction become important. They offer an important diagnostic for the variation of the flux over the distorted solar disk in the lower stratosphere and upper troposphere.

The MAESTRO spectrograph greatly extends the wavelength coverage of the ACE mission. Except for a gap between 1 and 2.4 microns, we will have continuous coverage from 0.28 to 13 microns from the FTS and spectrograph. MAESTRO will measure primarily ozone, nitrogen dioxide and atmospheric extinction, but with a higher signal-to-noise ratio than is possible in the infrared. Moreover, the vertical spatial resolution (~1 km) will be more than 3 times higher than the FTS. This high vertical resolution can be important because there are atmospheric layers (e.g., due to PSCs) that are of order 1 km thick. The wavelength dependence of the atmospheric extinction is useful for determining the aerosol size distribution and particle density. The oxygen A-band at 762 nm (as well as the B- and γ-bands) will be used by MAESTRO to make an independent determination of atmospheric temperature and pressure. It may even prove possible to detect BrO and OClO in polar regions undergoing perturbed chemistry. The FTS does not have the sensitivity to measure these molecules.

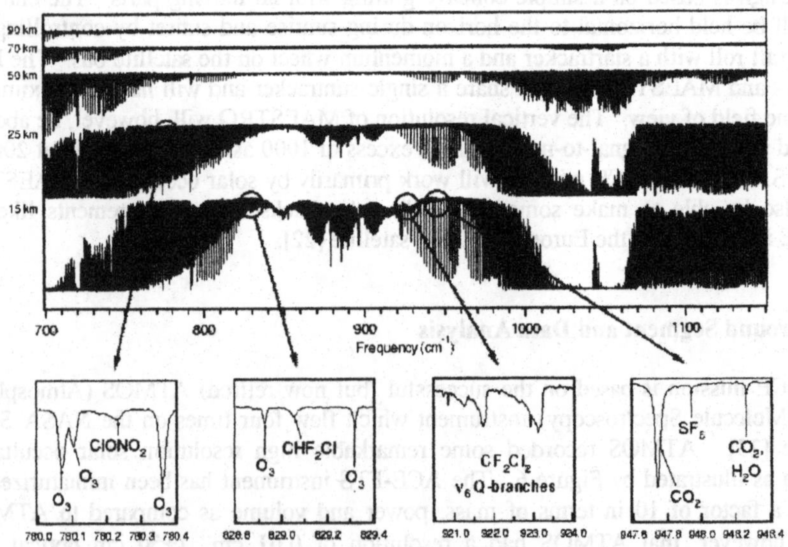

Figure 6. An ATMOS spectrum from C. Rinsland.

The raw ACE data will be sent to ground using at least 2 ground stations. The data volume is about 1 Gbyte per day. These data will be transferred from the Mission Operations Centre operated by Canadian Space Agency in St. Hubert near Montreal to the Science Operations Centre at the University of Waterloo. At Waterloo the data will be archived and transformed into data products for distribution to the science team members. In the case of the FTS, the raw interferograms (level 0) need to be transformed into corrected atmospheric spectra (level 1) by software to be supplied by the instrument contractor, ABB-Bomem.

The atmospheric temperature will be determined as a function of pressure by starting with *a priori* data from a weather forecast model and using a fixed CO_2 volume mixing ratio for the FTS and O_2 A-, B- and γ-bands for MAESTRO. FTS, MAESTRO and imager data will then be converted to height profiles (level 2) of atmospheric species in near real time. In the case of the ACE-FTS, we have inherited the ATMOS code [25] based on an onion-peeling algorithm for the level $1\rightarrow2$ transformation but we will likely use a global fit approach [26]. The MAESTRO algorithms will draw on the previous work by McElroy and co-workers with a variety of UV/visible spectrographs that have been deployed, for example, on the NASA ER-2 aircraft [27]. In the case of the imagers the data products will be SAGE-like extinction profiles [24] as a function of height. Some global data tools (level 3) will be provided for distribution of data over the Web to science team members and the general public.

The ACE mission is completely dependent on the availability of spectroscopic data for the required retrievals of atmospheric molecules and particles. These retrievals are based on the use of a forward model to calculate the radiance incident on the suntracker mirror. Our starting point is the ATMOS linelist [28] which will be up-dated and augmented by the required near UV, visible and infrared cross-sections. This is an enormous task for the members of the ACE science team but is crucial to the success of the mission.

7 Acknowledgements

Some of the material in this review paper was taken from the Phase A report prepared by the ACE Science team (see www.ace.uwaterloo.ca for a list). Many people made important contributions to the Phase A report but J. McConnell (York University) was responsible for much of the material in Section 2. We thank Bomem (M.-A. Soucy) for permission to reproduce Figure 5 and some of the material in Section 4.

8 References

1. D.I. Wardle, J.B. Kerr, C.T. McElroy, and D.R. Francis, eds., (1997) Ozone Science: A Canadian Perspective on the Changing Ozone Layer, Environment Canada.
2. Newman, P.A. *et al.*, (1997) Anomalously low ozone over the Arctic, *Geophys. Res. Lett.*, **24**, 2689-2692.
3. Müller, R., Crutzen, P.J., Grooss, J.-U., Brühl, C., Russell III, J.M., Gernandt, H., McKenna, D.S. and Tuck, A., (1997) Severe chemical ozone loss in the Arctic during the winter of 1995-96, *Nature*, **389**,

709-712.

4. Knudsen, B.M. *et al.*, (1998) Ozone depletion in and below the Arctic vortex for 1997, *Geophys. Res. Lett.*, **25**, 627-630.
5. Lefevre, F., Figarol, F., Carslaw, K.S. and Peter, T., (1998) The 1997 Arctic ozone depletion quantified from three-dimensional model simulations, *Geophys. Res. Lett.*, **25**, 2425-2429.
6. Farman, J. C., Gardiner, B.G., and Shanklin, J.D., (1985) Large losses of total ozone in Antarctica reveal seasonal ClO_x/NO_x interaction, *Nature* **315**, 207-210.
7. Brasseur, G.P., (1992) *Planet. Space Sci.*, **40**, 403.
8. Hofmann, D. J. and S. Solomon, (1989) Ozone destruction through heterogeneous chemistry following the eruption of El Chichon, *J. Geophys. Res.*, **94**, 5029--5041.
9. Martin, S.T., Salcedo, D., Molina, L.T., and Molina, M.J., (1998) Deliquescence of sulfuric acid tetrahydrate following volcanic eruptions or denitrification, *Geophys. Res. Lett.* **25**, 31-34.
10. Koop, T., and Carslaw, K.S., (1996) Melting of $H_2SO_4 \cdot 4H_2O$ particles upon cooling: Implications for polar stratospheric clouds. *Geophys. Res. Lett.*, **25**, 3747-3750.
11. Tolbert, M.A., (1996) Polar clouds and sulfate aerosols, *Science*, **272**, 1597.
12. Molina, L.T. and Molina, M.J., (1987) Production of Cl_2O_2 from the self reaction of the ClO radical, *J. Phys. Chem.*, **91**, 433.
13. Clyne, M.A.A., and Watson, R.T., (1977) Kinetic studies of diatomic free radicals using mass spectrometry, *J. Chem. Phys.*, **73**, 1169-1187.
14. Yung, Y.L., Pinto, J.P., Watson, R.T., and Sander, S.P., (1980) Atmospheric bromine and ozone perturbations in the lower stratosphere, *J. Atmos. Sci.*, **37**, 339-353.
15. Edouard, S.B., Legras, B., Lefevre, F. and Eymard, R., (1996) The effect of small scale inhomogeneities on ozone depletion in the Arctic, *Nature*, **384**, 444-447.
16. Carslaw, K.S. *et al.*, (1998) Increased stratospheric ozone depletion due to mountain-induced atmospheric waves, *Nature*, **391**, 675-678.
17. Santer, B.D. *et al.*, (1996) A search for human influences on the thermal structure of the atmosphere, *Nature*, **382**, 39-46.
18. Austin, J., *et al.*, (1992) Possibility of an Arctic ozone hole in a doubled-CO_2 climate, *Nature*, **360**, 221-225.
19. Shindell, D.T., Rind, D. and Lonergan, P., (1998) Increased polar stratospheric ozone losses and delayed eventual recovery owing to increased greenhouse-gas concentrations, *Nature*, **392**, 589.
20. Moreau, L., M.-A. Soucy, S. Fortin and J. Giroux, (Feb. 2001) in Technical Digest, Fourier Transform Spectroscopy: New Methods and Applications, Optical Society of America, Washington, D.C.
21. Mauldin, L.E., N.H. Zaun, M.P. McCormick, J.H. Guy and W.R. Vaughn, (1985) SAGE II Instrument: A Functional Description, *Opt. Eng.*, **24**, 307.
22. Burrows, J.P. et al., (1999) The Global Ozone Monitoring Experiment (GOME): Mission Concept and First Scientific Results, *J. Atmos. Sci.* **56**, 151-175.
23. Gunson, M.R. *et al.*, (1996) The Atmospheric Trace Molecule Spectroscopy (ATMOS) Experiment: Deployment on the ATLAS Space Shuttle missions, *Geophys. Res. Lett.* **23**, 2333-2336.
24. Kent, G.S. *et al.*, (1993) A model for the separation of cloud and aerosol in SAGE II occultation data, *J. Geophys. Res.*, **98**, 20,725-20,735.
25. Abrams, M.C. *et al.*, (1996) Remote sensing of the earth's atmosphere from space with high-resolution Fourier transform spectroscopy: Development of a methodology of data processing for the Atmospheric Trace Molecule Spectroscopy Experiment, *Appl. Opt.*, **35**, 2774-2786.
26. Carlotti, M., (1998) Global-fit approach to the analysis of limb-scanning atmospheric measurements, *Appl. Opt.* **27**, 3250-3254.
27. McElroy, C.T., (1995) A spectroradiometer for the measurement of direct and scattered solar irradiance from on-board the NASA ER-2 high-altitude research aircraft, *Geophys. Res. Lett.* **22**, 1361-1364.
28. Brown, L.R. *et al.*, (1996) The 1995 ATMOS linelist, *Appl. Opt.*, **35**, 2828-2848.

SPECTROSCOPIC MEASUREMENTS WITH MIPAS
(MICHELSON INTERFEROMETER FOR PASSIVE ATMOSPHERIC SOUNDING)

HERBERT FISCHER
Institut für Meteorologie und Klimaforschung
Forschungszentrum Karlsruhe / Universität Karlsruhe
Postfach 3640
76021 Karlsruhe
GERMANY
e-mail: herbert.fischer@imk.fzk.de

1. Introduction

The depletion of ozone in the stratosphere is not completely understood up to now. The different dynamical, microphysical and chemical processes in the stratosphere are supposed to be relatively well known but we are not able to simulate the decrease of ozone with numerical models with sufficient accuracy. Further problems in the atmosphere are connected with the exchange of air masses between the stratosphere and the troposphere and the pollution of the upper troposphere by subsonic aircrafts. The future development of the stratospheric ozone will depend on the decreasing abundances of ozone destroying industrial chemicals and the increasing abundances of the various greenhouse gases.

As a consequence, a better understanding of the various processes in the atmosphere is required. This can be achieved by simultaneous measurements of a considerable number of trace gases containing long-lived species for dynamical studies, reservoir species for microphysical and short-lived species for chemical investigations. Understanding of global changes requires measurements from polar orbiting satellites. In order to fulfil all these scientific requirements it is obvious to use a satellite-borne Fourier spectrometer in the limb-sounding mode detecting the mid-infrared (IR) spectrum as emitted by the atmosphere.

2. MIPAS experiments

Following the above-described arguments several MIPAS experiments have been designed, constructed and operated during the last 15 years [1].

An uncooled version of this interferometer was developed during the first half of the 1980s. This MIPAS-LM (laboratory model: spectral resolution ~0.03 cm^{-1}) has been used as a ground-based device for measuring column amounts of trace gases at polar

J. Demaison et al. (eds.), Spectroscopy from Space, 161–169.

stations by detecting the attenuated solar radiation. During different measurement campaigns in the time period between 1989 and 1993 time series of zenith column amounts of several trace gases (such as O_3, N_2O, CH_4, HNO_3, NO_2, HCI and HF) were derived from the recorded interferograms and analysed with respect to dynamic and chemical processes in the Arctic stratosphere.

A cooled version of MIPAS has been constructed to observe, with good vertical resolution, the composition and structure of the stratosphere and the corresponding diurnal variations from balloon-borne platforms. Field experiments have been conducted during balloon flights in France, Spain and Sweden in the time period between 1989 and 1999. Limb emission spectra were measured at heights between 5 and 40 km [2]. A considerable number of atmospheric trace species have been identified in the spectra, namely O_3, CO_2, H_2O, HDO, CH_4, N_2O, CCl_4, CF_2Cl_2, $CFCl_3$, CHF_2Cl, $ClONO_2$, HOCl, HNO_3, HNO_4, N_2O_5 and C_2H_6.

Besides the MIPAS ground-based and balloon experiments two cooled MIPAS aircraft instruments were flown in the nineties. The atmospheric sounder MIPAS-FT (Flugzeug Transall) detects infrared radiation emitted by various atmospheric trace constituents in the spectral range 750-1250 cm^{-1}. Due to the relatively low flight level of the Transall aircraft only column amounts of stratospheric trace gases have been determined [3]. The second version of our MIPAS aircraft instrument has been mounted on the high-flying air plane GEOPHYSICA and already successfully operated during the APE-GAIA campaign in fall 1999. By using the limb sounding technique two-dimensional cross sections of trace gases in the atmosphere are derived recently.

The MIPAS space experiment is a cooled high-resolution mid-infrared (4.1-14.6 µm) Fourier Transform Spectrometer that will be launched aboard ESA's ENVISAT satellite [4]. The instrument will scan across the earth limb in order to detect vertical profiles of temperature and more than 20 atmospheric trace species. Further important parameters to be derived from MIPAS observations are the distribution of aerosol optical depth, tropospheric cirrus clouds and polar stratospheric clouds.

The main specifications of the different cooled MIPAS experiments are summarized in Table 1. In this respect it has to be emphasized that the optics is cooled down to about 200 K and the measurement time for one interferogram depends on the type of instrument carrier.

The ENVISAT satellite will be launched in June 2001. The special azimuth scan of the MIPAS space experiment allows to cover the whole earth from pole to pole.

TABLE 1. Main specifications of different cooled MIPAS experiments

Spectral range	4 - 15 μm
Spectral resolution	0.1 - 0.025 cm^{-1}
Measurement time per interferogram	4 -10 s
Aperture diameter	10 - 30 cm
Data rate	200 - 600 kbit/s
Detectors	He-cooled Si:Ga/Si:As Mech. Cooler: HgCdTe
Cooled optics	~200 K

3. Information content and data processing

The MIPAS experiments are detecting broad-band spectra in the middle infrared region. The information content of these spectra is high due to ro-vibrational bands of many trace gases in the mid-IR and is relatively well known due to previous measurements with the air-borne MIPAS instruments.

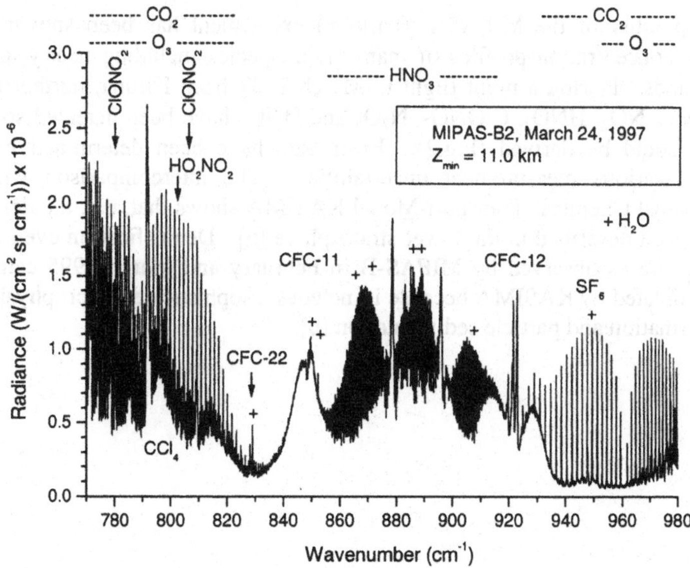

Figure 1. Limb emission spectrum as measured by the MIPAS balloon experiment at a tangent altitude of 11 km in March 1997

Fig.1 shows a spectrum measured with the MIPAS balloon experiment covering the mid-infrared window between 10 and 13 μm. The many spectral features can be

assigned to a considerable number of different trace species like CO_2, H_2O, O_3, HNO_3, CFC-11, CFC-12, CFC-22, $ClONO_2$ and others. Taking into account that the MIPAS instruments detect the mid infrared between 4 and 15µm, the information content of the measured spectra is tremendous.
As a consequence, operational processing will not include the complete measured spectra but will rely on selected, comparatively small spectral intervals called microwindows [12]. The process of selecting these microwindows is complicated because many criteria have to be considered as

a) a prominent feature of the target gas has to be within the microwindow,
b) the influence of other trace gases including line mixing should be small,
c) effects of non-LTE should be negligible,
d) uncertainties in modelling gaseous and other continua have to be small and so on.

An optimization method for the microwindow selection has been developed and applied. It leads for the target gas ozone to 30 or 162 microwindows respectively dependent on the available computer power. Up to now there are still possibilities for improving the proposed method, e.g. the combination of optimized microwindows is not necessarily optimal.

4. Results from MIPAS measurements

The capability of the MIPAS-B (balloon) experiment has been shown by obtaining vertical concentration profiles of many trace species including nearly all of the NO_y compounds. During a night flight in March 1997 from Kiruna, northern Sweden, the profiles of NO_2, HNO_3, $ClONO_2$, N_2O_5 and HNO_4 have been detected so that the NO_y budget could be derived (Fig.2). Error bars have been determined by taking into account various measurement uncertainties. The intercomparison with the three-dimensional Chemical Transport-Model KASIMA shows that, e.g. the denoxification is not yet well described in the lower stratosphere [5]. Denitrification events in the lower stratosphere as observed by MIPAS-B in February and March 1995 can be relatively well simulated by KASIMA because it includes a sophisticated microphysical model for PSC formation and particle sedimentation [6].

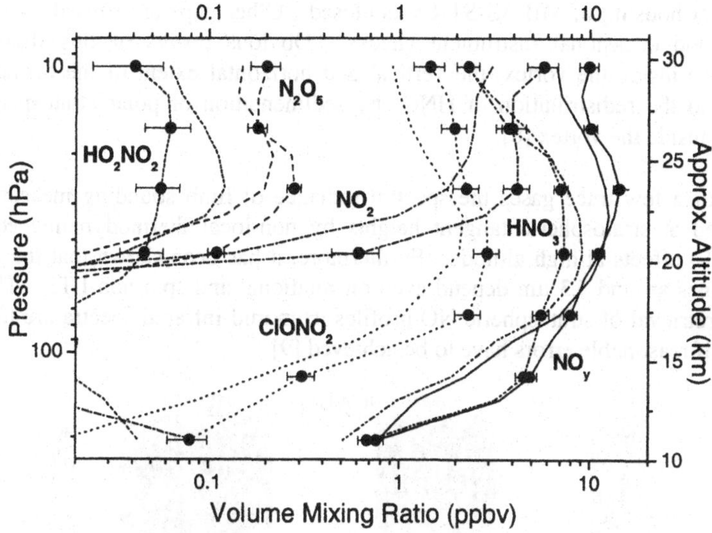

Figure 2. Vertical profiles of NO2, HNO3, ClONO2, N2O5 and HNO4 as derived from MIPAS-B measurements (lines with dots and error bars) compared with corresponding profiles as calculated with the CTM model KASIMA.

In 1999, for the first time vertical profiles of HDO inside the Arctic vortex along with H_2O were retrieved from MIPAS-B limb emission spectra. The deuterium to hydrogen ratio of water vapour shows a strong depletion in the lower stratosphere in February 1995 in comparison to that of Standard Mean Ocean Water. This strong depletion indicates an isotopic effect due to dehydration of the lower stratosphere by forming particles of Polar Stratospheric Clouds [7].
The inspection of the limb sounding sequences of MIPAS-B leads to the discovery of an interesting spectral feature at the tropopause region. The spectrum of the mid-infrared window between 10 and 12.5 μm for a tanget height of about 10 km (below the tropopause) exhibits a considerable number of prominent H_2O lines while these lines are missing in the corresponding spectrum for a tanget height above the tropopause. These weak H_2O lines can only be seen in tropospheric spectra due to the high water vapour amount. As a consequence, the H_2O concentration in the upper troposphere can be determined with high sensitivity in the mid-infrared window. As will be seen later this leads to special requirements for spectroscopic laboratory measurements.

Another quality of results has been derived more recently from the measurements of the MIPAS-STR (STRatospheric aircraft) during the APE-GAIA campaign at high southern latitudes. Upward looking measurements and limb sounding from the GEOPHYSICA flight level at about 17.5 km yield two-dimensional cross sections of trace gas concentrations along the flight route in the height region between 10 and 20 km. On September 23rd, 1999 the flight path of the GEOPHYSICA crossed the edge of the polar vortex and entered the ozone-depleted region. Figure 3 shows the corresponding distributions of temperature, ozone and HNO_3 mixing ratio. The data gap in the middle part of the plot is due to the dive of the aircraft during which the shutter of the instrument housing of MIPAS-STR was closed. Other gaps are caused by instrument failures and occasional instrument checks. Obvious results of this flight are the subsidence inside the vortex, the vertical and horizontal extent of the ozone-depleted region and the redistribution of HNO_3 by sedimentation of polar stratospheric cloud particles inside the vortex [8].

In case of a few trace gases the spectral radiance of limb sounding measurements is influenced at stratospheric tangent heights by non-local thermodynamic equilibrium (non-LTE) effects at high altitudes. Furthermore, it has been shown that the intensities of NO lines around 5.3 µm depend even on rotational and spin non-LTE. That means that the retrieval of stratospheric NO profiles from mid-infrared spectra are much more difficult if reasonable errors have to be achieved [9].

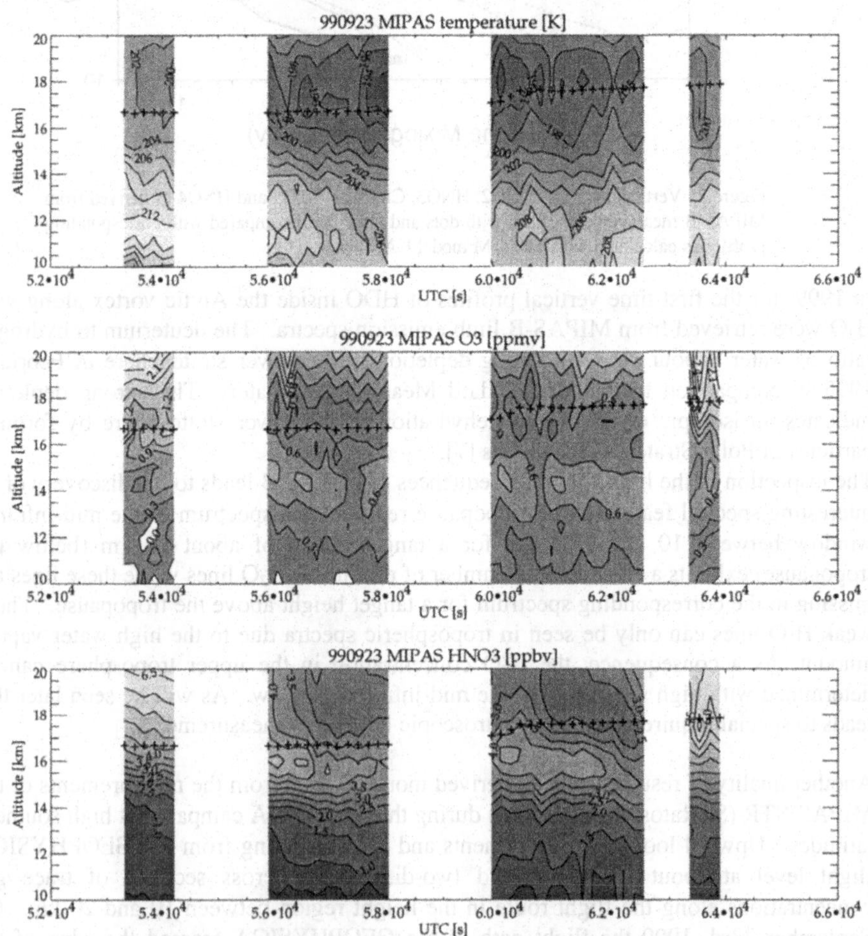

Figure 3. Cross sections of temprerature, ozone and HNO_3 along the flight path of the GEOPHYSICA from southern America to Antarctica and back (see also text).

5. Simultaneous derivation of the concentration of trace gases and particles

The MIPAS experiments offer the specific possibility to measure simultaneously trace species and particles (aerosol and cloud particles) in the atmosphere as long as the optical thickness of the particle ensemble is small. The particles cause a quasi-continuum in the mid-IR broad-band spectrum which is slowly changing with wavelength depending on the composition of the particles. Höpfner et al. [10] have recently shown that MIPAS limb emission spectra are sensitive to Polar Stratospheric Cloud (PSC) layers if the corresponding particle volume density is not too small. Under the assumption of a mean radius of larger than about 1 μm the number density of the particles and their mean radius can be determined from the spectra. In addition, the wavelength dependent quasi-continuum contains valuable information about the composition of the PSC-particles.

PSC clouds were observed in January 2000 in Kiruna (northern Sweden) by ground-based solar absorption mid-IR Fourier spectroscopy. A broad band analysis of the spectra was made by using the radiative transfer model KOPRA which is linked to a Mie model and provides analytical derivatives of the spectrum with respect to microphysical retrieval parameters. The agreement between calculated and measured spectra was best in case of applying the refractive index of water ice [11]. Furthermore, the mode-radius of a lognormal particle size distribution and the number density of the PSC particles could be retrieved. The results of simultaneous LIDAR observations at different wavelengths were in good accordance.

This capability of mid-IR Fourier spectrometers is very promising for future observations of the atmospheric state.

6. Conclusions

Mid-infrared Fourier spectrometers with high spectral resolution as used in atmospheric research yield a large amount of information on temperature and trace constituents simultaneously. The MIPAS space experiment is intended to provide global data sets for about five years after its launch in June 2001. In order to be able to derive global distributions of atmospheric parameters with high quality it is necessary to rely on accurate spectroscopic data. While the catalogues of spectroscopic data have been improved continuously during the past years the use of the new high-resolving spectrometers have lead to additional specific requirements for laboratory measurements. In the following some of these requirements for analysing corresponding atmospheric broad-band spectra will be listed:

1. The accurate line strength of weak lines of some molecules (e.g. H_2O, O_3) is needed at least for selected spectral intervals.
2. For the determination of pressure and temperature profiles the relative accuracy of CO_2-bands in the spectral range betweeen 4 and 16 μm has to be about 1%.
3. For non-LTE investigations missing transitions for several trace species (e.g. O_3 levels with vibrational energies between 4000 and 6550 cm^{-1}, highly excited

rotational levels of NO) have to be complemented in the current spectroscopic data bases.

4. Accurate spectroscopic parameters are needed for some trace species (like CH_3CCl_3, H_2CO, CH_3COCH_3) which are detectable in the upper troposphere with limb measurements.
5. Better laboratory data for complex refractive indices of ternary solution aerosols and nitric acid trihydrate at stratospheric temperatures is desired.

7. Acknowledgements

The work of many scientists of the Institut für Meteorologie und Klimaforschung (IMK) for the design, development, construction and operation of various MIPAS experiments is gratefully acknowledged. I also thank the members of the Scientific Advisory Group of the MIPAS space experiment for their support.

8. References

1. Fischer, H. (1993) Remote Sensing of Atmospheric Trace Gases, *Interdis. Sci. Rev.* **10**, 185.
2. Oelhaf H., T. von Clarmann, H. Fischer, F. Friedl-Vallon, C. Fritzsche, A. Linden, C. Piesch, M. Seefeldner and W. Völker (1994) Stratospheric $ClONO_2$, HNO_3 and O_3 Profiles Inside the Arctic Vortex from MIPAS-B Limb Emission Spectra during EASOE 1992, *Geophys. Res. Lett.* **21**, 1263.
3. Blom C., H. Fischer, N. Glatthor, T. Gulde, M. Höpfner, Ch. Piesch (1995) Spatial and Temporal Variation of ClONO2, HNO3 and O3 in the Arctic Winter 1992/93 as obtained by Airborne Infrared Emission Spectroscopy, *J. Geophys. Res.* **100**, 9101.
4. ESA 2000, ENVISAT: MIPAS, an instrument for atmospheric chemistry and climate research (March 2000), *ESA-report SP* **1229**.
5. Wetzel G., H. Oelhaf et al. (2000) Nitrogen partitioning in summer mid-latitudes and the late winter arctic vortex measured by MIPAS-B, *Proc. 5th European Workshop*, St. Jean de Luc, France, 384-387.
6. Ruhnke R., W. Kouker, Th. Reddmann, H. Oelhaf, H, Fischer, and Th. Peter (2000) Denitrification in Arctic winter 1994/1995 as calculated with KASIMA and the effect of the NOy partitioning, *Proceedings of the Quadrennial Ozone Symposium,* Hokkaido University Sapporo, Japan, pp. 617 - 618.
7. Stowasser M., H. Oelhaf, G. Wetzel, F. Friedl-Vallon, G. Maucher, M. Seefeldner, O. Trieschmann, T. von Clarmann, and H. Fischer (1999) Simultaneous Measurements of HDO, H20 and CH4 with MIPAS-B: Hydrogen Budget and Indication of Dehydration inside the Polar Vortex, *J. Geophys. Res.* **104**, 19213-19255.
8. Blom Cornelis E., Thomas Gulde, Michael Höpfner, Corneli Keim, Wolfgang Kimmig, Katrin Lessenisch, Christof Piesch, Christian Sartorius and H. Fischer (2000) MIPAS-STR measurements during APE-GAIA, *SOLVE-THESEO 2000 Science Meeting*, Palermo, Italy, 25-29 September 2000.

9. Funke B., M. Lopez-Puertas et al. (2000) A new NLTE retrieval method for atmospheric parameters from MIPAS-ENVISAT emission spectra, *Proceedings COPAR meeting July 2000*, Warsaw.
10. Höpfner M., Thomas von Clarman, Georg Echle, Alexandra Zimmermann (2000) Retrieval of PSC properties from MIPAS-ENVISAT measurements, *Proceedings of SPIE's 2nd International Asia-Pacific Symposium on Remote Sensing of the Atmosphere, Environment and Space*, 9-12 October 2000, Sendai, Japan.
11. Höpfner M., T. Blumenstock et al. (2001) Comparison of mountain wave PSC measurements over northern Scandinavia by ground-based FTIR and airborne LIDAR, *EGS meeting March 2001,* Nice, France.
12. Echle G., Thomas von Clarmann, A. Dudhia, J.-F. Flaud, B. Funke, N. Glatthor, B. Kerridge, M. López-Puertas, F.J. Martín-Torres and G.P. Stiller (2000) Optimized spectral microwindows for data analysis of the Michelson Interferometeer for Passive Atmospheric Sounding on ENVISAT, *Appl. Opt* **39**, 5531-5540.

INFRARED EMISSION SPECTROSCOPY

B. CARLI, U. CORTESI, L. PALCHETTI
Istituto di Ricerca sulle Onde Elettromagnetiche del CNR
Firenze - Italy

Abstract

A review is made of the instruments and of the observation modes that are used for the study of the Earth's atmosphere with emission spectroscopy. A short comparison is made between heterodyne and optical techniques and details are provided of the second technique with a few examples of real instruments. The role of Fourier transform spectroscopy is underlined and the rationale associated with the choice of the observation geometry and of the spectral resolution is discussed.

1. Introduction

Emission spectroscopy is a powerful tool for the monitoring of the Earth's atmosphere and is finding increasing applications in space observations. Techniques for the remote sensing of the atmosphere include active techniques (such as Radar, SAR, LIDAR), passive absorption spectroscopy (e.g. with solar occultation) and passive emission spectroscopy. Among these, emission spectroscopy, that uses the spontaneous thermal emission of the gas, has important advantages because it does not use energy demanding instruments as in the case of active techniques and can be performed continuously (both day and night) and in all directions without depending on external sources as in the case of absorption spectroscopy. On the other hand emission measurements are only possible in those spectral regions in which the atmosphere has an emission of detectable intensity (millimetre, submillimetre, far infrared and middle infrared spectral regions).
The different techniques that can be used for emission spectroscopy are compared in Sect. 2. In Sect. 3 a short description is made of Fourier transform spectroscopy (FTS) which is the most widely used technique in the case of incoherent detection. The options that are available for spectral resolution and geometry of observation are discussed in Sect. 4 and 5. In Sect. 6 a few examples are made of FTS instruments operating on space borne platforms. In the case of far infrared spectroscopy, for which a space application is not yet possible, examples are given in Sect. 7 of airborne observations. Finally, in Sect. 8 an instrument is described in which broad band rather high spectral resolution is the main instrument requirement. The discussion will be focused on applications for observations of the Earth's atmosphere.

J. Demaison et al. (eds.), Spectroscopy from Space, 171–186.

2. Incoherent and coherent detection

For the detection of signals due to thermal emission two techniques can be used: the heterodyne technique with coherent detection and the optical technique with incoherent detection. The heterodyne technique can be used at long wavelengths, typically in the microwave, millimetre and submillimetre spectral regions, while the optical technique can be used at shorter wavelengths, typically in the near- middle- and far infrared. In the spectral region at the boundary between the range of operation of these two techniques, where there was once a gap, there is now an overlap and a choice can be made between the two detection techniques.

The main characteristics of coherent detection are:

- detection of a single mode of the source, that implies a limited throughput of the instrument;
- a very selective frequency separation that allows a very high spectral resolution;
- detection noise proportional to the square root of the width of the resolved spectral element, that further favours the use of high spectral resolution;
- on the other hand, the spectral bandwidth and the number of independent spectral elements simultaneously observed with a single instrument are limited by technical constraints.

The main characteristics of incoherent detection are:

- simultaneous detection of several modes, that allows the exploitation of a large throughput in the case of extended sources;
- possibility of attaining high spectral resolution;
- in the case of cooled photon-noise-limited detectors, a measurements noise proportional to the square root of the width of the instantaneous spectral band;
- within the limits of the above property, possibility of observing a very broad spectral interval with a single instrument.

The comparison is further complicated by the cooling requirements of the detectors and in some case of the instrument as well. The cooling is an option for improved performances, but for some detectors the improvements are so large that it becomes almost a necessary feature of the instrument.

These general characteristics determine better performances of heterodyne technique at longer wavelengths and better performances of optical techniques at shorter wavelengths. More generally, the evolving technology and the specific requirements determine which technique provides the most efficient measurement in the different applications. The following discussion will be focused on emission spectroscopy performed with incoherent detection. In the field of incoherent detection the spectrometers that are more widely used for the measurement of the emission of the Earth's atmosphere are the Fourier Transform (FT) spectrometers, that will be discussed in the following section.

3. Fourier transform spectroscopy

Figure 1 shows the schematics of a FT spectrometer. The spectrometer splits the radiation of the source into two beams that travel different optical paths. The two beams are subsequently recombined and interfere with an amplitude that depends on their path difference. The interferogram, that is the amplitude of the interference as a function of the path difference, is the measured quantity. The FT of the interferogram is equal to the spectral distribution of the source.

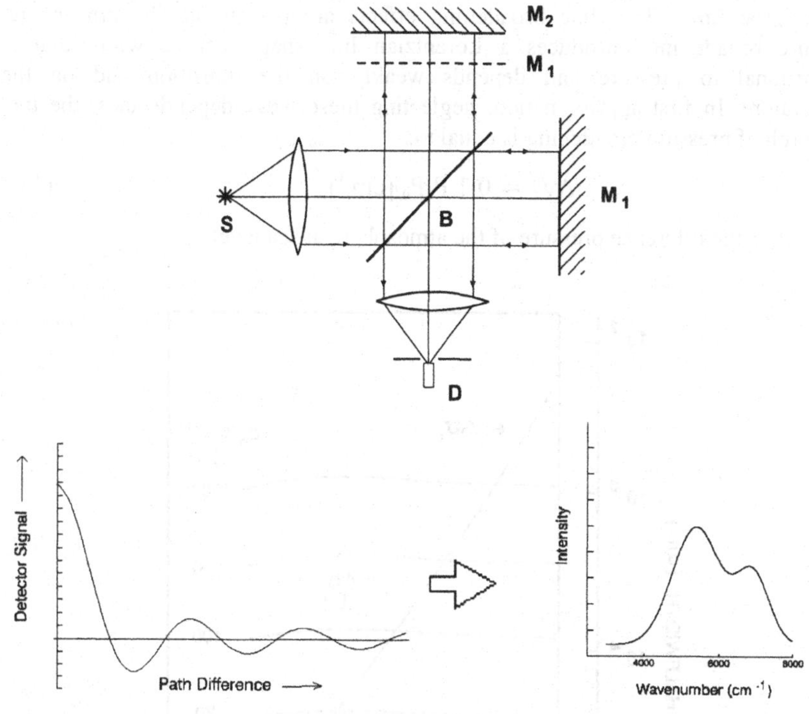

Figure 1
Schematics of a Fourier transform spectrometer

An extended literature can be found on FT spectroscopy (see for instance [1] and [2]), we briefly recall herewith its main properties. In summary:

- an indirect measurement of the spectral distribution is obtained;
- a broad spectral interval is observed with a single detector and a multiplex advantage is obtained in signal-to-noise ratio;
- a large throughput and a high spectral resolution are attained with a relatively small instrument;
- a very accurate frequency scale is obtained;
- moving optical components are needed;
- spectra are recorded with a relatively short measurement time.

4. Spectral resolution

Spectral resolution is a qualifying property of a spectroscopic measurement and must be chosen according to the specific objectives of the measurement.
In the case of spectroscopy of the Earth's atmosphere the value of the spectral resolution is chosen with the objective either of resolving the features of the studied atmospheric constituents from other interfering species or of measuring the line shape induced by the radiative transfer performed in an inhomogeneous medium. In both cases the actual resolution requirement depends on the linewidth of the atmospheric lines. Two line broadening effects are present in the atmosphere. Pressure broadening introduces a Lorentzian line shape with a width that is proportional to pressure and depends weakly on the transition and on the temperature. In first approximation, neglecting these weak dependencies, the line halfwidth of pressure broadening is equal to:

$$\Delta\nu_L \approx 0.1\ P/P_0\ [cm^{-1}] \tag{1}$$

where P_o is the reference pressure of the atmosphere at sea level.

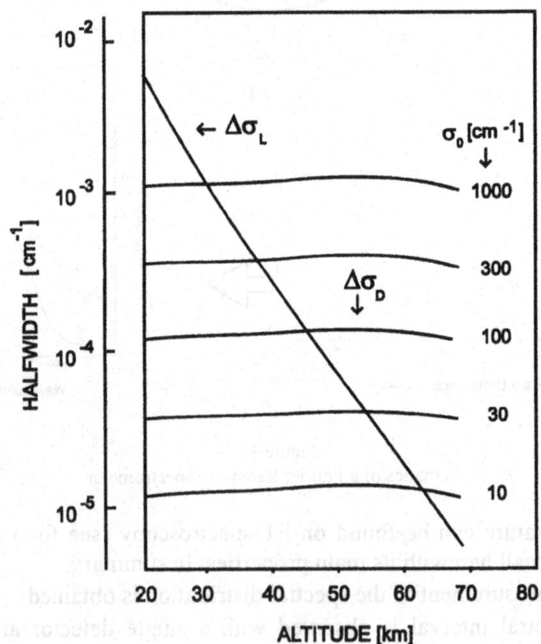

Figure 2
Altitude dependence of the pressure broadening halfwidth $\Delta\nu_L$ and of the Doppler broadening halfwidth $\Delta\nu_D$ for five different spectral frequencies

Temperature broadening introduces a Gaussian line shape with a width that is proportional to the frequency ν of the transition and depends on temperature and on the molecular weight of the molecule. In first approximation, neglecting the dependency on the molecular weight, the line halfwidth of temperature broadening is equal to:

$$\Delta\nu_D \approx 10^{-6}\,(T/T_0)^{1/2}\,\nu \qquad (2)$$

where T_o is a reference temperature of 300 Kelvin.

Figure 2 shows how the halfwidth of these two broadening effects varies as a function of frequency and altitude for a standard atmosphere.
The line profile of an atmospheric feature is equal to the convolution of the Lorentzian profile with the Gaussian profile and is called Voigt profile. The spectral distribution of the atmospheric spectrum is further complicated by the radiative transfer in an inhomogeneous medium and by the behaviour of strong transitions, which saturate at the central frequencies and increase their width as a result of the growth of the line wings. Finally, the observed spectral distribution is modified by the instrumental line shape.

5. Observation geometry

The geometry of observation is an important feature of atmospheric remote sensing since different information can be attained depending on this choice. Figure 3 shows an example of the different observation geometries.

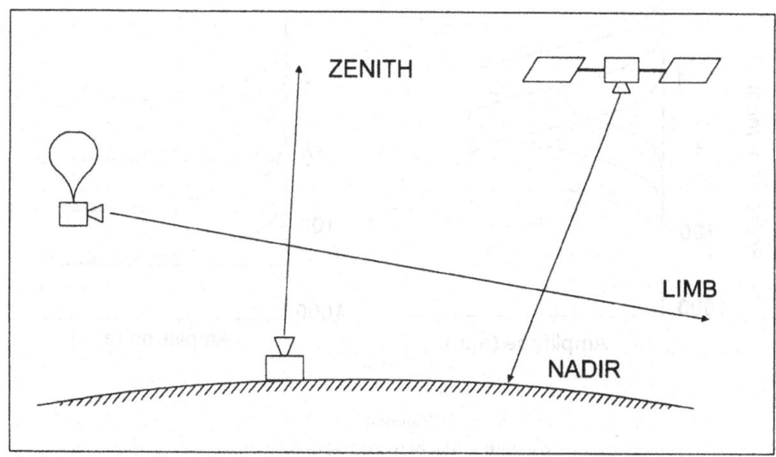

Figure 3
Graphical representation of the nadir sounding, zenith sounding and limb sounding observation geometries

Emission measurements are possible in all directions and all observation geometries can be used. Zenith sounding is made looking above in the vertical direction or in a slant direction. Nadir sounding is made looking below from some high altitude platform. Limb sounding is also performed from an high altitude platform and is made looking in the horizontal direction at the limb of the atmosphere so that the line of sight penetrates into lower layers of the atmosphere and emerges again towards space because of the Earth's curvature. From the ground only vertical sounding is possible and from space only nadir sounding and limb sounding are possible. From inside the atmosphere all the three geometries of observation are possible.

The objectives of remote sensing are the physical and chemical properties of the atmosphere (pressure, temperature and composition) with the best possible spatial resolution. Therefore, the aim of the observation geometry is not only to have a selective line of sight that discriminates either a particular geographical pixel or a particular direction, but also to resolve the different signals originated along the line of sight. An interesting example is provided by the different performances of nadir and limb sounding from space. The horizontal resolution of nadir sounding is directly determined by the extent of the solid angle with which the atmosphere is observed. Typically resolutions of about 50 km or less are obtained, which is a high resolution considering the horizontal variability of the atmosphere. The vertical resolution requires the separation of the different contributions along the line of sight. In the previous section we have seen that the line width depends on pressure, therefore the resulting line shape depends on the contributions at different altitudes.

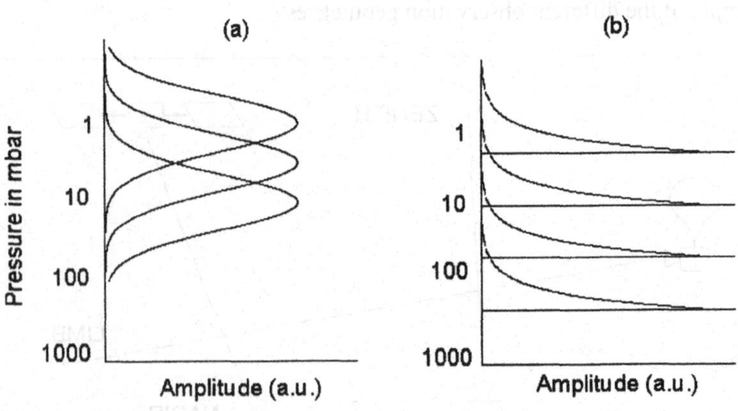

Figure 4
Weighting functions corresponding to
the vertical sounding (a) and limb sounding (b) observation modes

Figure 4a shows for a set of frequencies in the wing of a line the typical distribution of the contributions from different altitudes. Some discrimination of the different altitudes is present and an inversion of the measurements for the retrieval of the vertical distribution is possible.

The broad character of the distributions shown in Fig. 4a limits, however, the attainable vertical resolution. Typically a vertical resolution of about 10 km is obtained, which is a low resolution considering the vertical variability of the atmosphere.
In the case of a limb sounding measurement the distribution of the contributions from different altitudes is very peaked around the so-called tangent altitude of the measurement (i.e. the minimum altitude reached by the line of sight). Figure 4b shows the typical distributions for a set of measurements made with different tangent altitudes. The discrimination is very selective and a good vertical resolution (typically of 2 - 3 km) is attained. On the other hand, the line of sight travels very long distances in the horizontal direction and its horizontal resolution is poor (typically about 500 km).
Efforts are being made in order to overcome the individual limits of the two techniques. The vertical resolution of nadir measurements can be improved using redundant information (see Sect. 6.2). The horizontal resolution of limb measurements can be improved using a sequence of limb soundings made along the orbit to perform a two-dimensional retrieval [3].

6. Examples of some space borne FT spectrometers

The optical properties of the Earth's atmosphere and its emission spectrum depend on both the temperature and the composition of the atmosphere. Each molecular species present in the atmosphere has characteristic transitions due to its rotational and vibrational spectrum. Best information is obtained from measurements made in a broad spectral interval, so that numerous features are present, and at high spectral resolution, so that they are individually resolved. FT spectrometers combine these capabilities with an instrument of relatively small size and weight. This makes FT spectrometers the most common instruments for emission remote sensing of the Earth's atmosphere from space. In the following two subsections the examples are made of two instruments that will soon be launched.

6.1. MIPAS

MIPAS (Michelson Interferometer for Passive Atmospheric Sounding) is an example of an FT spectrometer that operates in a limb sounding mode. MIPAS will fly on the European satellite ENVISAT that is scheduled for a launch in June 2001 Table 1 summarises the main features of the instrument. Further details are given in [4]. Typical features of this limb sounding FT spectrometer are the capability of detecting simultaneously a large number of minor atmospheric constituents, a good vertical resolution and a medium horizontal resolution.

TABLE 1 – MIPAS	
Objective	Atmospheric chemistry and dynamics, global climatology
Products	T, H_2O, O_3, CH_4, HNO_3, N_2O and other minor constituents
Geometry	Limb sounding
Spectral interval	685 – 2410 cm^{-1}
Spectral resolution	0.025 cm^{-1}
Vertical resolution	3 km
Horizontal resolution	500 km

6.2. IASI

IASI (Infrared Atmospheric Sounding Interferometer) is an example of a FT spectrometer that operates in a nadir sounding mode. IASI will fly on the meteorological satellite METOP 1 that is scheduled for launch in 2002. Table 2 summarises the main features of the instrument [5].

TABLE 2 – IASI	
Objective	Operational sounding
Products	T and humidity, cloud properties, CH_4, N_2O, SO_2, and CO
Geometry	Nadir sounding
Spectral interval	645 – 2760 cm^{-1}
Spectral resolution	0.25 cm^{-1}
Vertical resolution	~ 2 km for T and humidity, columns for other minor constituents
Horizontal resolution	25 km spacing , 12 km pixel size

An important feature of IASI is the high spectral resolution of pressure and humidity sounding despite the fact that nadir measurements are not best suited for vertical resolution. Figure 5 shows the vertical resolution obtained in sensitivity tests of the IASI performances and its comparison with a previous instrument.

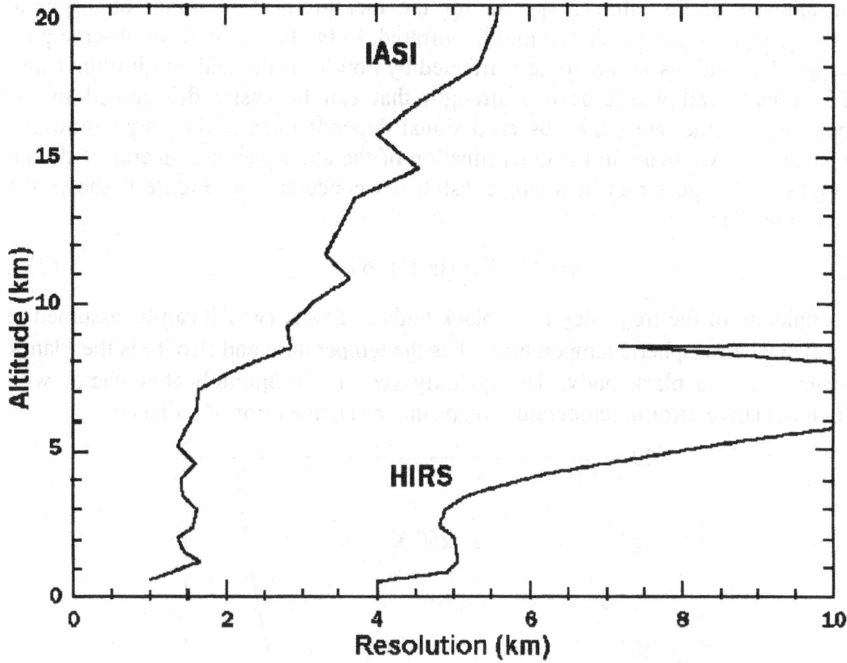

Figure 5
Calculated accuracy and vertical resolution of temperature measured by IASI compared with that of HIRS instrument

This result is made possible by the large number of observed spectral elements that depend on temperature and water vapour concentration. This data redundancy can be used to reduce the uncertainty of the retrieval and extract information with an improved vertical resolution. It is important to note that the same is not possible for the minor atmospheric constituents (CH_4, N_2O, SO_2, and CO), of which only a few spectral features are observed and for which only a column measurement is possible. In order to retrieve a vertical profile for these constituents a better spectral resolution would be necessary. Typical features of this nadir sounding FT spectrometer are therefore a good horizontal resolution, detection capability limited to the major atmospheric constituents, poor vertical resolution with the exception of quantities for which large redundancy of measurements is available.

7. Far-infrared measurements

Measurements at long wavelengths (20 – 250 cm^{-1}), typically the far infrared, depend upon the use of cooled detectors (at liquid helium temperatures). This is a technological difficulty that up to now has prevented the use of far infrared FT spectroscopy from space. Nevertheless, the far infrared spectral region has several interesting properties with respect to the middle infrared spectral region.

The analysis of far infrared spectra for the measurement of minor atmospheric constituents is easier than in the middle infrared. In the far infrared we observe pure rotational transitions which are less affected by non-local-thermal-equilibrium (non-LTE) effects and which have a strength that can be easily determined in the laboratory. Furthermore, the observed signal depends linearly on temperature and the errors that we make in the determination of the atmospheric temperature do not propagate as large errors in minor constituents concentration. Figure 6 shows the plot of the quantity:

$$k(\sigma) = T\, \partial\, (\ln B)/\, \delta T \tag{3}$$

as a function of the frequency for a black body at 250 K (which can be assumed to be a typical atmospheric temperature). T is the temperature and B(σ,T) is the Planck distribution of a black body. The quantity k(σ) is the multiplicative factor with which a relative error in temperature maps into a relative error of radiance.

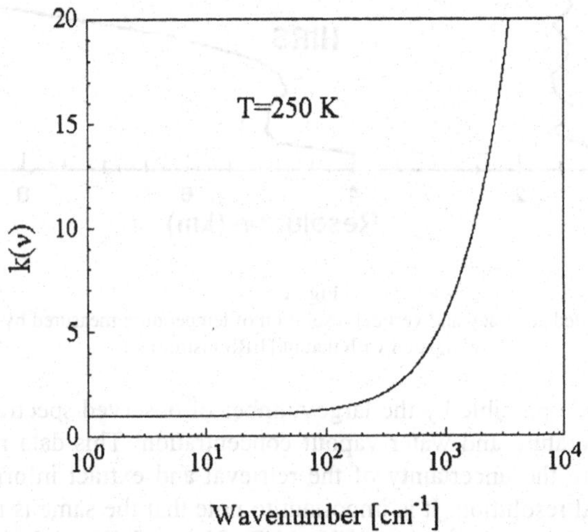

Figure 6
Plot of the ratio between the relative error in radiance ΔB(σ,T)/B(σ,T) and the relative error in temperature ΔT/T

Many important minor atmospheric constituents can be observed in the far infrared. In particular linear molecules (such as OH, HCl, HF, HBr, CO, ClO, and NO) have strong signatures in their rotational spectrum, and for some of these molecules the only measurement opportunity in emission occurs in the far infrared.
In the far infrared CO_2 and CH_4 have no dipole moment and no rotational spectrum. As a consequence no spectral clashing is occurring with these molecules, that are present in the atmosphere at a relatively high concentration (at the ppm level). On the other hand the other atmospheric molecular constituents that have a concentration at the ppm level, i.e. water vapour and ozone, have a rather strong rotational spectrum. Several far infrared instruments were flown on balloon platforms (see e.g. [6] and [7]) and recently the SAFIRE instrument was operated from a stratospheric aircraft [8].

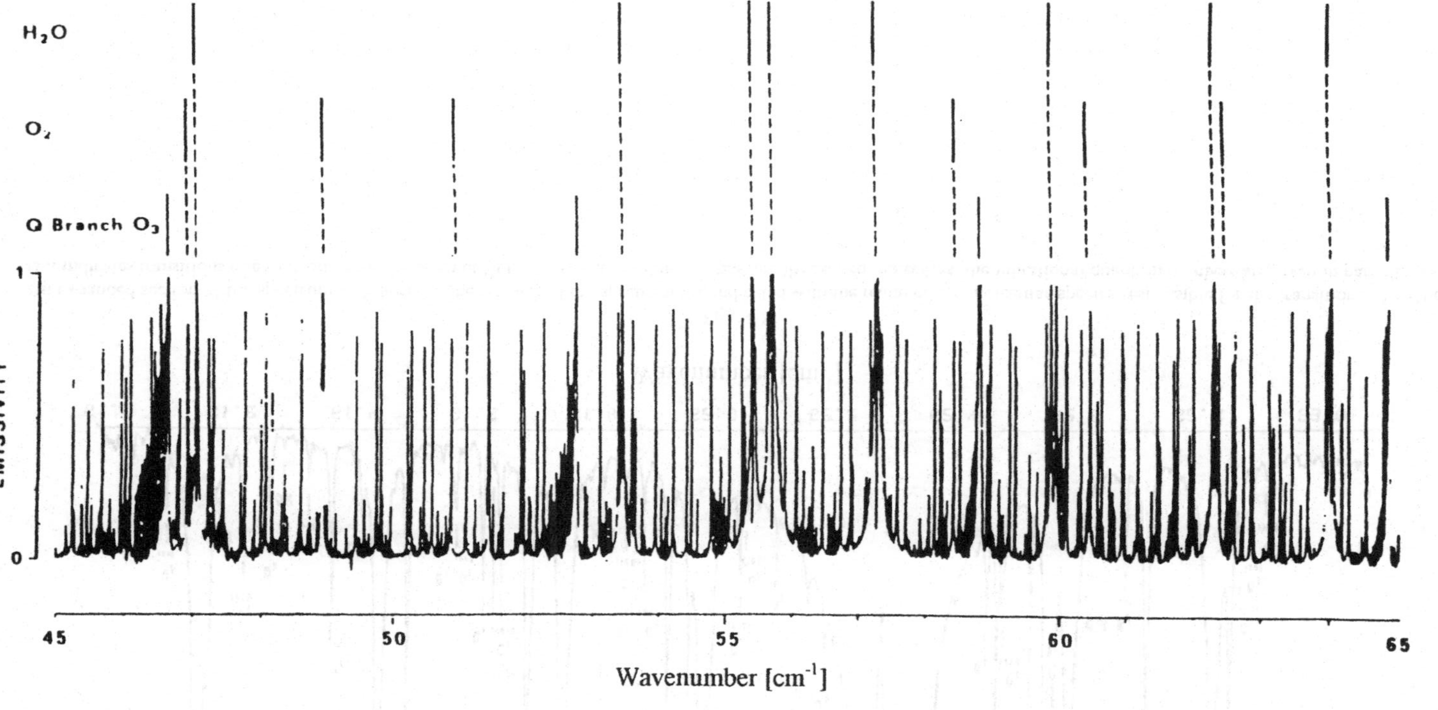

Figure 7
A sample of the atmospheric emission spectrum in the 45-65 cm-[1] spectral region. The main transitions of water vapour, molecular oxygen and ozone Q-type branches are marked.

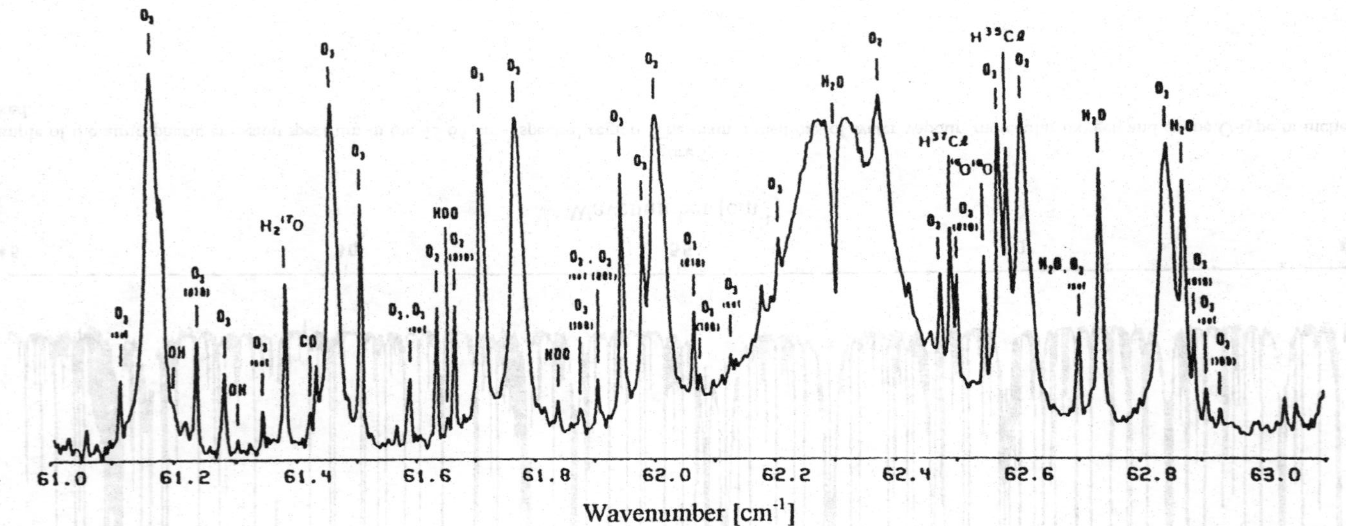

Figure 8

An expanded section of the spectrum of Figure 7. The features of the spectrum are indicated with the name of the molecular species responsible for the transition. The label *isot* indicates transitions due to ozone with one atom of ^{18}O. For transitions due to vibrationally excited molecules, the vibrational quantum numbers are given in parentheses.

7.1. BALLOON MEASUREMENTS

Figure 7 shows a sample of a far infrared atmospheric emission spectrum in the 45 – 65 cm^{-1} interval measured from a balloon altitude of 38 km in the limb direction with a tangent altitude of 32 km. An almost continuous sequence of lines can be observed. A closer look shows, as marked on the top of the figure, several strong water vapour lines, equispaced Q-branches of ozone and triplets of the rotational spectrum of molecular oxygen due to its magnetic dipole. Figure 8 shows an expanded view of the same spectrum in the interval 61 - 63 cm^{-1}. In between the strong features already observed in Fig. 7, many individual line of the ubiquitous ozone spectrum can be observed as well as some weaker features of other constituents such as water vapour isotopes, ozone isotopes, CO and OH.

Stronger features of OH [9] can be seen in Figure 9 where a less crowded spectral interval in the 118 cm^{-1} spectral region is shown. Other important measurements, that are typical of this spectral region, are those of ClO at the 23 cm^{-1} [10] and of HBr at 50 cm^{-1} [11].

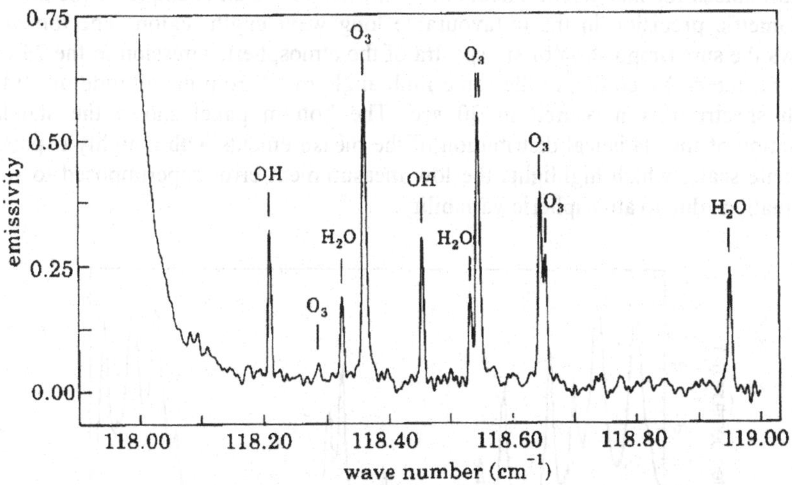

Figure 9
The atmospheric emissivity in the 118 cm^{-1} region, where two strong and well-resolved features of OH transitions are observed

7.2. AIRCRAFT MEASUREMENTS

SAFIRE (Spectroscopy of the Atmosphere using Far InfraRed Emission) is a far-infrared FT spectrometer that operates on board of a stratospheric aircraft. The aircraft is the Russian M-55 aircraft capable of reaching a maximum altitude of 21 km. Table 3 summarises the main features of the instrument [12].

TABLE 3 – SAFIRE	
Objective	Chemistry and dynamics at 10 – 20 km altitude
Products	Ozone, ClO, HNO_3, N_2O
Geometry	Limb sounding
Spectral interval	10 - 250 cm^{-1}
Spectral resolution	0.004 cm^{-1}
Vertical resolution	2 km
Horizontal resolution	100 km (along flight direction), 500 km (perpendicular to flight direction)

The instrument uses photon noise limited detectors that provide almost theoretical performances for incoherent detection. Figure 10 shows an example of the attained radiometric precision in the unfavourable long wavelength region. The top panel shows the superimposition of six spectra of the atmospheric emission in the 23 cm^{-1} spectral interval recorded at the same limb angle of 0° from the altitude of 20 km. Each spectrum is measured in 30 sec. The bottom panel shows the standard deviation of the statistical distribution of the measurements with a highly expanded ordinate scale, which highlights the low measurement error superimposed to some fluctuations due to atmospheric variability.

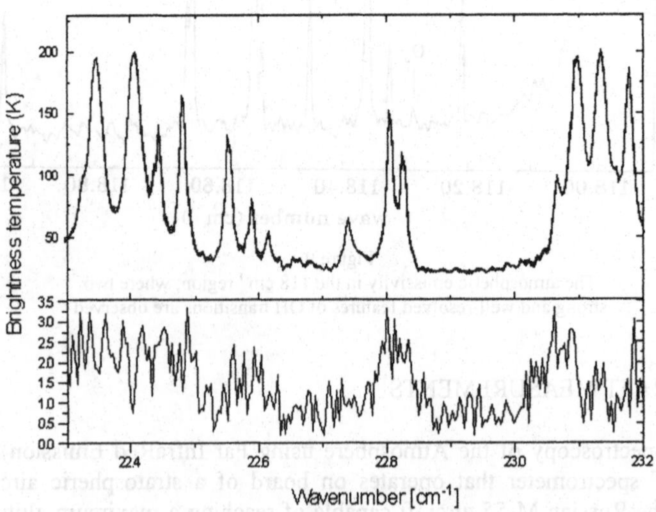

Figure 10

SAFIRE measurement performance: reproducibility for a sample of six calibrated spectra of the atmospheric emission measured at the same limb angle.

8. The Earth radiation budget

Most applications of emission spectroscopy from space exploit the resolved structure of molecular spectroscopy for the study of the atmospheric composition. However FT spectrometers have also the important property of broad band measurements. This property allows the observation with a single instrument of the full black body distribution of the thermal emission of the atmosphere and the measurement of the Earth radiation budget. Respect to radiometers (which do not resolve the atmospheric windows) FT spectrometers can resolve the signal inside and outside the atmospheric windows and determine the contributions of the different altitudes to the outgoing radiation flux. Theoretical calculations of the atmospheric cooling rate as a function of frequency and of altitude have shown that the cooling of the atmosphere occurs at frequency dependent altitude ranges determined by the absorption spectrum of water vapour [13]. Furthermore a significant fraction of the radiated power is emitted from the upper troposphere in a spectral region (frequencies lower than 500 cm^{-1}) in which measurements are lacking.

An FT spectrometer, called REFIR (Radiation Explorer in the Far InfraRed), was designed as part of an EU funded project for the broad band measurement of the radiative cooling of the Earth's atmosphere. Table 4 summarises the main features of the instrument [14]. The main features of REFIR are an optical design fully compensated for alignment error, broad band detectors, ambient temperature detectors and high optical efficiency.

TABLE 4 – REFIR	
Objective	Broad band Earth's radiation budget
Products	Radiance, T and H_2O
Geometry	Nadir sounding
Spectral interval	100 - 1000 cm^{-1}
Spectral resolution	0.5 cm^{-1}
Vertical resolution	~ 3 km for T and H_2O
Horizontal resolution	50 km spacing and 10 km pixel

9. Conclusions

Emission spectroscopy is a powerful tool for the passive study of the Earth's atmosphere. Among the instrument that use incoherent detection techniques FT spectrometers are the most efficient and productive instruments as they provide the best performances in terms of spectral resolution and band coverage with relatively compact instruments.

Among the numerous applications that are in progress, the MIPAS, IASI, SAFIRE and REFIR instruments have been recalled as they provide examples of instruments operating with nadir and limb geometry of observation, with low and high spectral resolution, from space and air borne platforms and in middle and far infrared spectral regions. This wide range of instruments and applications characterises the use of emission spectroscopy for the study of the atmosphere.

10. References

1. Bell, R.J., (1972) Introductory Fourier Transform Spectroscopy}, Academic Press, New York and London.
2. Chamberlain, J. (1978) The Principles of Interferometric Spectroscopy, John Wiles & Sons, New York.
3. Carlotti, M., Dinelli, B.M., Raspollini, P., Ridolfi, M. (2000) Geo-fit approach to the analysis of limb-scanning satellite measurements, in press.
4. Fisher, H. (2000) MIPAS: Michelson Interferometer for Passive Atmospheric Sounding, this volume.
5. Javelle, P. and Cayla, F. (1994) Infrared atmospheric sounding interferometer - instrument overview, *Space Optics 1994: Earth Observation and Astronomy*, G. Cerrutti-Maori and P. Roussel eds., Proc. SPIE 2209, 14-23.
6. Carli, B., Mencaraglia, F., Bonetti, A. (1984) Applied Optics, 23, 2594.
7. Traub, W.A., and Chance, K.V., (1981) Geophysical Research letters, 8, 1075.
8. Carli, B., Ade, P., Carlotti, M., Cortesi, U., Gignoli, A., Hamilton, P., Lanfranchi, M., Lee, C., MacKenzie, R., Phillips A., (1999) Minor Constituent Concentrations Measured from a High Altitude Aircraft using High Resolution Far-Infrared Fourier Transform Spectroscopy, Journal of Atmospheric Chemistry, JOCH 1363, pp. 1-21
9. Carli, B, Carlotti, M , Dinelli, B.M., Mencaraglia, F. and Park, H.J. (1989) The mixing ratio of the stratospheric hydroxyl radical from far infrared emission measurements. Journal of Geophysical Research. 94, 11049.
10. Carli, B. et al. (1988) Submillimetric measurement of stratospheric chlorine monoxide, Journal of Geophysical Research, 93, 7063 - 7068.
11. Carlotti, M., Ade, P., Carli, B., Ciarpallini, P., Cortesi, U., Griffin, M.J., Lepri, G., Mencaraglia, F., Murray, A.G., Nolt, I.G., Park, J.H., Radostitz J.V., (1995) Measurements of stratospheric HBr using high resolution far infrared spectroscopy, Geophysical Research Letters. Vol. 22 N°. 23, 3207.
12. Carli, B., Peter A., Cortesi, U., Dickinson, P., Epifani, M.,.Gannaway, F., Gignoli, A., Keim, C., Lee, C., Leotin, J., Mencaraglia, F., Murray, A., Nolt, I., Ridolfi, M. (1999) SAFIRE-A: Spectroscopy of the Atmosphere using Far-InfraRed mission /Airborne, *Journal of Atmospheric and Oceanic Technology*, 1313 – 1328.
13. Clough, S.A., Iacono, M.J and Moncet, J.L.(1992) Line-by-line calculations of atmospheric fluxes and cooling rates: application to water vapor, Journal of Geophysical Research, 97, pp. 15761-15785.
14. Palchetti, L., Barbis, A., Harries, J.E., Lastrucci, D., (1999) Design and mathematical modelling of the space-borne far-infrared Fourier transform spectrometer for REFIR experiment, Infrared Physics & Technology 40, pp. 367-377.

INFRARED SPECTROSCOPY IN THE ATMOSPHERE

J.-M. FLAUD
Laboratoire de Photophysique Moléculaire
Bâtiment 210, Université Paris-Sud, 91405 ORSAY Cedex, France

1. Introduction

Remote sensing techniques are widely used to probe the terrestrial atmosphere and retrieve P, T profiles as well as concentration profiles of a number of atmospheric constituents. These techniques cover the whole spectral range from the microwave to the visible-UV but is it worth noticing that the infrared domain is particularly suited to retrieve the concentration profiles of a large number of species in the atmosphere since almost all the atmospheric species exhibit strong fundamental vibration-rotation bands in the infrared spectral domain. The remote sensing techniques working in the infrared are numerous using various instruments with different resolutions (radiometers, spectrometers, Fourier transform interferometers, diode laser spectrometers, …) and performing measurements from different platforms (ground, aircraft, balloons, satellites) with different viewing geometries (limb, nadir, ...). However in all cases a precise knowledge of the spectral parameters (line positions, intensities, widths, …) is required for an accurate retrieval of the atmospheric profiles and this is true, not only for the species of interest, but also for the interfering species (i.e. species absorbing in the same spectral interval as the species of interest). Moreover it is worth noticing that, given the recent progress in the atmospheric instruments (better resolution, better signal/noise ratio, …), the spectral parameters need to be of better and better quality. For the spectroscopic point of view, this represents a significant amount of work which needs in order to be performed properly to combine in the laboratory modern experimental techniques and theoretical methods. It is indeed not always possible to observe and measure some unstable species and in some cases only a portion of the spectrum of the species of interest can be measured (For example, in many cases it is difficult to realize in the laboratory absorption path length equivalent to those encountered in atmospheric measurements). In such cases it is necessary to rely on theoretical methods. They can be “effective” i.e. using an Hamiltonian whose constants are determined from the fit of the available experimental data or “fundamental” (ab-initio calculations, …) or also “mixed” by combining the previous methods.

The goal of this paper is to show on a few selected examples the impact of the spectral parameters on the atmospheric retrieval and the difficulties which are faced to generate accurate spectral parameters.

J. Demaison et al. (eds.), Spectroscopy from Space, 187–200.

2. The $C\ell ONO_2$ molecule

$C\ell ONO_2$ is a reservoir molecule for the $C\ell O_x$ ($C\ell$, $C\ell O$, ...) as well as the NO_x (NO, NO_2, ...) species which play an important catalytic role in the chemical processes leading to the destruction of the ozone molecule.

From the spectroscopic point of view $C\ell ONO_2$ is a heavy molecule which possesses low frequency vibrational modes and its infrared spectrum exhibits, even at high resolution, a pseudo continuum on which more or less wide Q branches stand out. Figure 1 presents the absorption spectrum of $C\ell ONO_2$ in the 500 – 1800 cm^{-1} spectral range where appear some fundamental bands of this molecule.

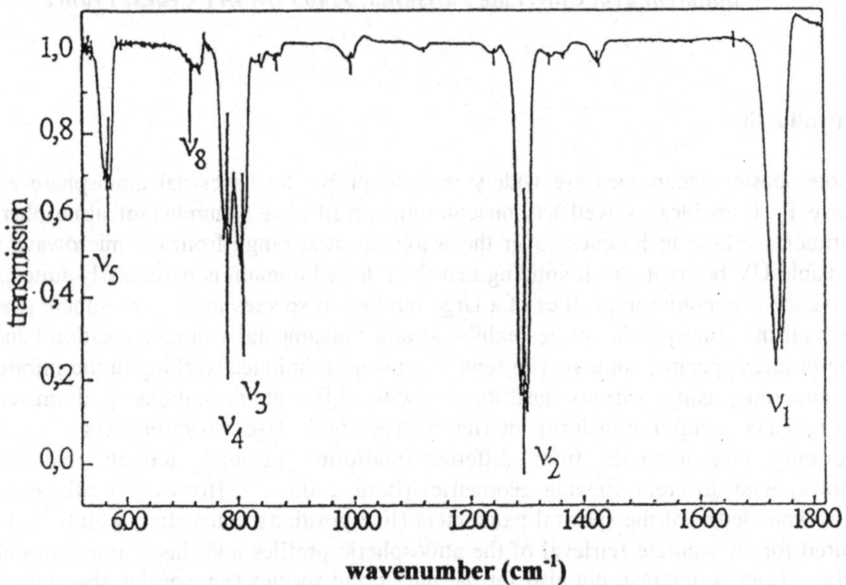

Figure 1. Absorption spectrum of $C\ell ONO_2$ between 20 and 5.5 μm. The Q branches of some vibration-rotation are clearly observable. [From ref. 1].

As a consequence and given the fact that it was not possible until recently to observe the rotational structure of the vibration-rotation bands of this molecule, only absorption cross-sections were measured in the laboratory [1-4] and no theoretical model was developed. Accordingly, retrievals of $C\ell ONO_2$ concentration profiles in the atmosphere [5-8] were performed using these absorption-cross sections. Unfortunately these laboratory data were suffering from some deficiencies, indeed :

The absorption cross sections were mainly measured for two temperatures (around 220 and 296 K) and this is not sufficient for the atmospheric applications : The atmospheric temperatures span indeed the range 200–300 K and the laboratory measurements do not allow to interpolate and/or extrapolate reliably.

No systematic study of pressure effects had been done. Figure 2 which presents a fit of the Q branch of the ν_4 bands of $C\ell ONO_2$ located around 780 cm^{-1} for an atmospheric spectrum recorded by the MIPAS experiment [9] shows clearly that the absorption cross-section recorded in the laboratory do not allow one to reproduce satisfactorily the atmospheric spectrum.

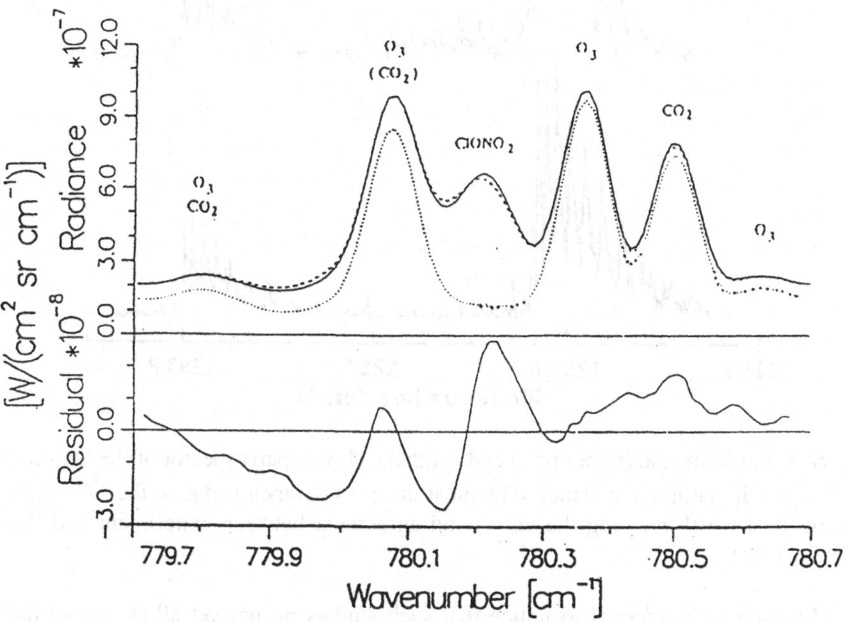

Figure 2. Fit of an emission atmospheric spectrum from the MIPAS balloon experiment.__ Experimental spectrum, --- or ... synthetic spectra including or not the $C\ell ONO_2$ emission. The difference between the observed and calculated spectra shows clearly that the laboratory absorption cross sections are not accurate enough (From ref. 7).

In order to improve the situation, a series of experiments at high spectral resolution using molecular beams combined with diode laser spectrometers were performed recently [9-11]. Indeed, given the low temperatures reached in a beam (T ≅ 30 K) the absorption due to hot bands is very weak allowing one to observe the rotational structure of the cold bands. As a consequence it is possible to perform the vibrorotational analysis of the spectrum. Figure 3 which presents the Q branches of the ν_2 bands of the $^{35}C\ell$ and $^{37}C\ell$ isotopic variants of $C\ell ONO_2$ in natural abundances shows clearly the quality of the theoretical analysis.

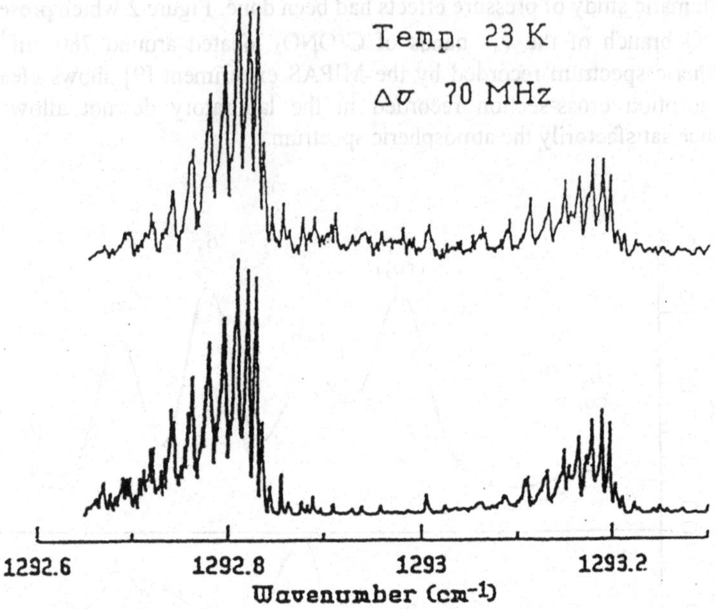

Figure 3. Experimental (upper part) and synthetic (lower part) spectra of the Q branches of $C\ell ONO_2$ in natural abundance. The most intense Q branch is due to the $^{35}C\ell$ isotopic variant. It is worth noticing the very good agreement between experiment and theory (From ref. 11).

However it is essential to notice that such studies do not yet allow one to model satisfactorily the atmospheric spectra. The reasons for that are numerous :

- Only transitions involving low and medium values of the rotational quantum numbers J and K_a are observed in molecular beam spectra because of the low temperature of the beam. As a consequence the extrapolation to transitions involving high J and K_a quantum numbers and which contribute to the atmospheric absorptions is tricky. Such an extrapolation, to be accurate, needs indeed to know the values of the high order terms in the Hamiltonian.
  Also because usually diode laser spectrometers do not cover a wide spectral range, it is possible that other bands (combination bands for example) which interact with the studied band are not observed. In such a case, the extrapolation is even trickier. Such a situation was encountered during a study of some fundamental bands of the N_2O_4 molecule [12] and it was shown that the analysis of the fundamental bands when considering them as isolated leads to "abnormal" values of the high order Hamiltonian constants and hence to rather poor extrapolations.
  As already said, the hot bands are not observed in the molecular beam spectra while in fact they contribute strongly to the atmospheric absorptions. Table 1 gives an estimation of the contribution of the first hot bands to the absorption at two

different temperatures. It is easy to see that, even at the lowest temperature, the contribution to the absorption of the fundamental band is less than 50%.

TABLE 1. Contribution of the hot bands to the $C\ell ONO_2$ absorption.

Band	T = 213 K	T = 296 K
ν_i $(i \neq 9,7)$	$\cong 0.43$	$\cong 0.25$
$\nu_i + \nu_9 - \nu_9$	$\cong 0.19$	$\cong 0.14$
$\nu_i + 2\nu_9 - 2\nu_9$	$\cong 0.09$	$\cong 0.08$
$\nu_i + \nu_7 - \nu_7$	$\cong 0.07$	$\cong 0.07$
$\nu_i + 3\nu_9 - 3\nu_9$	$\cong 0.04$	$\cong 0.04$

$\tilde{\nu}_0(\nu_9) \cong 121\,\text{cm}^{-1}$

$\tilde{\nu}_0(\nu_7) \cong 273\,\text{cm}^{-1}$

Fortunately, very recently high resolution Fourier transform spectra of $C\ell ONO_2$ have been recorded [13] at various pressures and temperatures. They have been used to produce absorption cross sections of better quality than the previous ones and their analysis is in progress.

From this example one can see that concomitant progress in experiment and theory are needed in order to generate the precise spectral parameters required for a reliable analysis of the atmospheric spectra.

3. The HOCℓ molecule

The chemical processes involving chlorine atoms are believed to be important as far as the depletion of ozone in the stratosphere is concerned [14]. Among the molecular species involved in these processes, hypochlorous acid is considered to be a temporary reservoir of chlorine atoms and has been detected in the atmosphere at mid-latitudes by far infrared measurements [15]. The ν_2 band located around 8.1 μm lies in an atmospheric window and was used to measure the column abundance of HOCℓ over the Antartic [16] or to estimate its concentration in the Artic stratosphere [7]. The crucial question here is the quality of the line intensities which were used for the retrieval since, in the linear region of the curve of growth an error of + x % on the line intensities will propagate as an error of – x % on the concentrations. The main difficulty indeed when aiming at the measurements of absolute intensities is that HOCℓ is an unstable species and that it is not at all easy to determine its concentration in the sample. Indeed HOCℓ always exists in the following equilibrium :

$$H_2O + C\ell_2O \leftrightarrows 2HOC\ell \qquad (1)$$

Various methods can be used to determine the partial pressure of HOCℓ in the sample. Niki et al [17] used two chemical titrations based on reaction 1 : the hydrogen mass balance $\Delta(H_2O) = 2\Delta(HOC\ell)$ and the rapid reaction of HCℓ with $C\ell_2O$. Su et al

[18] relied on the modeling of the UV photolysis of dilute mixtures of $C\ell_2$, H_2 and O_3 in excess of O_2. Finally Burkholder [19] used the UV absorption cross sections of HOCℓ derived from HOCℓ absorption spectra recorded after UV photolysis of equilibrium mixtures of $C\ell_2O$, H_2O and HOCℓ to eliminate $C\ell_2O$ and the chlorine mass balance method. Unfortunately, as shown in Table 2, the band intensities derived from these experiments show a rather large scatter preventing a precise determination of the HOCℓ concentration in the atmosphere.

To cope with this problem (Determination of the concentration of HOCℓ in the sample), a study [20] based on an optical determination of the HOCℓ partial pressure was performed. It consists in probing the same sample in the infrared and the far infrared ranges and to record the spectra **simultaneously** in both ranges. Indeed the far infrared intensities can be calculated accurately using the dipole moment deduced from Stark measurements [21]. As a consequence they can be used to derive the HOCℓ concentrations in the sample and it is these concentrations which are then used to derive the intensities of the lines of the infrared ν_2 band. In this way one gets rid of the problem of the estimation of the HOCℓ concentration through chemical methods.

Figure 4 gives a schematics of the experiment : the far infrared pure rotation lines were recorded using a high resolution Fourier transform interferometer whereas the ν_2 lines were measured using a tunable diode laser spectrometer.

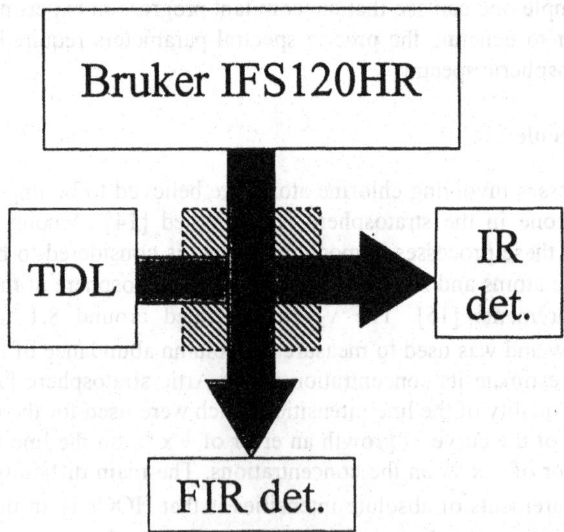

Simultaneous measurement

Figure 4. Schematics of the experimental set up used for a simultaneous measurement of the HOCℓ absorptions in the far infrared and the infrared.

It is outside the scope of this paper to give all the experimental details which are described in ref. [20] but it is worth noticing, as already stated, that the main difficulty comes from the fact that HOCℓ is unstable. Figure 5 shows the integrated absorption coefficients of a given line during four experiments (1 → 4) plotted as a function of the time delay from the start of the experiment. The solid curves represent the result of the adjustement of each set of data points to a polynomial expansion. The figure shows clearly the decrease of the temporal evolution of the HOCℓ amount from experiment 1 to experiment 4, leading to a stable sample for experiment 4.

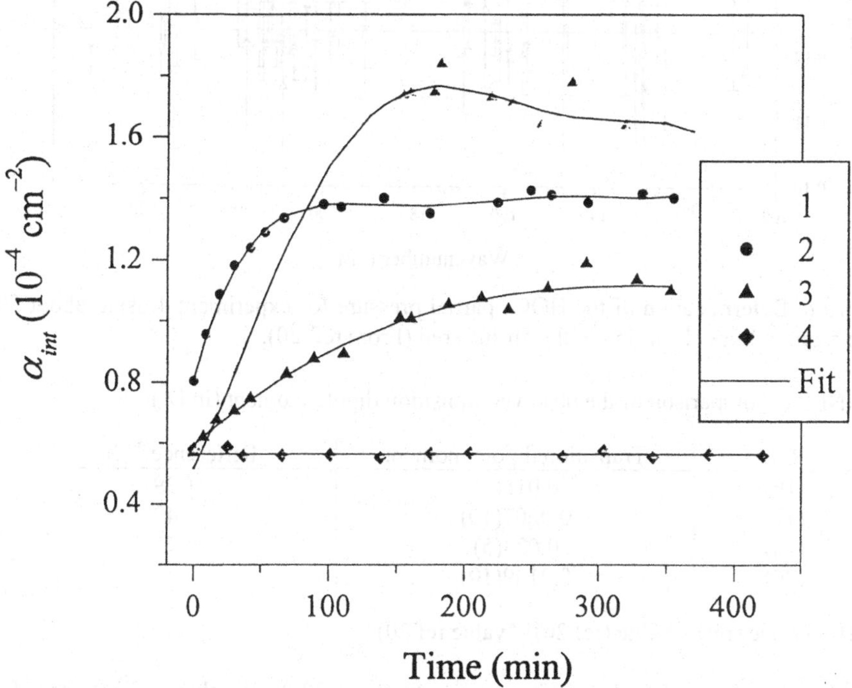

Figure 5. Evolution of the HOCℓ concentration in the sample. It is worth noticing that the HOCℓ concentration is stable during experiment 4 (From ref. 20).

Figure 6 presents the partial pressure of HOCℓ determined from the pure rotation spectrum measured in experiment 4 and the far infrared lines [22] calculated using the dipole moment deduced from Stark experiments [21]. It is worth noticing that about 50 lines with different quantum numbers were used. In this way it was possible to determine the partial pressures during the various experiments to within an average uncertainty of about 3 %. These partial pressures were then used to derive the intensities of the infrared lines and we give in Table 2 the corresponding transition dipole moment.

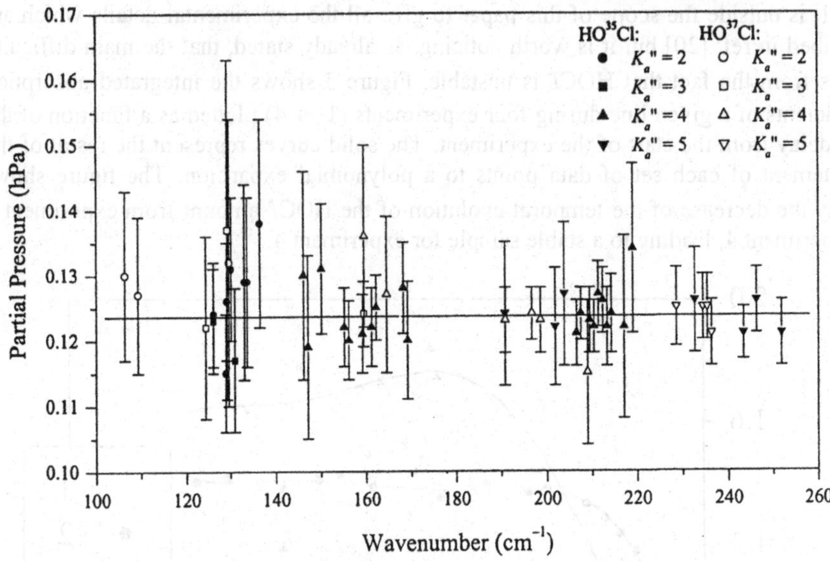

Figure 6. Determination of the HOCℓ partial pressure for experiment 4 using about 50 pure rotation lines absorbing in the far infrared (From ref. 20).

TABLE 2 Comparison of the observed transition dipole moment (in D^2)

Ref.	Transition dipole moment	Difference [a] (%)
(19)	0.018(3)	+ 29
(17)	0.0207(10)	+ 49
(18)	0.024(5)	+ 72
(20)	0.0139(10)	

[a] Diff = (value (ref) – value (ref 20)) / value ref 20)

It is worth noticing that differences up to about 70 % are observed between the previous measurements [17-19] and the results of ref. [20] showing clearly the extreme difficulty to get precise intensities when dealing with an unstable molecule. Again, one can see that to generate accurate spectral parameters it is necessary to rely both on sophisticated experiments and theoretical methods.

4. The ozone molecule

The decrease of stratospheric ozone (See for example14, 23) has led to the development of numerous techniques aiming at an accurate determination of the concentration profiles of this molecule in the atmosphere. It has also led to a large number of spectroscopic studies aiming at the generation of accurate spectroscopic parameters ; indeed :

- the ozone molecule, contrary to other less abundant atmospheric constituents, exhibits an atmospheric absorption on a very wide spectral domain ranging typically from the microwave to about 2.5 μm and in the ultraviolet. As a consequence the atmospheric measurements are performed, depending on the instruments, in different spectral regions and the comparison of the results of these different instruments, requires to be significant that the spectral parameters are consistent in the various spectral domains.
- It proved necessary to study the spectroscopic properties not only for the main isotopic species $^{16}O_3$ but also for the isotopic variants $^{16}O^{18}O^{16}O$, $^{16}O^{16}O^{18}O$, $^{16}O^{17}O^{16}O$ and $^{16}O^{16}O^{17}O$. Indeed the transitions of these less abundant isotopic variants appear clearly in the atmospheric spectra.
- In the upper stratosphere and in the mesosphere the O_3 molecule is in a Non Local Thermodynamic Equilibrium (non-LTE). Because of that the populations of the levels do not follow the Boltzmann law and the highly excited vibrational levels are "abnormally" populated. As a consequence, one observes in the atmospheric spectra the appearance of transitions involving high energy levels which are not (or with great difficulty) observed in the laboratory. It is then necessary to rely on sophisticated theoretical methods in order to model them.

It is outside the scope of this paper to discuss in great details the spectroscopic properties of the ozone molecule as well as the theoretical methods which have been developed and we refer the reader to refs. [24-26] for more details. We will then give here only the relevant details necessary to understand the problems which are faced when trying to generate spectral parameters for this molecule.

In its electronic ground state the ozone molecule is an asymmetric rotor with three normal modes ν_1, ν_2 and ν_3 located approximately at 1103, 701 and 1042 cm^{-1}. They are such that $3\omega_2 \cong 2\omega_1 \cong 2\omega_3$ and this complicates seriously the spectroscopic study of the molecule. Indeed the previous relations lead to the fact that the states with the same pseudo quantum number $s = v_1 + v_3$ are nearby in energy and can resonate. More precisely, to calculate the energy levels of ozone, one has to consider at least two types of interactions.

- The Darling Dennison interaction coupling the states $(v_1\ v_2\ v_3)$ and $(v_1 \pm 2\ v_2\ v_3 \mp 2)$
- The Coriolis interaction coupling the states $(v_1\ v_2\ v_3)$ and $(v_1 \pm 1\ v_2\ v_3 \mp 1)$

This implies that one has to consider polyads of interacting states and the lowest ones are given in Table 3. It is clear that the size of the polyads increases with increasing energies and consequently that it is necessary to use more and more complicated Hamiltonians. Moreover it is worth stressing that the resonance pattern presented here is less and less valid as the energies increase. Local resonances between states belonging to different polyads are indeed to be accounted for as was shown in various papers (See for example 27, 28). Finally it is worth noticing that in their very interesting paper on the potential energy surface of the ozone molecule Tyuterev and coworkers (29) have demonstrated that at high energies neither a normal mode model nor a local mode one are really suitable to reproduce the energy levels of O_3.

TABLE 3 First polyads of interacting states for $^{16}O_3$

	$v_2 = 0$			$v_2 = 1$		
	v_1	v_3	Band center (cm^{-1})	v_1	v_3	Band center (cm^{-1})
s = 3	3	0	3290	3	0	3967
	2	1	3186	2	1	3850
	1	2	3084	1	2	3739
	0	3	3046	0	3	3698
s = 2	2	0	2201	2	0	2886
	1	1	2110	1	1	2785
	0	2	2058	0	2	2726
s = 1	1	0	1103	1	0	1796
	0	1	1042	0	1	1727
s = 0	0	0	0	0	0	701

Nevertheless, despite these difficulties, it is important, as already stated, to generate spectral parameters concerning hot bands involving highly excited energy levels. Indeed in the upper atmosphere the ozone molecule is no longer in thermodynamic equilibrium. More precisely, the most important source of production of excited vibrational states is the reaction :

$$O_2 + O + M \rightarrow O_3 (v_1\ v_2\ v_3) + M$$

where v_1, v_2, v_3 can reach high values ($\cong 7$).

Then the excited states can relax either through collisions or spontaneous emission. It is outside the scope of this paper to describe in a detailed manner the kinetics of these excited states (See for example 30) but it is easy to understand that the result of these non-LTE phenomena is the appearance in the atmospheric spectra of transitions involving highly excited ozone states : these transitions do not appear in LTE situations since, as already noticed, the populations of the levels follow the Boltzman law. A number of studies have dealt with the non-LTE phenomena (See for example 30-34) using satellite measurements such as those of LIMS (Limb Infrared Monitor of the Stratosphere) or CLAES (Cryogenic Limb Array Etalon Spectrometer). However these atmospheric measurements were performed at medium and low resolutions (0.25 cm^{-1} for CLAES, radiometer for LIMS) preventing very precise analyses. The launch in the near future of the MIPAS (Michelson Interferometer for Passive Atmospheric Sounding) experiment [35] which is a high resolution (0.03 cm^{-1}) Fourier transform spectrometer covering a large spectral range (14.6 – 4.15 µm) has stimulated new non-LTE ozone studies since the high spectral resolution should allow one to perform much more precise analyses. Also these studies are necessary to assess the influence of the non-LTE effects on the retrievals of the ozone concentration profiles.

As already stated, from a spectroscopic point of view, this implies generating spectral parameters for hot bands involving highly excited vibrational states. This has been done [36] allowing to model non-LTE emissions in the spectral domains at 14.6, 10 and 4.8 μm covered by MIPAS. It was shown that the non-LTE effects are different from one spectral region to another : whereas they are rather weak at 14.6 μm they are significant at 10 μm and 4.8 μm. Figure 7 shows the influence of non-LTE phenomena on the emission spectrum of the atmosphere at the limb around 9 μm as it should be observed by MIPAS. It can be clearly seen that, for O_3 as well as for CO_2, the differences between spectra taking into account non-LTE phenomena and assuming LTE are well above the noise. One can then expect that **i)** retrievals of ozone profiles using the MIPAS spectra should be performed either accounting for non-LTE effects or by using spectral intervals where these effects are negligible, and **ii)** that MIPAS will be an excellent opportunity to study in great details non-LTE phenomena.

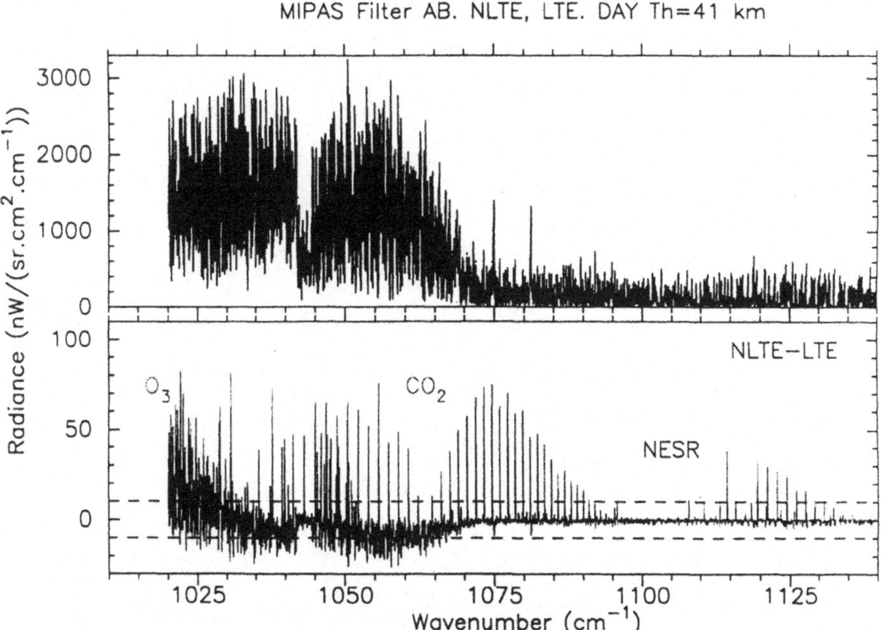

Figure 7. Synthetic limb radiances at a tangent height of 41 km for MIPAS. The differences between non-LTE and LTE radiances are well above the noise equivalent spectral radiance (NESR) for CO_2 as well as for O_3.

5. Conclusion

It is hoped that the few examples given during this course are a good illustration of the many problems one has to face when trying to derive from optical remote sensing experiments precise concentration profiles of atmospheric constituents. In particular one can emphasize that the accuracy of the retrieved profiles is directly linked to the quality of the spectral parameters used in the retrieval process. Finally, from a spectroscopic point of view, it should be clear that to generate high quality spectroscopic data it is necessary to rely on both modern experimental techniques and sophisticated theoretical models.

6. References

1. Orphal J., (1995) "Infrared Spectroscopy at High Resolution of Unstable Molecules of Atmospheric Interest : $C\ell ONO_2$, $C\ell NO_2$ and $BrNO_2$", *Thesis n° 3510, Université Paris XI, France.*
2. Orphal J., Morillon-Chapey M. and Guelachvili G., (1994), *J. Geophys. Res.*, **99**, 14549-14555.
3. Ballard J., Johnston J., Gunson M.R. and Wassel P.T., (1988) *J. Geophys. Res.*, **93D2**, 1659-1665.
4. Davidson J.D., Canttrel C.A., Shetter R.E., Daniel Mc and Calvert J.G. (1987) *J. Geophys. Res.*, **92D**, 10921-1925.
5. Rinsland C.P., Zander R., Demoulin P. and Mahieu E., (1966), *J. Geophys. Res.*, **101**, 3891-3899.
6. Johnson D.G., Orphal J., Toon G.C., Chana K.V., Traub W.A., Jucks K.W., Guelachvili G. and Morillon-Chapey M., (1996), *Geophys. Res. Lett.*, **23**, 1745-1748.
7. von Clarmann T., Wetzel G., Oelhaf J., Friedl-Vallon F., Linden A., Moucher G., Seefeldner M., Trieschmann O. and Lefèvre F., (1997), *J. Geophys. Res.*, **102**, 16157-16168.
8. Zou D.K., Bingham G.E., Rezai B.K., Anderson G.P., Smith D.R. and Nadile R.M., (1997) *J. Geophys. Res.*, **D102**, 3559-3573.
9. Xu S., Blake T.A. and Sharpe S.W., (1996), *J. Mol. Spectrosc.*, **175**, 303-314.
10. Mueller H.S., Helminger P. and Young S.H., (1997), *J. Mol. Spectrosc.*, **181**, 363-378.
11. Domenech J.L. Flaud J.-M., Fraser G.T., Andrews A.M., Lafferty W.J. and Watson P.L., (1997), *J. Mol. Spectrosc.*, **183**, 228-233.
12. Herp M., Georges R., Herman M., Flaud J.-M. and Lafferty W.J., (2000), *J. Mol. Structure,* **517-518**, 171-180.
13. Birk M. and Wagner G., (2000) *Private communication.*
14. World Meteorological Organisation, Scientific assessment of ozone depletion : (1994), *in Global Ozone Research and Monitoring Project,* Report n° 37, Geneva, Switzerland, 1995.
15. Chance K.V., Johnson D.G. and Traub W.A. (1989), *J. Geophys. Res.*, **94**, 11059-11069.
16. Toon G.C. and Farmer C.B., (1989), *Geophys. Res. Lett.*, **16**, 1375-1377.
17. Niki H., Maker P.D., Savage C.M. and Breitenbach P.L., (1979), *Chem. Phys. Lett.*, **66**, 325-328.
18. Su F., Calvert J.G., Lindley C.R., Uselman W.M. and Shaw J.H., (1979), *J. Phys. Chem.*, **83**, 912-920.
19. Burkholder J.B., (1993), *J. Geophys. Res.*, **98**, 2963-2974.
20. Vander Auwera J., Kleffmann J., Flaud J.-M., Pawelke G., Bürger H., Hurtmans D. and Pétrisse R., J. Mol. Spectrosc., in press.
21. Singbeil H.E.C., Anderson W.D., Davis R.W., Gerry M.C.L., Cohen E.A., Rickett H.M., Lovas F. and Suenram R.D., (1984), *J. Mol. Spectrosc.*, **103**, 466-485.

22. Flaud J.-M., Birk M., Wagner G., Orphal J., Klee S. and Lafferty W.J., (1998), *J. Mol. Spectrosc.*,**191**, 62-367.
23. Abbatt J.P.D. and Molina M.J., (1993),*Annu. Rev. Energy Environ.*, **18**, 1.
24. Flaud J.M., Camy-Peyret C., Rinsland C.P., Smith M.A.H. and Malathy Devi V., (1990), *Atlas of ozone line parameters from microwave to medium infrared,* Academic press, New York.
25. Flaud J.-M. and R. Bacis R.,(1998), *Spectrochimica Acta, part A*, **54**, 3-16.
26. Bacis R., Bouvier A.J. and Flaud J.-M.,(1998), *Spectrochimica Acta, part A*, **54**, 16-35.
27. Flaud J.-M., Barbe A., Camy-Peyret C. and Plateaux J.J., (1996), *J. Mol. Spectrosc.*, **177**, 34-39.
28. Mikhailenko S.N., Barbe A., Tyuterev Vl.G. and Chichery A., (1999), *Atmos. Oceanic Opt.*, vol. **12**, n° 9, 771-785.
29. Tyuterev Vl.G., Taskkun S., Jensen P., Barbe A. and Cours T., (1999), *J. Mol. Spectrosc.*, **198**, 57-76.
30. Manuilova R.O., Gusev O.A., Kutepov A.A., Von Clarmann T., Oelhaf H., Stiller G.P., Wegner A., Lopez-Puertas M., JMartin-Tores J., Zaragoza G. and Flaud J.-M., (1998), *J.Q.S.R.T.*, **3-5**, 405-422.
31. Solomon S., Kiehl J.T., Kerridge B.J., Remsberg E.E. and Russel III, J.M., (1986), *J. Geophys. Res.*, **91**, 9865.
32. Mlynczak M.G. and Drayson S.R., (1990), *J. Geophys. Res.*, **95**, 16497.
33. Connor B.J., Suskind D.E., Tsou J.J., Parish A. and Remsberg E.E., (1994), *J. Geophys. Res.*, **99**, 16757.
34. Edwards D.P., Kumer J.B., Lopez-Puertas M., Mlynczak M.G., Gopolan A., Gille J.C. and Roche A., (1996), *J. Geophys. Res.*, **101**, 26577.
35. Endemann M. and Fischer H., (1994), *ESA Bull.*, **76**, 47.
36. Lopez-Puertas M., Lopez-Valverde M.A., Martin-Torres F.J., Zaragoza G., Dudhia A., von Clarmann T., Kerridge B.J., Koutoulaki K. and Flaud J.-M., (1999), *European Symposium on Atmospheric Measurements from Space,* ESTEC, Noordwijk, The Netherlands.

HIGH RESOLUTION INFRARED SPECTROSCOPY

A contribution from ULB, Belgium

M. HERMAN, D. HURTMANS, and J. VANDER AUWERA
Laboratoire de Chimie Physique Moléculaire C. P. 160/09
Université Libre de Bruxelles
50, avenue F. D. Roosevelt, B-1050 Brussels, Belgium

1. Introduction

Chemistry at the molecular scale played a major role in recent exciting scientific developments such as computing the architecture of molecules, watching atoms dance, and preventing the crises of atmospheric chemistry. The activities of the ULB-infrared spectroscopy research group is aimed at developing high-resolution type instrumental and methodological means to investigate the internal structure of complex molecules, undress reaction mechanisms and, simultaneously, provide reference information required to quantitatively probe the composition of atmospheres. The common ground supporting our activities is the investigation, at high spectral resolution, of the energy, shape and intensity of spectral lines and bands, from the far- to the near-infrared spectral ranges. High resolution Fourier transform spectroscopy is extensively applied to unravel the spectra of ever more complex and highly vibrationally excited molecules, helped by a variety of means including near infrared laser investigations.

In this paper, we focus on the expertise we have developed in these areas, highlighting atmospheric-oriented studies, in particular. Such investigations concern vibration-rotation bands in mid-infrared atmospheric windows and also, stimulated by recent diode laser-based development, in the near infrared region of the spectrum. Emphasis is actually set on this higher energy range, given the lack of published reference data. The experimental means we are using and the homemade software are described in sections 2 and 3, respectively. Recent research activities and results are illustrated in section 4. An overview of all species investigated by the group over the last five years is presented in section 5. Literature references around the results presented in this paper are not cited before section 5, except in the legend of figures, and for instrumental and review papers. All references are listed in section 6. Finally, we have to acknowledge most efficient and continuous partnership with *ab initio* and other theoreticians at ULB (Profs/Drs Godefroid, Liévin and Vaeck), from the same laboratory.

J. Demaison et al. (eds.), Spectroscopy from Space, 201–218.

2. Experimental means

2.1. FOURIER TRANSFORM SPECTROMETERS

A commercial high-resolution Fourier transform spectrometer (FT) Bruker IFS120HR, having a programmable resolution up to 0.0018 cm^{-1} and operating from the far infrared (FIR) to the ultraviolet (UV), is available in our laboratory (FT1). A second Bruker IFS120HR instrument, with a programmable resolution up to 0.004 cm^{-1}, operating from the FIR to the visible (VIS) and equipped with emission and time resolved spectroscopy facilities, will be available early 2001 (FT2). Both FT1 and FT2 are equipped with a variety of absorption sources, beamsplitters, optical filters and detectors, to cover the above-mentioned spectral ranges. The two instruments can be evacuated using two-stage rotary pumps.

Figure 1 presents the optical scheme of FT1. Short path absorption cells are placed at the sample position, under vacuum. Long path cells require the use of transfer optics installed at the same position (section 2.2). The supersonic jet expansion set-up (section 2.3) makes use of the parallel beam going out of the interferometer through the exit port beside the sources compartment. Future activities are planned, which involve (*i*) the use of a laser beam (see section 2.4) as the source of one FT spectrometer and (*ii*) the use of the two interferometers to measure simultaneously in different spectral ranges (for instance, to conduct absolute intensities measurements similar to the one described in section 4.1.2).

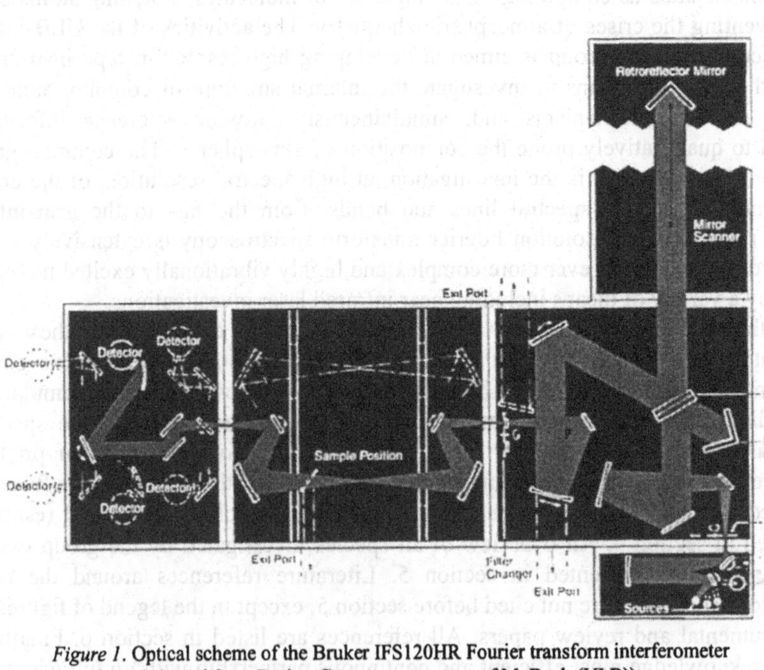

Figure 1. **Optical scheme of the Bruker IFS120HR Fourier transform interferometer (adapted from Bruker Company, courtesy of M. Bach (ULB)).**

2.2. ABSORPTION CELLS

2.2.1. Long path cells

Two commercial multiple-reflections absorption cells are available in our laboratory, providing total absorption paths ranging from 60 cm to 50 m. They allow weak vibration-rotation transitions to be probed from the mid-infrared to the visible. The absorption spectrum of acetylene presented in Fig. 2 has been recorded in the visible using one of these multiple-reflections cells.

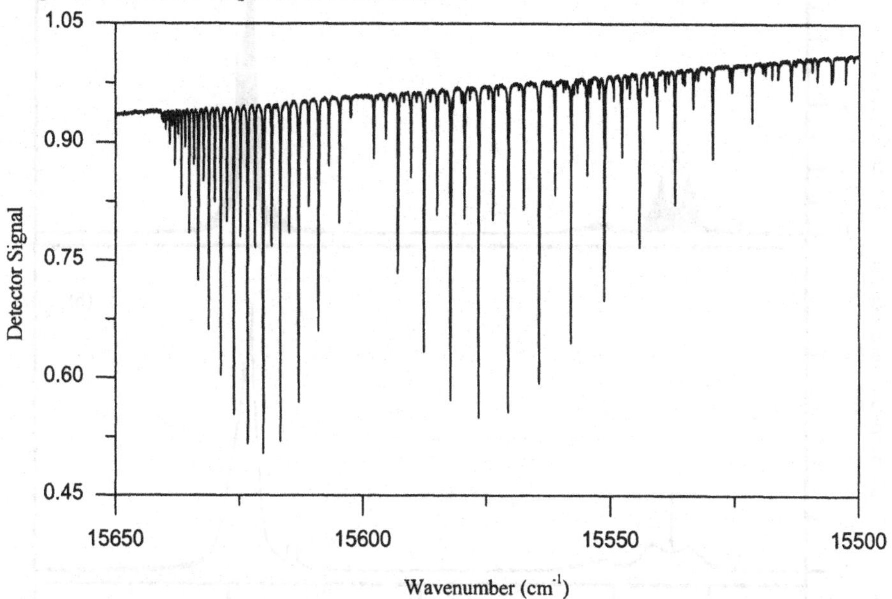

Figure 2. $2\nu_1+3\nu_3$ band observed in the visible absorption spectrum of acetylene ($^{12}C_2H_2$), recorded using a Fourier transform spectrometer (path = 49 m, pressure = 250 mbar, room temperature; adapted from [1]).

2.2.2. Thermostatic cells

Homemade thermostatic absorption cells are also available, with nominal path lengths ranging from 1.5 to 25 cm and designed to operate at any temperature between 200 and 300 K. As an example of the use of these cells, Fig. 3 presents the absorption cross-sections of HFC-152a, which we have measured between 600 and 1700 cm^{-1} for the pure gas and mixtures with dry air.

2.3. SUPERSONIC JET EXPANSION

A supersonic jet system was built in the laboratory, operating with the high-resolution FT spectrometer. It is described, with its performances compared to other similar equipment, in [2]. It consists in a slit or a multi-nozzle system, both 16 cm long and heated up if required [3], pumped by a Roots system (4000 m^3/h). A new system with a 29-cm slit is presently being built. The overall design of the jet system is presented in Fig. 4. Typically, rotational temperatures around 30 K are obtained with Ar containing mixtures. The set-up was used to achieve spectral investigations of a number of larger species (see section 4.3.2).

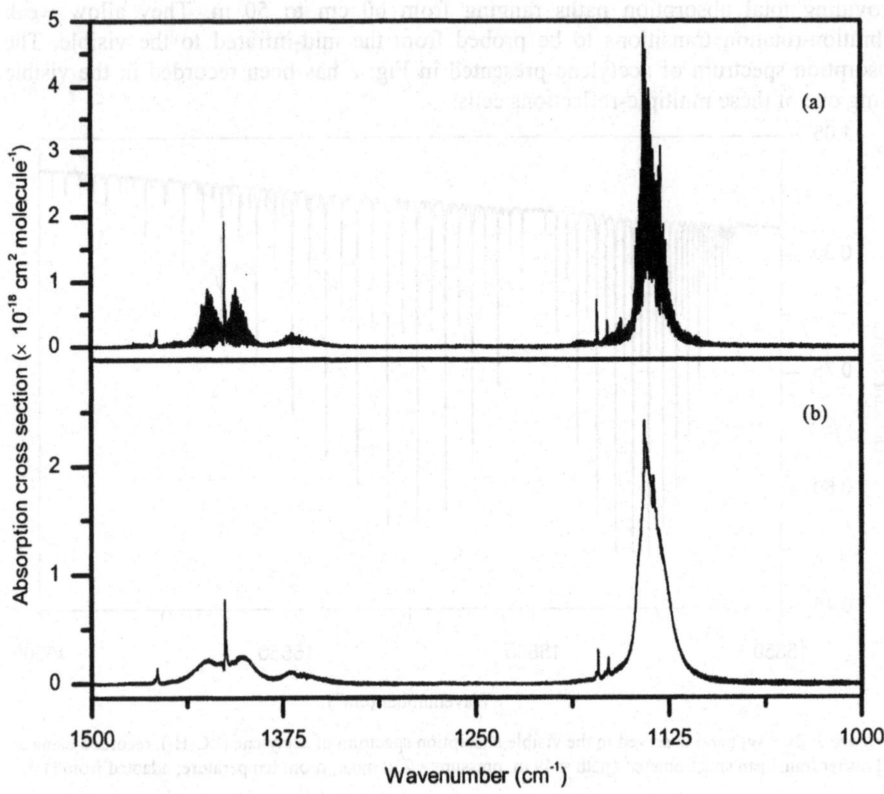

Figure 3. Absorption cross-sections of HFC-152a: pure sample at 203 K (a) and mixture with dry air at a total pressure of 800 mbar at 293 K (b) (adapted from [4]).

2.4. NEAR INFRARED LASER

An autoscan Ti:Sa laser Coherent model 899-29 is also available in our laboratory. It is equipped with a multiple-reflections absorption cell with coated optics, providing total absorption paths up to 100 m, and optoacoustic cells (OA). The overall set-up is presented in Figure 5. Such a system was recently used to measure pressure broadening and shift of vibration-rotation lines in the near infrared range, between roughly 12000 and 14000 cm^{-1} (see section 4.2). The laser is presently being transformed to achieve optical coupling with the FT facilities available.

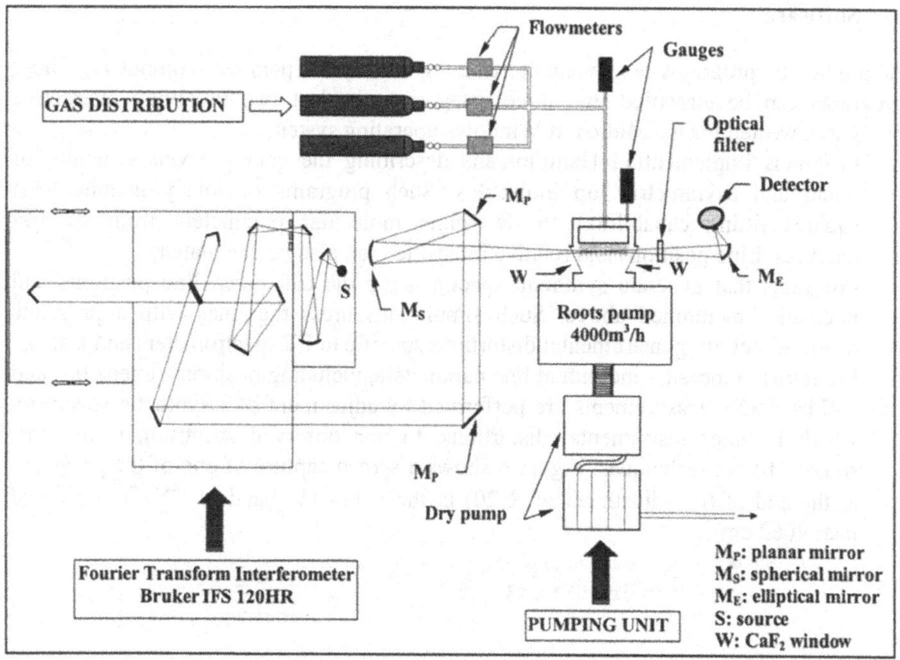

Figure 4. Schematic design of the jet-FT set-up at ULB (adapted from [5]).

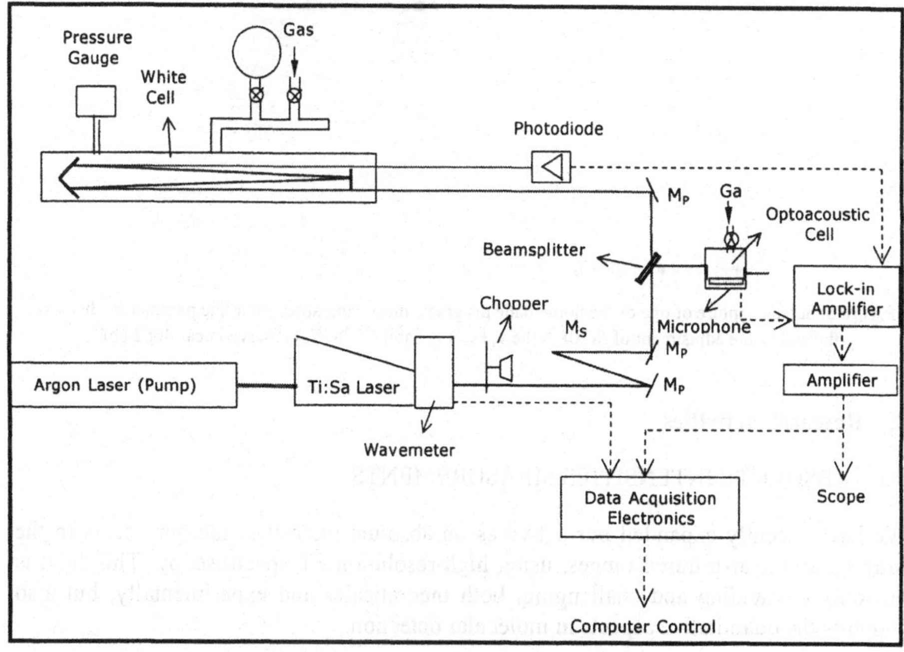

Figure 5. Schematic view of the Ti:Sa laser spectrometer available at ULB (adapted from [6]).

3. Software

A number of programs are available, running mostly on personal computers. These programs can be classified into three categories, the last two involving interactive programs, written for the Microsoft Windows operating system:

- Programs implementing Hamiltonians describing the energy levels structure of linear and asymmetric top molecules. Such programs obviously include least squares fitting capabilities to determine molecular parameters from spectral analyses. Line positions and relative intensities can also be calculated;
- Programs that generate synthetic spectra using the calculated line positions and intensities mentioned above. Such simulations dress the lines with appropriate profiles, including instrumental distortions specific to FT spectrometers and lasers;
- Programs to measure individual line parameters, including positions, intensities and widths. Such measurements are performed by adjustment of a synthetic spectrum, which includes instrumental distortions, to one observed spectrum, using least squares fitting techniques. Figure 6 shows a screen capture of one of the programs at the end of the adjustment of R(20) in the $\nu_1+\nu_2^1+\nu_3$ band of $^{14}N_2{}^{16}O$ observed near 4062 cm^{-1}.

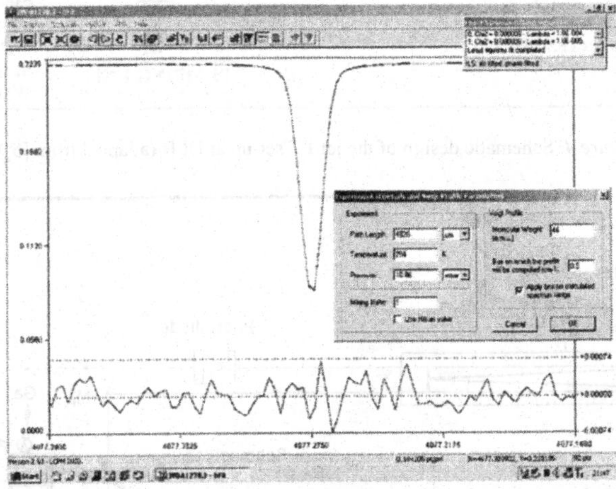

Figure 6. Screen capture of one of the homemade programs measuring absorption line parameters. It shows the end of the adjustment of *R*(20) in the $\nu_1+\nu_2^1+\nu_3$ band of $^{14}N_2{}^{16}O$ observed near 4062 cm^{-1}.

4. Research activities

4.1. ABSOLUTE INTENSITIES MEASUREMENTS

We have recently expanded our activities on absolute intensities measurements in the infrared and near-infrared ranges, using high-resolution FT spectroscopy. This field is not only demanding and challenging, both theoretically and experimentally, but also supports the quantitative aspects in molecular detection.

4.1.1. Stable species

We measure absolute intensity information for chemically stable compounds with accuracy better than 5%. Our focus is on trace atmospheric species and molecules of astrophysical interest. In this context, OCS, HFCs and hydrocarbons such as C_2H_2 and C_2H_6 are or have been considered. As an example, Fig. 7 presents the square of the vibrational transition dipole moment measured for vibration-rotation lines in the $3\nu_1+2\nu_2^0$ band of $^{14}N_2{}^{16}O$ near 5026 cm^{-1}. Some absolute line intensities measurements were also performed in the near-infrared spectral range using the laser spectrometer described in section 2.4 (see section 4.2).

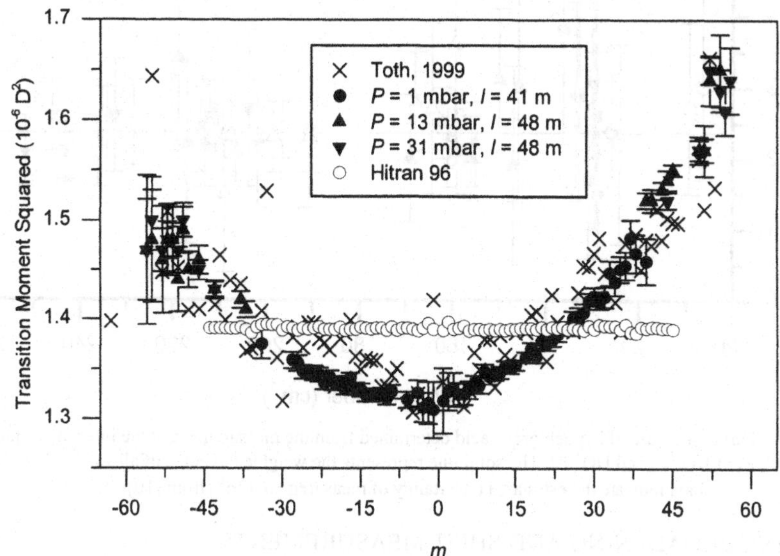

Figure 7. Square of the vibrational transition dipole moment measured for lines in the $3\nu_1+2\nu_2^0$ band of $^{14}N_2{}^{16}O$ near 5026 cm^{-1}, plotted versus m ($-J''$ for the P-branch lines and $J''+1$ for the R-branch lines) [7]. They are compared with previous measurements [8] and Hitran 96 [9]. The error bars indicate the estimated uncertainty of measurement (3σ).

4.1.2. Unstable species

We have recently extended absolute line intensities measurements to chemically unstable and reactive species, such as acidic compounds found in the troposphere and stratosphere of the Earth. Such measurements are most challenging because the required concentration of the species is not directly accessible, for instance through the measurement of the total gas pressure. These unstable compounds indeed usually exist only in equilibrium mixtures. Furthermore, their concentration may evolve while their spectra are being recorded, as a result of their reactivity.

In that frame, we have measured absolute line intensities in the ν_2 band near 1238 cm^{-1} of hypochlorous acid, with accuracy estimated to 7 and 9% for $HO^{35}Cl$ and $HO^{37}Cl$ respectively. To determine accurately the partial pressure of the species in the sample mixture involving HOCl, H_2O and Cl_2O, we relied on known absolute line intensities in the pure rotational far infrared (FIR) spectrum, recorded simultaneously with the IR spectrum. Figure 8 presents the partial pressure determined for various pure rotational lines of $HO^{35}Cl$ and $HO^{37}Cl$.

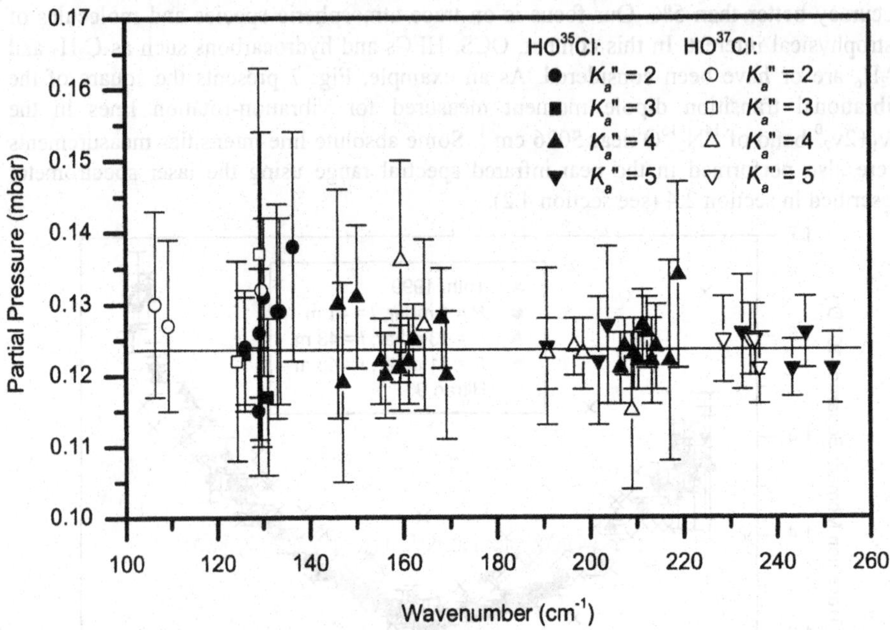

Figure 8. Partial pressure of hypochlorous acid determined from the measurement of the intensity of pure rotation lines of $HO^{35}Cl$ and $HO^{37}Cl$. The solid line represents the weighted average of all values. The error bars indicate the estimated uncertainty of measurement (3σ) (from [10]).

4.2. LINE BROADENING AND SHIFT MEASUREMENTS

We investigated line self-, Ar- and Xe-broadening and shift effects in acetylene, in the $\nu_1+3\nu_3$ and $3\nu_1+\nu_3$ bands and also, previously, in the $5\nu_3$ band near 15676, 13033 and 15946 cm^{-1} respectively. Among the many results, including the observation of line narrowing effects, we highlighted some unexpected and significant relationship between intra- and inter-molecular couplings. Such an effect is probably directly related to the extent of the vibrational excitation and therefore only expected to occur in the overtone range. An example of the results of the fitting procedure applied to $R(9)$ of the $\nu_1+3\nu_3$ band is presented in Fig. 9.

4.3. INVESTIGATIONS OF LARGER SPECIES

We are investigating larger species, up to simple aromatic compounds, in the infrared range at high spectral resolution. Significant dynamical effects such as intramolecular vibrational relaxation (IVR) can be predicted to affect conventional lineshape effects, including pressure effects, in such species. The same comment applies to highly excited vibrations even in simpler species, supporting overtone spectroscopy research within the environmental domain, as described in section 4.4.

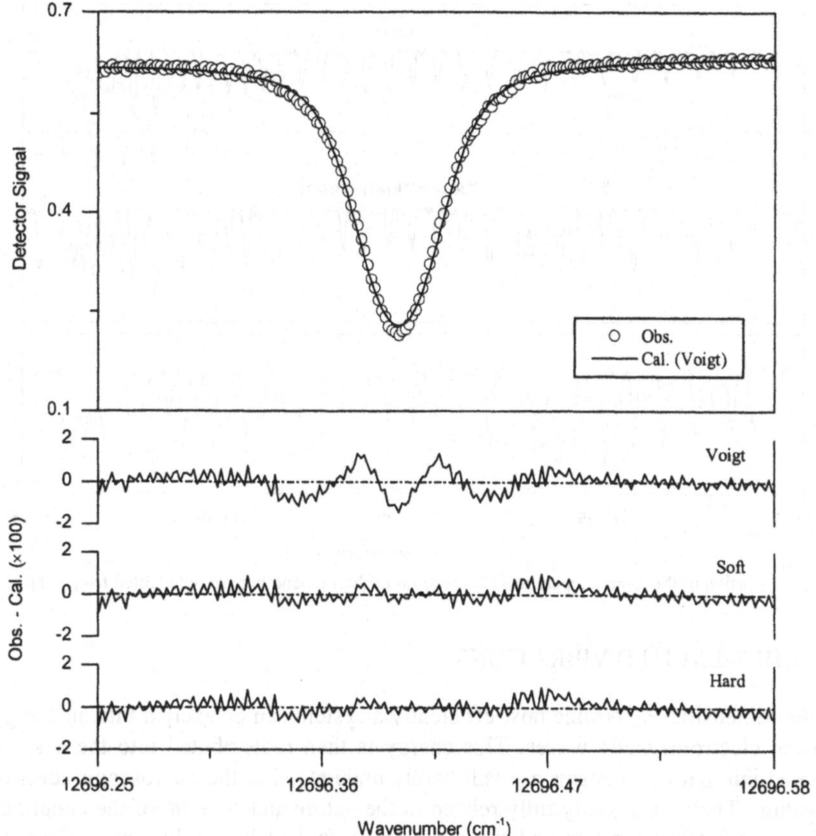

Figure 9. Example of the results of the fitting procedure applied to self-pressure broadened line shapes in acetylene ($^{12}C_2H_2$ at 76 mbar, adapted from [6]).

4.3.1. *Room temperature*

Among other species, we have recently recorded and analyzed vibration-rotation bands in pyrrole (C_4H_5N) and in furan (C_4H_4O). Figure 10 presents a portion of the room temperature FT spectrum of the ν_{22} band in pyrrole.

4.3.2. *Jet-cooled*

The jet-FT set-up was extensively used to investigate vibration-rotation bands of hydrocarbons and NO containing species in the mid- and near-infrared ranges, as recently reviewed in [2]. Special effort was devoted to methane, ethylene and ethane among the first category. NO, N_2O, N_2O_3 and N_2O_4 were studied, in the second family. As an example, Fig. 11 presents a jet-cooled spectrum of C_2H_6.

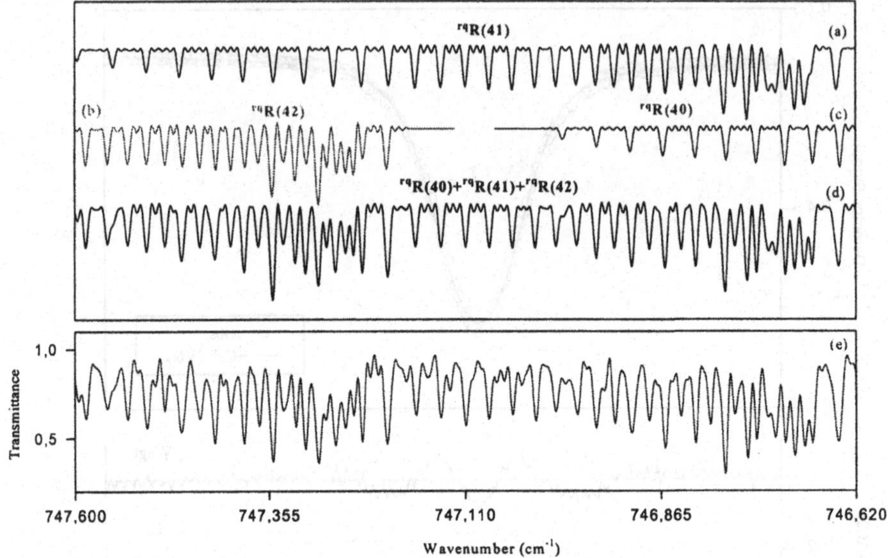

Figure 10. Portion of the room temperature FT spectrum of the ν_{22} band in pyrrole (adapted from [11]).

4.4. HIGHLY EXCITED VIBRATIONS

Absorption intensities determine how efficiently a system can be excited into an energy level using electromagnetic waves. This energy is then redistributed into the system. The related internal mechanisms are still barely understood at the microscopic level of investigation. They are actually fully related to the nature and strength of the couplings between the excited eigenstate and the nearby manifold of levels. In any of the case studies mentioned hereafter, the concepts of "resonances" and of "bright", "dark" and "doorway" states play a crucial role. We are investigating such problems in a variety of species – C_2H_2, C_2H_4, C_2H_6, in particular – introducing new quantum numbers related to the concept of polyads [12]. As pointed out in previous sections, such high resolution overtone investigations also help characterizing lineshape features that are relevant to atmospheric-type investigations, and they also provide reference data in spectral ranges today accessible *in situ* using supersensitive laser-based detection techniques. We also take advantage of overtone investigations to probe chemically relevant events, as recently experienced in *trans*-formic acid. The latter work stimulated our interest in intra- and inter-molecular hydrogen bonding. This field is likely to drive some forthcoming studies in our research group. We present in Fig. 12 an overview of the $2\nu_{OH}$ band in HCOOH.

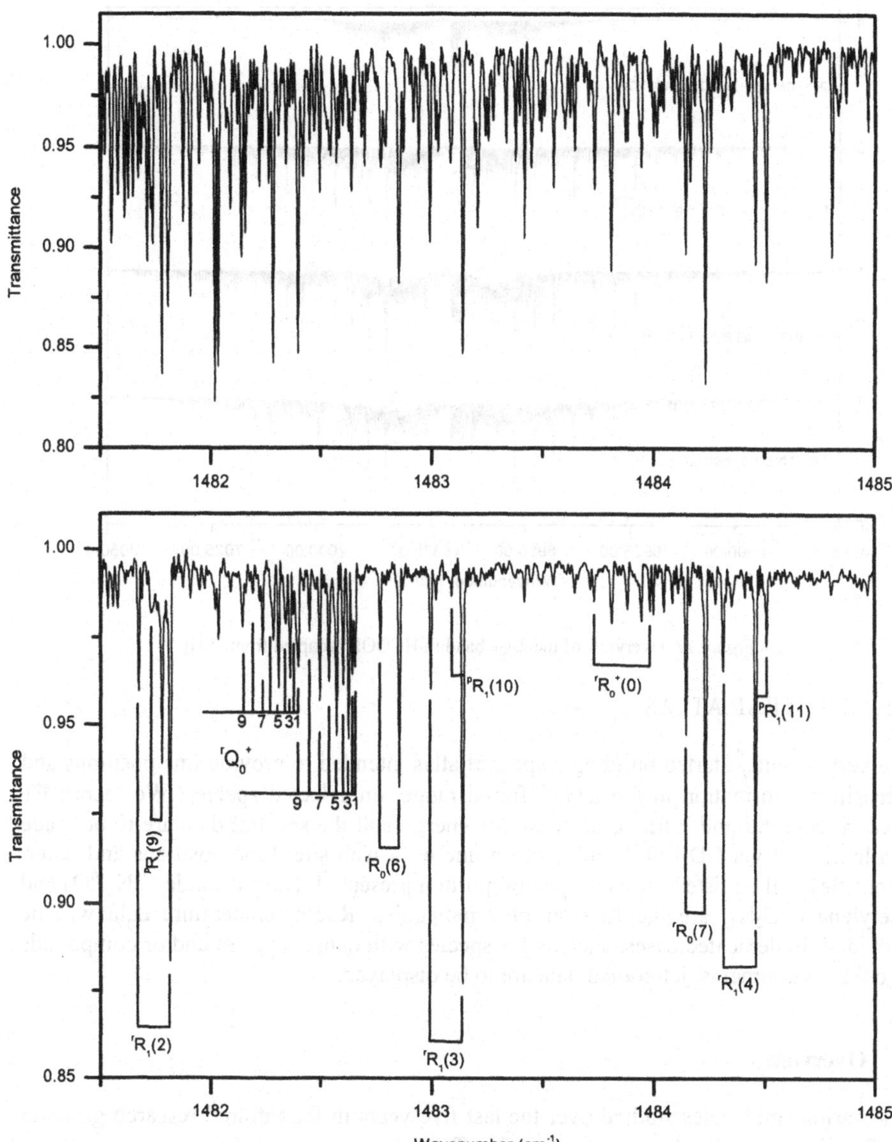

Figure 11. Jet cooled FT data of the ν_8 and $\nu_4+\nu_{12}$ bands in C_2H_6 (upper: room temperature; lower: jet-cooled) (adapted from [13]).

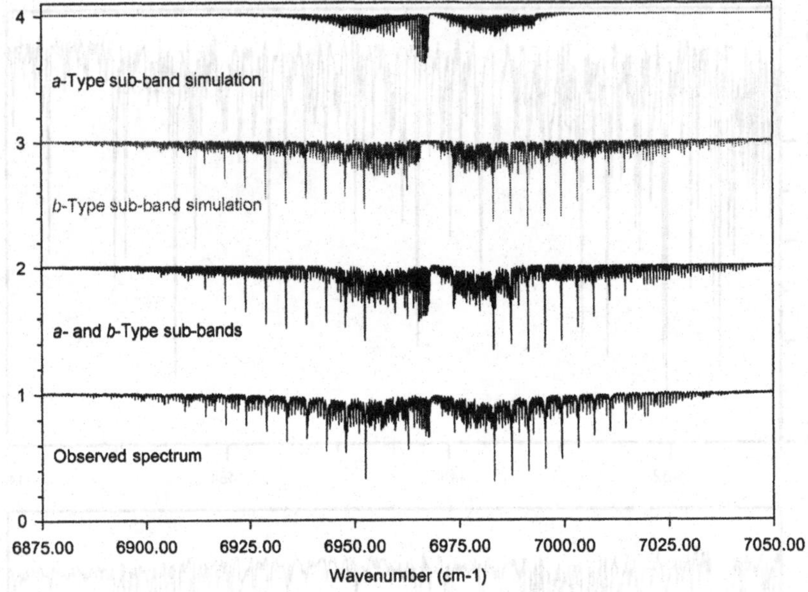

Figure 12. Overview of the $2\nu_{OH}$ band in HCOOH (adapted from [14]).

4.5. SPECTRAL ATLAS

We very recently started building a spectral atlas intended to provide line positions and intensities information in the near infrared range, on selected species. We record FT spectra from the mid-infrared up to higher energy and the spectral data are to be made available first via CD-ROM and then on-line on a web site. Line positions and, later, intensities will be listed for each spectral portion presented. Nitrous oxide ($^{14}N_2{}^{16}O$) and acetylene ($^{12}C_2H_2$) are the first targeted molecules. Room temperature data will be provided. In dedicated cases, such as for species with denser spectra and/or compounds in cold environments, jet-cooled data are to be displayed.

5. Overview

The various molecules studied over the last five years in the infrared research group at ULB and the related references are listed in Table 1.

Table 1. Molecules studied over the last five years by the authors of this paper and their co-workers (Except when indicated, the more abundant isotopomer is concerned).

Species	Range (cm^{-1})	Aim[(a)]	Collab.[(b)]	References
Triatom				
N_2O	4300 – 5200	Absolute line intensities measurements	LPMA	[7]
	6500 – 15000	Global vibrational analysis; vibration-rotation analysis	LSM	[15, 16]
NO_2	1500 – 3400	Vibration-rotation analysis, absolute line intensities measurements	PPM	[17]
OCS	2500 – 3100	Absolute line intensities measurements	LLN	[18]
	4800 – 8000	Vibration-rotation analysis	LLN	[19]
HOCl	21 – 103	Dipole moment measurement using Stark spectroscopy	LENS, BOL	[20]
	1200 – 1300	Absolute line intensities measurements	PPM, BUGH	[10]
Fouratom				
C_2H_2	7000 – 18500	Vibration-rotation analysis	LSM	[21-24]
	12000 – 15000	Pressure effects and absolute line intensity measurement		[6, 25-27]
	0 – 25000	Global vibrational analysis, intramolecular dynamics	LSM, LIT	[12, 28-32] [33]
	1240 – 1420	Absolute line intensities measurements		[34]
C_2HD	3000 – 16000	Global vibrational analysis		[35]
C_2D_2	0 – 14100	Global vibrational analysis; vibration-rotation analysis	LSM, BOL	[36]
$^{13}C_2H_2$	0 – 23700	Global vibrational analysis	BOL, NRC	[37]
Fiveatom				
CH_4	3000 – 6500	Vibration-rotation analysis (JC)	LPUB	[38]
N_2O_3	1800 – 1900	Vibration-rotation analysis (JC)	PPM	[39]
HCO_2H	1200 – 14000	Vibration-rotation analysis, global vibrational analysis	LSM	[14, 40]
$CHClF_2$	600 – 1700	Absorption cross sections measurements		[41]
Sixatom				
C_2H_4	0 – 17000	Global vibrational analysis		[42]
	700 – 8000	Vibration-rotation analysis	PPM	[43-45]
N_2O_4		Vibration-rotation analysis (JC)	NIST, PPM	[46]
		Global vibrational analysis		[47]
Sevenatom				
CH_3CHO	1300 – 14000	Vibration-rotation analysis (RT/JC)	PHLAM, PPM	[48, 49]
CH_3C_2H	3000 – 10000	Global vibrational analysis; vibration-rotation analysis	LSM	[50-52]
SF_6	800 – 1000	Vibration-rotation analysis (JC)	LPUB, UK	[53]
Eightatom				
C_2H_6	1200 – 4500	Vibration-rotation analysis (JC)		[13, 54-57]
1,2-$C_2H_4Cl_2$	0 – 1000	Global vibrational analysis; vibration-rotation analysis (JC)		[58-60]
	600 – 1700	Absorption cross sections measurements		[4]
1,1-$C_2H_4F_2$	600 – 1700	Absorption cross sections measurements		[4]

Species	Range (cm^{-1})	Aim[(a)]	Collab.[(b)]	References
Nineatom				
C_4H_4O	0 – 10000	Vibration-rotation analysis (RT/JC) and global vibrational analysis	PHLAM	[61, 62]
$C_3H_3F_3$	40 – 340	Vibrational analysis	PPM	[63]
Tenatom				
C_4H_5N	0 – 16500	Global vibrational analysis; vibration-rotation analysis	BG	[11, 64, 65]
C_4H_6	700 – 800	Vibration-rotation analysis and *ab initio* calculations	LMSU	[66]
Elevenatom				
C_3H_8	1200 – 1600	Vibration-rotation analysis (RT/JC)	NIST, PPM	[67]

(a) RT = Room temperature, JC = Jet-cooled.

(b) **BGSU**: Department of Chemistry, Bowling Green State University, OH, USA;
BUGH: Physikalische Chemie and Anorganische Chemie, Bergische Universität–Gesamthochschule Wuppertal, Wuppertal, Germany;
BOL: Dipartimento di Chimica Fisica e Inorganica, Università di Bologna, Italy;
LIT: MREID, Université du Littoral, Dunkerque, France;
LLN: Laboratoire de Spectroscopie Moléculaire, Université Catholique de Louvain, Belgium;
LMSU: Laboratory of Molecular Spectroscopy, Department of Chemistry, M. V. Lomonosov Moscow State University, Moscow, Russia;
LPUB: Laboratoire de Physique de l'Université de Bourgogne, France;
LSM: Laboratoire de Spectrométrie Physique, Université Fourier, Grenoble, France;
MIT: Department of Chemistry, Massachusettes Institute of Technology, Cambridge, MA, USA;
NIST: National Institute for Standards and Technology, Gaithersburg, MD, USA;
NRC: Steacie Institute for Molecular Sciences, National Research Council, Ottawa, Canada;
PHLAM: Laboratoire de Physique des Lasers, Atomes et Molécules, Lille, France;
PPM: Laboratoire de Photophysique Moléculaire, Université de Paris-Sud, Orsay, France;
UK: Physikalisches Institut, Universität zu Köln, Germany.

6. References

1. Herman, M., Abbouti Temsamani, M., Lemaitre, D., and Vander Auwera, J. (1991) The Fourier-transform vibrational spectrum of acetylene in the visible range, *Chem. Phys. Lett.* **185**, 220-224.
2. Herman, M., Georges, R., Hepp, M., and Hurtmans, D. (2000) High resolution Fourier transform spectroscopy of jet-cooled molecules, *Int. Rev. Phys. Chem.* **19**, 277-325.
3. Hepp, M., Herregodts, F., and Herman, M. (1998) FTIR jet spectroscopy using a heated slit source, *Chem. Phys. Lett.* **294**, 528-532.
4. Vander Auwera, J. (2000) Infrared absorption cross-sections for two substituted ethanes: 1,1-difluoroethane (HFC-152a) and 1,2-dichloroethane, *J. Quant. Spectrosc. Radiat. Transfer* **66**, 143-151.
5. Georges, R., Bach, M., and Herman, M. (1997) High resolution FTIR spectroscopy using a slit jet: sampling the overtone spectrum of $^{12}C_2H_4$, *Mol. Phys.* **90**, 381-387.
6. Herregodts, F., Hurtmans, D., Vander Auwera, J., and Herman, M. (1999) Laser spectroscopy of the ν_1+3ν_3 absorption band in $^{12}C_2H_2$(I): pressure broadening and absolute line intensity measurements, *J. Chem. Phys.* **111**, 7954-7960.

7. Daumont, L., Vander Auwera, J., Teffo, J-L., Perevalov, V.I., and Tashkun, S.A., Linestrengths measurements in $^{14}N_2{}^{16}O$ from 4300 to 5200 cm^{-1} and their treatment using the effective dipole moment approach, to be submitted.
8. Toth, R.A. (1999) Line positions and strengths of N_2O between 3515 and 7800 cm^{-1}, *J. Mol. Spectrosc.* **197**, 158-187.
9. Rothman, L.S., *et al.* (1998) The HITRAN molecular spectroscopic database and HAWKS (HITRAN Atmospheric WorKStation): 1996 edition, *J. Quant. Spectrosc. Radiat. Transfer* **60**, 665-710.
10. Vander Auwera, J., Kleffmann, J., Flaud, J-M., Pawelke, G., Bürger, H., and Pétrisse, R. (2000) Absolute ν_2 line intensities of HOCl by simultaneous measurements in the infrared with a tunable diode laser and far-infrared region using a Fourier transform spectrometer, *J. Mol. Spectrosc.* **204**, 36-47.
11. Mellouki, A., Vander Auwera, J., and Herman, M. (1999) Rotation-vibration constants for the ν_1, ν_{22}, ν_{24}, $\nu_{22}+\nu_{24}$ and ground states in pyrrole ($^{12}C_4H_5N$), *J. Mol. Spectrosc.* **193**, 195-203.
12. Herman, M., Liévin, J., Vander Auwera, J., and Campargue, A. (1999) Global and accurate vibration Hamiltonians from high resolution molecular spectroscopy, *Adv. Chem. Phys.* **108**, 1-431.
13. Hepp, M. and Herman, M. (1999) Effective rotation-vibration parameters for the ν_8 and $\nu_4+\nu_{12}$ bands of ethane, *J. Mol. Spectrosc.* **194**, 87-94.
14. Hurtmans, D., Herregodts, F., Herman, M., Liévin, J., Kachanov, A.A., and Campargue, A. (2000) Spectroscopic and *ab initio* investigation of the ν_{OH} overtone excitation in trans-formic acid, *J. Chem. Phys.* **113**, 1535-1545.
15. Campargue, A., Permogorov, D., Bach, M., Abbouti Temsamani, M., J., V.A., Herman, M., and Fujii, M. (1995) Overtone spectroscopy in nitrous oxide, *J. Chem. Phys.* **103**, 5931-5938.
16. Weirauch, G., Kachanov, A.A., Campargue, A., Bach, M., Herman, M., and Vander Auwera, J. (2000) Refined investigation of the overtone spectrum of nitrous oxide, *J. Mol. Spectrosc.* **202**, 98-106.
17. Perrin, A., Flaud, J-M., Camy-Peyret, C., Hurtmans, D., and Herman, M. (1996) The [2 ν_3, $4\nu_2$, $2\nu_2+\nu_3$] and 2 ν_3-ν_3 bands of $^{14}N^{16}O_2$: line positions and intensities, *J. Mol. Spectrosc.* **177**, 58-65.
18. Errera, Q., Vander Auwera, J., Belafhal, A., and Fayt, A. (1995) Absolute intensities in $^{16}O^{12}C^{32}S$: The 2500 – 3100 cm^{-1} region, *J. Mol. Spectrosc.* **173**, 347-369.
19. Rbaihi, E., Belafhal, A., Vander Auwera, J., Naïm, S., and Fayt, A. (1998) Fourier transform spectroscopy of OCS from 4800 to 8000 cm^{-1} and new global analysis of $^{16}O^{12}C^{32}S$, *J. Mol. Spectrosc.* **191**, 32-44.
20. Modugno, G., De Natale, P., Bellini, M., Inguscio, M., DiLonardo, G., Fusina, L., and Vander Auwera, J. (1996) Precise measurement of molecular dipole moments with a tunable far-infrared Stark spectrometer: application to HOCl, *J. Opt. Soc. Amer. B* **13**, 1645-1649.
21. Campargue, A., Abbouti Temsamani, M., and Herman, M. (1997) The absorption spectrum of $^{12}C_2H_2$ between 12800 and 18500 cm^{-1}. I. Vibrational assignments, *Mol. Phys.* **90**, 793-805.

22. Yang, S.F., Biennier, L., Campargue, A., Abbouti Temsamani, M., and Herman, M. (1997) The absorption spectrum of $^{12}C_2H_2$ between 12800 and 18500 cm^{-1}. II. Rotational analysis, *Mol. Phys.* **90**, 807-816.
23. Campargue, A., Biennier, L., and Herman, M. (1998) The visible absorption spectrum of $^{12}C_2H_2$. III. The region 14500 – 17000 cm^{-1}, *Mol. Phys.* **93**, 457-469.
24. Weirauch, G., Campargue, A., El Idrissi, M.I., and Herman, M. (2000) The absorption spectrum of $^{12}C_2H_2$ (IV): The regions 7600-9200 and 10600-11500 cm^{-1}, *Mol. Phys.*, in press.
25. Georges, R., Van Der Vorst, D., Herman, M., and Hurtmans, D. (1997) Ar and self pressure broadening coefficient of the *R*(11), $5\nu_3$ line of $^{12}C_2H_2$, *J. Mol. Spectrosc.* **185**, 187-188.
26. Herregodts, F., Hepp, M., Hurtmans, D., Vander Auwera, J., and M., H. (1999) Laser spectroscopy of the $\nu_1+3\nu_3$ absorption band in $^{12}C_2H_2$(II): self-collisional lineshift measurements, *J. Chem. Phys.* **111**, 7961-7966.
27. Herregodts, F., Hurtmans, D., Vander Auwera, J., and Herman, M. (2000) Ar-induced pressure effects in the $\nu_1+3\nu_3$ absorption band in $^{12}C_2H_2$, *Chem. Phys. Lett.* **316**, 460-464.
28. Campargue, A., Abbouti Temsamani, M., and Herman, M. (1995) The absorption spectrum of acetylene in the $2\nu_2+3\nu_3$ region. A test of the cluster model, *Chem. Phys. Lett.* **242**, 101-105.
29. Abbouti Temsamani, M. and Herman, M. (1995) The vibrational energy levels in acetylene $^{12}C_2H_2$: Towards a regular pattern at higher energies, *J. Chem. Phys.* **102**, 6371-6385.
30. Abbouti Temsamani, M., Herman, M., Solina, S.A.B., O'Brien, J.P., and Field, R.W. (1996) Highly vibrationally excited $^{12}C_2H_2$ in the X $^1\Sigma_g^+$ state: Complementarity of absorption and dispersed fluorescence spectra, *J. Chem. Phys.* **105**, 11357-11359.
31. Abbouti Temsamani, M. and Herman, M. (1996) The vibrational energy pattern in $^{12}C_2H_2$(II): Vibrational clustering and rotational structure, *J. Chem. Phys.* **105**, 1355-1362.
32. El Idrissi, M.I., Liévin, J., Campargue, A., and Herman, M. (1999) The vibrational energy pattern in acetylene. IV. Updated vibrational constants in $^{12}C_2H_2$, *J. Chem. Phys.* **110**, 2074-2086.
33. Zhilinskii, B.I., El Idrissi, M.I., and Herman, M. (2000) The vibrational energy pattern in acetylene (VI): Inter- and intra-polyad structures, *J. Chem. Phys.*, in press.
34. Vander Auwera, J. (2000) Absolute intensities measurements in the $\nu_4+\nu_5$ band of $^{12}C_2H_2$: analysis of Herman-Wallis effects and forbidden transitions, *J. Mol. Spectrosc.* **201**, 143-150.
35. Abbouti Temsamani, M. and Herman, M. (1996) Anharmonic resonances in monodeuteroacetylene ($^{12}C_2HD$), *Chem. Phys. Lett.* **264**, 556-557.
36. Herman, M., El Idrissi, M.I., Pisarchik, A., Campargue, A., Gaillot, A-C., Biennier, L., Di Lonardo, G., and Fusina, L. (1998) The vibrational energy levels in acetylene. III. $^{12}C_2D_2$, *J. Chem. Phys.* **108**, 1377-1389.

37. Di Lonardo, G., Fusina, L., Venuti, E., Johns, J.W.C., El Idrissi, M.I., Liévin, J., and Herman, M. (1999) The vibrational energy pattern in acetylene. V. $^{13}C_2H_2$, *J. Chem. Phys.* **111**, 1008-1016.
38. Georges, R., Herman, M., Hillico, J.C., and Robert, D. (1998) High-resolution FTIR spectroscopy using a jet: sampling the rovibrational spectrum of $^{12}CH_4$, *J. Mol. Spectrosc.* **187**, 13-20.
39. Georges, R., Liévin, J., Herman, M., and Perrin, A. (1996) The ν_1 band in N_2O_3, *Chem. Phys. Lett.* **256**, 675-678.
40. Hurtmans, D. and Herman, M. (2000) Analysis of vibration-rotation bands in formic acid, to be submitted.
41. Ballard, J., Knight, R.J., Vander Auwera, J., Herman, M., Di Lonardo, G., G., M., Nicolaisen, F.M., Beukes, J.A., Christensen, L.K., Mc Pheat, R., Duxbury, G., Freckleton, R., and Shine, K.P. (2000) An intercomparison of laboratory measurements of absorption cross-sections and integrated absorption intensities for HCFC-22, *J. Quant. Spectrosc. Radiat. Transfer* **66**, 109-128.
42. Georges, R., Bach, M., and Herman, M. (1999) The vibrational energy pattern in ethylene ($^{12}C_2H_4$), *Mol. Phys.* **97**, 279-292.
43. Bach, M., Georges, R., Hepp, M., and Herman, M. (1998) Slit-jet FTIR spectroscopy in $^{12}C_2H_4$: cold and hot bands near 3000 cm^{-1}, *Chem. Phys. Lett.* **294**, 533-537.
44. Bach, M., Georges, R., Herman, M., and Perrin, A. (1999) Investigation of the fine structure in overtone absorption bands of $^{12}C_2H_4$, *Mol. Phys.* **97**, 265-277.
45. Hurtmans, D., Rizopoulos, A., Herman, M., Salah, H., and Perrin, A. (2000) Vibration-rotation analysis of jet-cooled ν_{12}, $\nu_7+\nu_8$ and $\nu_1+\nu_{10}$ absorption bands of $^{12}C_2H_4$, *Mol. Phys.*, accepted.
46. Hepp, M., Georges, R., Herman, M., Flaud, J-M., and Lafferty, W.J. (2000) Striking anharmonic resonances in N_2O_4: supersonic jet Fourier transform spectra at 13.3, 7.9, 5.7 and 3.2 μm, *J. Mol. Struct.* **517-518**, 171-180.
47. El Youssoufi, Y., Herman, M., Liévin, J., and Kleiner, I. (1997) *Ab initio* and experimental investigation of the vibrational energy pattern in N_2O_4: the mid and near infrared ranges, *Spectrochim. Acta A* **53**, 877-890.
48. Herman, M., Herregodts, F., Georges, R., Hepp, M., Hadj Bachir, I., Lecoutre, M., and Kleiner, I. (1999) Spectroscopic investigation of vibration-rotation bands in acetaldehyde: focus on the $n\nu_3$ (n=1-5) aldehyde CH stretch bands, *Chem. Phys.* **246**, 433-444.
49. Kleiner, I., Georges, R., Hepp, M., and Herman, M. (1999) High resolution FT spectroscopic investigation of acetaldehyde around 7 μm, *J. Mol. Spectrosc.* **192**, 228-230.
50. Campargue, A., Biennier, L., Garnache, A., Kachanov, A., Romanini, D., and Herman, M. (1999) High resolution absorption spectroscopy of the ν_1=2-6 acetylenic overtone bands of propyne: spectroscopy and dynamics, *J. Chem. Phys.* **111**, 7888-7903.
51. Campargue, A., Bertseva, E., Graner, G., and Herman, M. (2000) High resolution absorption spectroscopy of the $3\nu_1$ and $3\nu_1+\nu_3$ bands of propyne, *J. Mol. Spectrosc.* **201**, 156-163.

52. El Idrissi, M.I., Herman, M., Campargue, A., and Graner, G. (2000) Vibrational energy pattern in propyne $^{12}CH_3{}^{12}C_2H$, to be submitted.
53. Boudon, V., Hepp, M., Herman, M., Pak, I., and Pierre, G. (1998) High resolution jet-cooled spectroscopy of SF_6: The $\nu_2 + \nu_6$ combination band of $^{32}SF_6$ and the ν_3 band of the rare isotopomers, *J. Mol. Spectrosc.* **192**, 359-367.
54. Hepp, M., Georges, R., and Herman, M. (1997) The $\nu_6 + \nu_{10}$ band of ethane, *Chem. Phys. Lett.* **275**, 513-518.
55. Hepp, M. and Herman, M. (1998) The jet cooled spectrum of ethane between 4000 and 4500 cm^{-1}, *Mol. Phys.* **94**, 829-838.
56. Hepp, M. and Herman, M. (1999) Weak Combination Bands in the 3μm Region of Ethane, *J. Mol. Spectrosc.* **197**, 56-63.
57. Hepp, M. and Herman, M. (2000) Vibration-rotation bands in ethane, *Mol. Phys.* **98**, 57-61.
58. El Youssoufi, Y., Herman, M., and Liévin, J. (1998) The ground electronic state of 1,2 dichloroethane: I. Ab initio investigation of the geometrical, vibrational and torsional structure, *Mol. Phys.* **94**, 461-472.
59. El Youssoufi, Y., Liévin, J., Vander Auwera, J., Herman, M., Federov, A., and Snavely, D.L. (1998) The ground electronic state of 1,2-dichloroethane. II. Experimental investigation of the fundamental and overtone vibrations, *Mol. Phys.* **94**, 473-484.
60. El Youssoufi, Y., Georges, R., Liévin, J., and Herman, M. (1998) High-Resolution spectroscopic investigation of ν_{16} in trans-1,2-dichloroethane, *J. Mol. Spectrosc.* **186**, 239-245.
61. Mellouki, A., Herman, M., Demaison, J., Lemoine, B., and Margulès, L. (1999) Rotational analysis of the ν_7 band in furan (C_4H_4O), *J. Mol. Spectrosc.* **198**, 348-357.
62. Mellouki, A., Vander Auwera, J., and Herman, M. (2000) Rotational analysis of the ν_6 band in furane, to be submitted.
63. Vander Auwera, J. and Kleiner, I. (1997) The far infrared spectrum of 3,3,3-trifluoropropene, *Spectrochim. Acta A* **53**, 1701-1703.
64. Held, A. and Herman, M. (1995) High resolution spectroscopic study of the first overtone of the N-H stretch and of the fundamentals of the C-H stretches in pyrrole, *Chem. Phys.* **190**, 407-417.
65. Mellouki, A., Georges, R., Herman, M., Snavely, D.L., and Leytner, S. (1997) Spectroscopic investigation of ground state pyrrole ($^{12}C_4H_5N$): the N-H stretch, *Chem. Phys.* **220**, 311-322.
66. De Maré, G.R., Panchenko, Y.N., and Vander Auwera, J. (1997) Structure of the high-energy conformer of 1,3-butadiene, *J. Phys. Chem.* **101**, 3998-4004.
67. Flaud, J-M., Lafferty, W.J., and Herman, M. (2000) First high-resolution analysis of the absorption spectrum of propane in the 6.7 – 7.5 μm region, *J. Chem. Phys.*, submitted.

High Resolution Infrared Laboratory Spectroscopy of Atmospheric Constituents at DLR

M. BIRK†, D. HAUSAMANN‡, F. SCHREIER†, G. WAGNER†
† *Remote Sensing Technology Institute*
‡ *German Remote Sensing Data Center*
DLR Oberpfaffenhofen, D-82234 Wessling, Germany

Abstract
High resolution spectroscopy utilising a Bruker IFS 120 HR spectrometer was established at DLR during the end of the 80s. Since then extensive infrastructure has been built up regarding laboratory equipment as well as software tools. This includes cooleable/heatable single and multi-pass absorption cells (temperature range 190 – 1000 K, length 0.15 – 100 m), a flow system for generation of unstable species, capability for atmospheric absorption and emission measurements. For data reduction, among other utilities, the flexible line parameter retrieval code FitMAS (Fit Molecular Absorption Spectra) was developed. The work is focused on the generation of a quality-assured spectroscopic database of atmospheric constituents. Within nationally and internationally funded projects numerous species were investigated in the region from 10 to 3000 cm^{-1}: CO, CO_2, O_3, ClO, Cl_2O_2, HOCl, $ClONO_2$, BrO, OH, H_2O, HO_2, H_2O_2, NO, NO_2, N_2O_5. Validated accuracies for line strengths of 1% and for pressure broadening parameters of 2% have been established.

1. Introduction

The spectroscopic database is a key input for remote sensing applications such as atmospheric and climate research, environmental monitoring, fire detection, gas phase analytics, and astronomy. Only an accurate knowledge of line positions, line strengths, pressure broadening parameters, and pressure induced line shifts or absorption cross sections (ACS) allows to determine distributions of the respective species from remote measurements.

The goal of the laboratory spectroscopy work at DLR is to determine spectroscopic parameters of species of atmospheric relevance with highest possible accuracy [1]. The work focuses on the needs for atmospheric remote sensing which requires high accuracy for all parameters. To achieve this goal it is necessary (i) to conduct the experiment in a way to avoid error sources which are hard to quantify or correct, (ii) to quantify all remaining error sources, and (iii) to do quality assurance of the results.

Existing spectroscopic data are collected in a number of different individual databases such as HITRAN [2], GEISA [3], JPL [4], ATMOS [5], SAO [6] for general atmospheric applications or in special collections such as HITEMP for high temperature applications. The main deficiency of the existing collections is the lack of a global reference database with standardised and characterised quality information.

J. Demaison et al. (eds.), Spectroscopy from Space, 219–233.

Although for most of the atmospheric constituents spectroscopic data are available from microwave to the ultraviolet, the database is far from being complete for all remote sensing applications. The targeted geolocation in the atmosphere combined with different observation geometries require different subsets of spectroscopic data.

Especially line strength, pressure broadening, and ACS data are susceptible to numerous systematic error sources which may arise anywhere in the chain from sample preparation over the instrument to the data reduction. A rigorous error analysis covering all these aspects is rare within the existing spectroscopic databases. Thus, errors of atmospheric trace gas distributions from remote sensing propagated from spectroscopic database errors cannot be quantified in most cases.

2. Laboratory infrastructure

2.1. HIGH-RESOLUTION FOURIER-TRANSFORM SPECTROMETER

For the conduction of laboratory spectroscopy a Bruker IFS 120 HR (maximum optical path difference (MOPD) 6 m, spectral range 10 – 10000 cm^{-1}) Fourier transform (FT) spectrometer was used. About 5 years ago the extension for double-sided measurements was installed. The spectrometer is operated with the OPUS software running under OS/2. Recently, the instrument was extended to allow atmospheric measurements in absorption and emission.

2.2. IR/FAR INFRARED (FIR) DETECTORS

A number of detectors is available to cover various spectral ranges:

- Custom made Si bolometer, Infrared Laboratories Inc., 1.4 K, 10 – 400 cm^{-1} (various cold filters).
- Standard Si bolometer, Infrared Laboratories Inc., 4.2 K, 20 – 600 cm^{-1} (various cold filters).
- Standard MCT, EG&G, 77 K, 500 – 2000 cm^{-1}, Dewar was modified to accommodate cold stops and filters, however, due to thermal excitation noise, there was only little gain in the noise figures.
- Standard InSb, EG&G, 77 K, 1800 – 10000 cm^{-1}.
- A new MCT detector optics has been developed. A standard EG&G MCT chip is mounted in an Infrared Laboratories Dewar. A new imaging concept was developed using raytracing software developed in our laboratory. The Dewar contains two imaging mirrors, a flat folding mirror, and a variable cold stop. Applying cold filters a factor of 8 improvement in signal to noise was obtained when compared to the standard MCT.

2.3. ABSORPTION CELLS

Absorption cells are subject to a number of requirements: (i) Absorption path must be well defined, (ii) channelling should be avoided, (iii) good temperature homogeneity above or below ambient temperature is required, (iv) the cell must be filled or emptied without repositioning to allow measurements of suitable reference spectra, (v) cell and window material must be chemically inert with respect to the gases to be investigated.

All cells with absorption path smaller than 30 cm were mounted inside the evacuable sample compartment of the spectrometer. One or two in/outlets allowed to fill the cell or to set up a steady flow from outside the spectrometer. To avoid channelling windows are wedged and inclined with larger angles in the FIR.

2.3.1. *Cooleable short cell*

This glas cell (absorption path 25 cm) is double jacketed to allow coolant flow. Temperature homogeneity is established by heat-sinking the windows combined with the insulating vacuum inside the sample compartment. Temperature range is 190 to 300 K. Temperature homogeneity has been tested by comparison of retrieved CO temperatures from relative line strengths of pure rotational transitions with readings of Pt 100 sensors attached to the cell body. The retrieved gas temperature of 190.7(3) K was in perfect agreement with the sensor readings of 190.8(5) K, indicating the window temperature to be identical to that of the glas body. For corrosive gases (e.g. N_2O_5 or $ClONO_2$) AgCl windows were applied.

2.3.2. *Hot cell*

For the purpose of high temperature database up to 1000 K an electrically heated quartz cell (absorption path 16 cm) was developed featuring radiatively heated CaF_2 windows, equipped for steady state flow and mounted in an evacuated housing.

2.3.3. *Cooleable White-type cell*

This cell was optimised for FIR spectroscopy where large optical throughput and diffraction are an issue. Since unstable species like OH, HO_2, ClO, and BrO were to be investigated, the cell volume was minimised and a 250 m^3/h roots blower was installed to allow for flow experiments with short residence times. A raytracing program was developed in order to specify transfer optics as well as T- and D-type mirrors. A four-row spot design on the T-type mirror was selected. The baselength is 80 cm, the diameter 20 cm, and the maximum absorption path about 100 m. The cell body is double jacketed glas and can be cooled down to 200 K. White-cell mirrors were thermally isolated from the ambient and heat-sinked to the cooled glas body. The transmittance of the cell is close to theory as calculated from the reflectance of the gold mirrors.

2.4. SAMPLE PREPARATION

The infrastructure for sample preparation and handling includes a gas supply system with various MKS flow controllers, a microwave discharge for producing atomic species, a number of glas vacuum lines, various temperature stabilised MKS Baratron pressure gauges, vacuum pumps ranging from 8 to 250 m^3/h, a closed cycle cooler for temperatures down to 190 K, and two gas mixing chambers (4 and 150 l) with stirrers.

3. Data Processing

3.1. LEVEL 0-1 PROCESSING: FROM INTERFEROGRAM TO TRANSMITTANCE SPECTRA

Several corrections have to be applied to the raw interferogram and spectral data:

3.1.1. *Detector non-linearities*

Detector non-linearities cause a small part of the spectral signal to be redistributed, thus causing radiometric errors. If the signals are redistributed to regions where the signal originally is zero they can be used for correcting higher order terms in the detector's response to the incident photon flux. The disappearance of these out-of-band contributions indicate an appropriate non-linearity correction. Usually a quadratic correction term is sufficient.

3.1.2. *Sample and window thermal radiation*

In case of two-port interferometers the thermal emission arising from sample, cell windows, etc. causes systematic radiometric errors. The effect was corrected with a novel method utilising low resolution source-on and source-off reference spectra. The standard correction requires to measure sample source-off spectra at sample source-on spectral resolution [1]. To take source-off spectra with the same signal to noise ratio at source-on spectral resolution requires two times more total measurement time. The subtraction of source-on and source-off spectra increases the noise by $\sqrt{2}$. Thus, applying the usual correction scheme, the total measurement time for a corrected transmittance spectrum with the same signal to noise ratio as an uncorrected one is four times longer.

3.1.3. *Channelling*

In some cases periodic baseline variations (channelling) dominate the noise in the transmittance spectra, caused by slightly different standing waves in the spectrometer optics in sample and reference spectra due to drifts. The channel spectra appear as a kind of small copy of the centerburst in the interferogram (sideburst). In order to avoid this problem, usually reference spectra are taken before and after the sample spectra and a linear combination is applied, yielding an almost channelling-free transmittance spectrum. Alternatively, contaminated interferogram regions were cut, which, however, introduces instrumental line shape distortions and can only be applied in interferogram regions, mainly at higher optical path difference, where modulations are rather small. In case of low path difference the following method was found useful. In some cases, the incompatible channelling in reference and sample spectra can be explained by a simple shift of the sideburst between sample and reference measurements. In case of strong modulations the shift cannot be determined from the measured interferograms themselves. To determine the shift spectral regions containing high resolution spectral information were cut out in the sample spectrum, and the same regions were cut out in the reference spectrum. The inverse FT was applied to both spectra to reveal the shift. Applying this shift to the reference interferogram yielded a more compatible reference spectrum.

3.2. LEVEL 1-2 PROCESSING: FROM TRANSMITTANCE SPECTRA TO LINE PARAMETERS

3.2.1. *Absorption cross sections vs. line data*

The connection of ACSs to the line database is given by:

$$\alpha(\sigma) = -\frac{1}{nl}\ln(\tau) = \sum_i S_i f_i(\sigma - \sigma_{0i}, \gamma_i, \ldots)$$

where S_i is the line strength of an individual transition and f_i the line shape at the transition wavenumber σ_{0i}. Line data to be retrieved are transition wavenumber, line strength (at 296 K), broadening parameter γ_i (at 296 K), temperature exponent, and pressure-induced line shift.

As can be seen from the equation above line parameters are more fundamental when compared to ACSs since line data allow to calculate ACSs for any desired condition i.e., pressure and temperature. However, at high pressures or high line density individual transitions are not resolvable in the experimental spectra and thus line data cannot be obtained. The alternative is the measurement of ACSs directly without attempting a further data reduction to line data. However, since no physical model allows the conversion to temperatures and pressures different from the experiment measurements have to be taken with a temperature and pressure grid allowing interpolation to any condition within the targeted range. Of course, the measurement effort for obtaining ACSs directly is substantially higher. Further constraints for ACS measurement are caused by the instrumental line shape (ILS) distorting the observed features. This is no problem in case of the line data because the ILS can be modelled and separated from the retrieved line data allowing for a slight instrumental distortion of the observed spectra. In case of ACSs it must be assured that the instrument has a negligible effect on the observed spectrum. Simulations have been carried out which showed that distortions are indeed negligible when compared with the noise level for an instrumental line width being four times smaller than the smallest Lorentzian width. The 100% transmittance must be accurate over the entire spectrum. In case of line data baseline accuracy requirements are relaxed since baseline offset and slope are fitted in the retrieval within each microwindow.

3.2.2. *Retrieval methodology*

The retrieval of spectral line parameters (i.e., positions, strengths, widths, etc.) from measured FT spectra is done with the FitMAS (Fit Molecular Absorption Spectra) code [7]. FitMAS performs a non-linear least squares fit using a damped Gauss-Newton type iteration or an implementation of the Levenberg-Marquardt trust region algorithm essentially based on the MINPACK routine [8]. The model function to be adjusted to the observed spectrum is the convolution of the monochromatic transmission and the instrumental line shape function (ILS). A variety of line shapes can be used for the monochromatic transmission, e.g., Lorentz, Voigt, correlated and uncorrelated Rautian profiles [9]. For the ILS either a simple sinc function (accounting for the finite optical path difference of the FTS), a convolution of sinc and box (additionally accounting for the interferometer field stop) or a phase distorted sinc are possible options. Groups of dense lines can be treated as multiplets, i.e., relative line position and/or strengths can be frozen and only the multiplet position and total line strength will be fitted. Furthermore, baseline effects can be considered by adjusting the coefficients of an appropriate polynomial in wavenumber. It should be noted that the Jacobian, i.e., the derivatives of the model function with respect to the parameters to be fitted, is computed fully analytically, hence avoiding the numerical delicacies of finite differences and resulting in a considerable computational efficiency gain [10]. The convolution of monochromatic transmission and instrument line shape function is performed in the wavenumber domain; truncation errors in the quadrature are avoided by using asymptotic expansions of the integrand and adding the corresponding contributions to the numerically evaluated contribution in the microwindow region.

Initial and frozen line parameters can be read from HITRAN [2] or JPL [4] type databases or simple peaklists. FitMAS can be used in batch mode or interactively. The user has complete flexibility in his choice of line parameters to be fitted, e.g., some or all parameters of weak lines can be held fixed at pre-selected values. This choice can be specified in the initial guess line parameter list or can be given interactively by simple keystrokes in the plot of the observed spectrum.

FitMAS has been written in FORTRAN77 and utilises the SLATEC (http://www.netlib.org/) and LAPACK [11] libraries. It has been extensively tested and validated, e.g. with J. Johns' INTBAT/GMET program [1], [12], and has been used in the analysis of the large variety of spectroscopic investigations listed within this paper.

4. Systematic error sources

4.1. PARTICLE DENSITY ERRORS

Errors in particle density are proportional to line strength and absorption cross section errors. The sample preparation process is, therefore, extremely important.

Pure samples of stable species: The main issue is an accurate pressure measurement.

Gas mixtures: The preparation of gas mixtures requires special attention with respect to the determination of partial and total pressure. Adsorption effects of polar species on walls have to be considered.

Unstable species: In this case pressure measurements are not sufficient. Particle densities of such species can be derived by simultaneous intensity measurements in another spectral region such as UV (where absorption cross sections are often available in the literature) or in the FIR (where knowledge of the dipole moment allows to calculate the sample concentration) or titration reactions leading to stable measurable species. In the UV the radiometric accuracy as well as the spectroscopic database are error sources. The titration may have systematic error sources that are hard to quantify.

4.2. INSTRUMENTAL ERRORS

Numerous systematic error sources can affect the accuracy of typical Fourier transform intensity measurements [1] and thus retrieved line strengths, such as detector non-linearities, thermal radiation of sample and windows, and in some cases ILS errors. In case of laser spectrometers other systematic error sources exist (e.g. baseline reproducibility).

Pressure broadening parameters are affected by ILS-induced errors which become important when the ILS begins to influence the observed line shape. Additionally, channelling is a severe problem when line width and channel period are about equal.

4.2.1. *ILS errors*

The ILS of an ideal FTS with an extended circular source is the convolution of the sin(x)/x function with a box. ILS errors affect the spectrum of the interferometer field stop (box) and can be caused by vignetting, by poor alignment, and by illumination errors. These errors always cause the box to be asymmetric, however, the area under the spectral feature is not affected. Resulting errors in the retrieved line parameters can be avoided by using the actual (distorted) ILS for the line fitting.

Vignetting: The vignetting error originates from divergence of the collimated beam in the interferometer and causes radiation passing the interferometer field stop close to its edge to be increasingly attenuated with increasing optical path difference. The upper wavenumber edge of the box remains unchanged while the values towards the lower wavenumber edge are decreased.

Alignment error: Poor alignment of the optical axis in the interferometer with respect to the mirror displacement path can cause substantial deterioration of the box and gives rise to an asymmetric and broadened ILS.

Illumination error: If the interferometer field stop is not illuminated homogeneously or the combination of optics and detector have a variable response depending on the position and angle of rays within the interferometer field stop, the intensity distribution within the box will vary and almost certainly result in an asymmetry. Another type of illumination error occurs when the detector can see an interferogram of thermal radiation arising from the back side of the interferometer field stop. Since this radiation has a larger angle of divergence it gives rise to an additional wider box extending towards lower wavenumber with a much smaller height.

Influence of ILS errors on retrieved line strength: Since the area under the spectral feature is not influenced by ILS errors, line strength determinations based on the equivalent width method are not affected, provided only optically thin lines are considered. In the least squares based line fit approach, the results will be correct if the ILS function used represents the observed data within the observed noise level. Since the observed spectrum is a convolution of the monochromatic spectrum with the ILS function, ILS errors can be reduced by ensuring that the lines in the monochromatic spectrum are wider than the ILS function.

4.2.2. *Residual phase errors*

An ideal interferogram is symmetric with respect to zero-path-difference and has cosine contributions only. In a real interferogram, however, phase differences occur, arising essentially from a wavenumber dependence of the exact position of zero-path-difference caused by refraction, electronic signal delays, and thermal radiation from the beamsplitter/substrate/compensator. Furthermore, additional phase errors are caused by the finite sampling grid, i.e. by not matching the region of zero-path-difference. Despite phase corrections, residual phase errors may remain. These errors may result in an asymmetry but will in any case decrease the area under a spectral feature, thus causing the line strength to be systematically too small.

4.2.3. *Further radiometric errors*

Systematic errors caused by detector non-linearities, thermal sample emission, and channelling have been discussed above (Ch. 3.)

5. Measurement Strategies for Data Quality Assurance

One major focus of the spectroscopic work at DLR was to avoid systematic errors, especially when they are hard to quantify, and to quality assure our results. In-depth investigations of these issues have been carried out by means of FIR CO measurements [1]. Line strength accuracies better than 1% have been achieved. In case of ozone also line strength accuracies better than 1% were obtained [13]. Pressure broadening parameters of CO and O_2 have been measured with an accuracy of 2% as validated by independent millimeter-wave measurements [14].

5.1. SAMPLE PREPARATION

Especially for line strength and ACS measurement the knowledge of sample particle densities is essential. The simplest case are stable species because it only requires pure samples together with accurate pressure measurements. In case of gas mixtures the mixing chambers mentioned in §2.4 are used. Gases are consecutively filled and the pressures are measured. For non wall adhesive gases accuracies better than 1% were achieved.

We have developed a method to provide defined particle densities of pure ozone and ozone gas mixtures in an absorption cell based on pressure measurements [15]. By means of FIR pure rotational spectra an accuracy better than 1% has been proven [13].

In case of unstable species a new method was developed. Since line strengths for pure rotational transitions in the FIR can be calculated theoretically from the permanent electric dipole moment, the measurement of FIR line intensities is suitable for determining the average particle density within the optical path in a flow experiment. This method has been successfully applied in case of ClO [16] and ClOOCl [17].

5.2. OPTIMISATION OF THE INSTRUMENT

5.2.1. *Optical alignment*

In case of high resolution FT spectroscopy the alignment of the optical axis with respect to the interferometer axis is critical since even small deviations well below 1 mrad can cause ILS distortions. We have developed adjustment procedures to obtain an ILS very close to theory even at very high resolution (0.002 cm^{-1}).

5.2.2. *Optimisation of instrumental parameters*

Resolution: ILS errors are, of course, less critical when the observed spectrum is only slightly distorted by the ILS. Thus, a sufficient instrumental resolution helps to avoid such errors.

Measurement time: High spectral resolution in turn requires a longer measurement time, good alignment and very good detectors. In order to allow long measurement times the laboratory was thermostated to minimise drifts. Measurement time was selected to avoid deteriorations in the sample but still obtain sufficient signal to noise ratio.

Optical throughput: The maximum throughput for optimum information content is determined by the maximum wavenumber and the MOPD.

Spectral range: The sampling frequency in the interferogram was chosen to avoid folding noise and out of band contributions from detector non-linearity into the alias of interest.

Filters: Optical filters are required to reduce the data amount, reduce detector non-linearities, and, if cooled, reduce background photon flux.

Scanner velocity: In case of 1/f noise, microphonics, and other modulations higher scanner velocity helps to reduce or change spectral perturbations. Furthermore, higher velocity reduces the dynamic range in the interferogram.

5.3. OPTIMISATION OF SAMPLE PARAMETERS

Sample pressure and absorption path should be selected in order to avoid transmittances of target lines smaller 5%. At such high optical depth baseline errors, noise, and ILS errors have a stronger impact on retrieved spectroscopic parameters. In case of pressure broadening parameter and line strength retrieval the broadening gas pressure should be as high as possible without introducing blendings of adjacent lines. This reduces the amount of measurement time substantially.

5.4. RETRIEVAL PROCEDURES

Two independent least squares line parameter retrieval codes exist at DLR, which have been cross checked and validated with synthetic spectra [1],[12]. For further details see §3.2 and §5.5.

5.5. CONSISTENCY CHECKS AND VALIDATION

Numerous procedures to assure the quality and to validate the data were elaborated.

5.5.1. *Redundancy*

Measurements under different conditions yielding redundant information are an important tool for quality assurance.

Lambert-Beer check: Repeat measurements with various target gas pressures. Assessment of radiometric accuracy, pressure measurement, gas mixing procedure.

Apply various broadening gas pressures: For line strengths similar to Lambert-Beer check, additionally different influence of ILS induced errors. For pressure broadening ILS errors and channelling induced errors can be detected.

Area under ACS spectrum (ACS database): The area under ACS of an entire infrared band system is not dependent on the broadening gas pressure. Furthermore, we found that for at least several molecules the area is almost (<1%) independent of temperature. Assessment of radiometric accuracy, pressure measurement, gas mixing procedure, as well as temperature homogeneity.

Fit of particle density and temperature from retrieved line strengths for individual measurements (line database): Analysis tool for Lambert-Beer and broadening gas pressure variation check. Particle density and temperature are fitted by analysing retrieved products of line strength and column density utilising modelled reference line strength data (e.g. HITRAN). In case of correct relative reference line strength data the average gas temperature can be obtained, allowing for temperature homogeneity checks or temperature measurement in absence of suitable temperature sensors.

Apply different spectral resolution/field stop sizes: This helps to detect errors introduced by the ILS.

5.6. OTHER QUALITY ASSURANCE METHODS

Fitting procedures involving line parameters: Any further data reduction or test procedure based on line parameters involving a fitting procedure allows to (i) calculate standard deviation from input data uncertainties and (ii) calculate standard deviation from residuals. Differences indicate systematic errors or model deficiencies.

Comparison of model and experimental spectra: Calculate model spectra from final database and compare with experimental spectrum. This checks the adequacy of the data reduction. If the experimental spectrum was not used for the data reduction, also interpolation or extrapolation capabilities of the model can be checked.

Spectral residuals: (i) Line parameter fitting: If the residuals show random noise only this indicates the adequacy of the line model and the absence of line shape related instrumental errors. (ii) Level 2 processing of atmospheric spectra: The absence of residuals or the improvement when applying the new database indicates validity of the new database.

FIR line strength: Absolute line strengths can be calculated from the permanent electric dipole moment and serve as a calibration standard. This allows to check the entire chain from sample preparation to data reduction.

Comparison of measurements by different groups and instrument techniques: This is only useful if most points above have been considered. The comparison of pressure broadening parameters [14] mentioned at the beginning of this section may serve as an example.

5.7. ERROR ANALYSIS

It is attempted to quantify as many error sources as possible and to propagate these into the spectroscopic parameters.

5.8. FURTHER DATA REDUCTION

Quantum mechanical data reduction is a standard approach in case of line positions and line strengths. In case of pressure broadening parameters we have successfully applied polynomial representations. The data reductions have two advantages: (i) A spectroscopic database produced from the reduced data has a better precision when compared to the original data and is more complete since interpolation and, to some extent, extrapolation is possible. (ii) The data reduction usually involves a fitting procedure which produces residuals (obs-calc) allowing to calculate a standard deviation. Since the errors of the spectroscopic parameters also allow to calculate an a priori standard deviation differences in these numbers help to detect systematic error sources.

6. Results

Several species have been investigated for different purposes and applications. An overview is given in the following table, details are described in the subsequent sections.

#	species	FIR	MIR	purpose/application	remark	ref.
1	O_3	S, γ(T)	σ, S, γ(T), α(T,p)	database improvement		19, 17, 26, 14, 13, 18, 15
2	$ClONO_2$		α(T,p)	database improvement, MIPAS	difficult synthesis	17, 20, 21
3	N_2O_5		α(T,p)	database improvement, MIPAS		17
4	OH/HO_2	σ		new methodology	extremely unstable	22
5	BrO	σ, γ(T)		database improvement, MASTER/SOPRANO	extremely unstable	23, 14
6	ClO	σ, γ(T)	σ, S	database improvement, MASTER/SOPRANO	unstable	14, 16
7	ClOOCl		α(T,p)	detectability with remote sensing	sample preparation difficult	17
8	HOCl	σ		FIR database		24
9	CO	S, γ(T)	S, γ(T)	error characterisation, high temperature database, Q/A	<1% radiometric accuracy	25, 26, 14, 1
10	CO_2		α(T,p)	high temperature database		25
11	H_2O		σ, S, γ(T)	high temperature database improvement, climate, MIPAS, IASI	sample preparation difficult	25

12	NO		σ,S, γ(T)	high temperature data-base, engine emissions		25
13	NO_2		α(T,p)	high temperature data-base, engine emissions		25

6.1. OZONE (O_3)

A patented method was developed for generation of defined ozone/inert gas mixtures [15]. FIR line strengths with an accuracy of 1% have been measured [13]. The weak difference band ν_3-ν_2 was investigated regarding line strength and line positions [18]. A new mid infrared spectroscopic database for the three fundamentals focused on line strengths and N_2/O_2 broadening parameters with temperature exponents was determined [17], [19]. Independent millimeter wave and FT measurements of air broadening parameters and temperature exponents of ozone around 500 GHz were carried out yielding agreement better than 2% [14].

6.2. CHLORINE NITRATE ($CLONO_2$)

Various measurements of $ClONO_2$/air mixtures covering atmospheric conditions were conducted to obtain ACSs as function of temperature and air pressure [17], [20], [21]. Marked differences to HITRAN96 were observed. The accuracy of the new data is in the range 3-5%.

6.3. DINITROGEN PENTOXIDE (N_2O_5)

For N_2O_5 an MIR database was produced with similar quality as for $ClONO_2$. Main results are that air pressure dependence is negligible and the band structure remains unresolved [17].

6.4. HYDROPEROXYL (HO_2) AND HYDROXYL (OH)

Since the lifetime of the hydroperoxyl radical (HO_2) is controlled by its self-disproportion, its spectroscopy required a flow synthesis (reaction of purified H_2O_2 with atomic fluorine) in combination with the White-type cell described in §2.3.3. About 366 line positions of the electronic and vibrational ground state could be identified in the FIR between 50 and 200 cm^{-1} covering the branches r_{R0} until r_{R4} and r_{Q1} until r_{Q4}. Rotational constants in A- and S-reduction have been derived with high accuracy. High-order centrifugal distortion terms could be determined for the first time. With these improved molecular constants a better prediction of line positions and, finally, of the line intensities is available. Furthermore, for the hydroperoxyl radical (OH) a number of spin-forbidden rotational transitions in the FIR spectral range have been measured with high accuracy [22].

6.5. BROMINE OXIDE (BrO)

The FIR pure rotational spectrum of BrO between 12 and 28 cm^{-1} has been measured to determine the air broadening parameter and temperature exponent of the 499.6 GHz transition of ^{79}BrO. BrO was formed in a steady state flow experiment from atomic bromine and ozone utilising the White-type cell [14], [23].

6.6. CHLORINE OXIDE (ClO)

ClO was formed in a flow reaction from Cl_2O and atomic chlorine and measured in the White-type cell. Air broadening parameters and temperature dependence were determined for pure rotational transitions between 12 and 28 cm^{-1} [14]. To determine ClO fundamental line strengths a series of subsequent FIR, MIR and, again, FIR spectra was recorded within one experiment. From the FIR transitions and their calculated line strengths the particle density could be determined [16].

6.7. DICHLORINE DIOXIDE (ClOOCl)

Motivation for measuring ACSs of ClOOCl in the region 500 – 800 cm^{-1} was to explore the capability of FT limb sounders for remote sensing of this species relevant to perturbed "ozone hole" chemistry. In similarity to ClO infrared fundamental line strength determination FIR and MIR measurements within the same experiment were applied. ClOOCl was generated in a flow reaction from ClO in the White-type cell [17].

6.8. HYPOCHLOROUS ACID (HOCl)

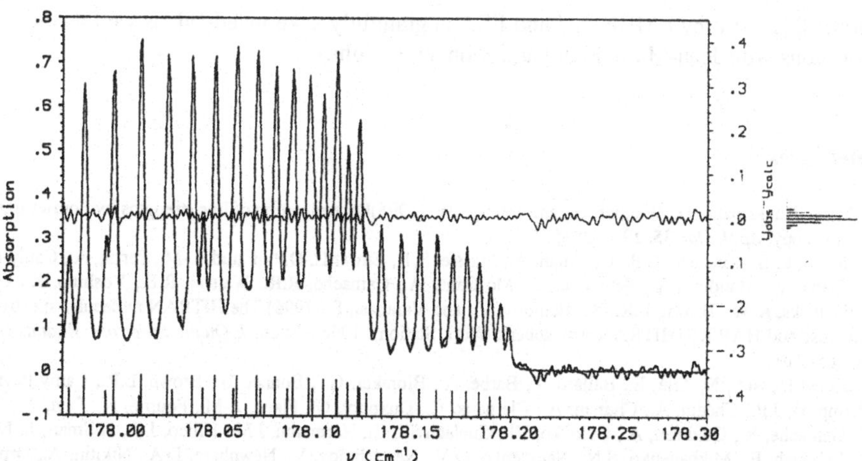

Figure 1. Part of the multi-line HOCl spectrum at 178 cm^{-1}: Line parameters of 39 lines, indicated by vertical impulses, have been retrieved with FitMAS (cf. subsection 3.2.2). Small differences between the observed and modelled spectra only show up in the regions between the lines and in the tail of the band ($\sigma > 178.2$ cm^{-1}). Residuals (difference of observed and modeled spectrum, right scale) are essentially due to noise.

The FIR spectrum of HOCl has been recorded between 20 and 360 cm^{-1}. As an example a Q branch is shown in Figure 1. The measured lines, combined with microwave and tunable FIR data available in the literature were least squares fitted [24].

6.9. CARBON MONOXIDE (CO)

A rigorous test of the radiometric accuracy of commercial high resolution FTSs has been made by measuring line strengths in the pure rotational spectrum of carbon monoxide with a Bruker IFS 120 HR (transitions: J" = 7...23) and a Bomem DA3.002 (J" = 27...35). The experimental results show that radiometric errors affected the retrieved line strengths by less than 1% [1].

Consistent air broadening coefficients and temperature dependencies have been obtained with the very different experimental techniques of coherent millimeter-wave spectroscopy (University of Lille group) and incoherent Fourier transform spectroscopy (DLR). The results show that reliable results with an accuracy of 2% can be achieved by both techniques [26].

6.10. CARBON MONOXIDE (CO), CARBON DIOXIDE (CO_2), WATER (H_2O), NITROGEN OXIDE (NO), NITROGEN DIOXIDE (NO_2)

A high temperature database (300 – 750 K) was generated for support of non-intrusive engine exhaust gas analysis. In case of CO, NO, H_2O line parameters were determined, whereas for CO_2 and NO_2 ACSs were derived [25].

7. Acknowledgement

Support of this work by BMBF, EU and ESA is gratefully acknowledged, as well as the collaborations with Jean-Marie Flaud and John W. C. Johns.

8. References

1 Birk, M., Hausamann, D., Wagner, G., and Johns, J.W. (1996)Determination of line strengths by Fourier transform spectroscopy, *Appl. Opt.* **35**, 2971–2985.

2 Rothman, L. S., Rinsland, C.P., Goldman, A., Massie, S.T., Edwards, D.P., Flaud, J.-M., Perrin, A., Camy-Peyret, C., Dana, V., Mandin, J.Y., Schroeder, J., Mc Cann, A., Gamache, R.R., Watson, R.B., Yoshino, K., Chance, K.V., Jucks, K.W., Brown, L.R., Nemtchinov, V., and Varanasi, P. (1996) The HITRAN molecular spectroscopic data base and HAWKS (HITRAN atmospheric Work Station): 1996 edition, *J. Quant. Spectrosc. Radiat. Transfer* **60**, 665–710.

3 Jacquinet-Husson, N, Arié, E., Ballard, J., Barbe, A., Bjoraker, G. , Bonnet, B., Brown, L.R., Camy-Peyret, C., Champion, J.P., Chédin, A., Chursin, A., Clerbaux, C., Duxbury, G., Flaud, J.-M., Fourrié, N., Fayt, A., Graner, G., Gamache, R., Goldman, A., Golovko, Vl., Guelachvili, G., Hartmann, J.M., Hilico, J.C., Hillman, J., Lefèvre, G., Lellouch, E., Mikhailenko, S.N., Naumenko, O.V., Nemtchinov, V., Newnham, D.A., Nikitin, A., Orphal, J., Perrin, A., Reuter, D.C., Rinsland, C.P., Rosenmann, L., Rothman, L.S., Scott, N.A., Selby, J., Sinitsa, L.N., Sirota, J.M., Smith, A.M., Smith, K.M., Tyuterev, Vl.G., Tipping, R.H., Urban, S., Varanasi, P., and Weber, M. (1999) The 1997 spectroscopic GEISA databank. *J. Quant. Spectrosc. Radiat. Transfer* **62**, 205-254.

4 Pickett, H.M., Poynter, R.L., and Cohen, E.A. (1992) Submillimeter, Millimeter, and Microwave Spectral Line Catalogue, *JPL Publication* **80-23**, Rev. 3, Jet Propulsion Laboratory, Pasadena, CA, USA.

5 Brown, L.R., Gunson, M.M., Toth, R.A., Irion, F.W., Rinsland, C.P., and Goldman, A. (1996) 1995 Atmospheric Trace Molecule Spectroscopy (ATMOS) Linelist, *Appl. Opt.* **35**,2828.

6 Chance, K.V., Jucks, K.W., Johnson, D.G., and Traub, W.A. (1994) The Smithsonian Astrophysical Observatory Database 1992, *J. Quant. Spectrosc. Radiat. Transfer* **52**, 447 - 457.

7 Schreier, F., Schimpf, B., and Birk, M. (1993) FitMAS -- Least-squares fitting of molecular line parameters from high resolution Fourier transform spectra (1993) In *13. Colloquium on High Resolution Molecular Spectroscopy*, Riccione, Italy.

8 Dennis, J.E. and Schnabel, R.B. (1983) Numerical Methods for Unconstrained Optimization and Nonlinear Equations Prentice-Hall.

9 Varghese, P.L. and Hanson, R.K. (1984) Collisional narrowing effects on spectral line shapes measured at high resolution, *Appl. Opt.* **23**, 2376-2385.

10 Schreier, F. (1992) The Voigt and Complex Error Function: A Comparison of Computational Methods, *J. Quant. Spectrosc. Radiat. Transfer* **48**, 743-762.

11 Anderson, E., Bai, Z., Bischof, C., Demmel, J., Dongarra, J.J., Du Croz, J., Greenbaum, A., Hammarling, S., McKenney, A., Ostrouchov, S., and Sorensen, D. (1995) LAPACK Users' Guide (2nd edition), SIAM.

12 Johns, J.W.C., Schreier, F., Schimpf, B., and Birk, M. (1994) Analysis of High-Resolution Fourier Transform Molecular Spectra, In *The Future of Spectroscopy: From Astrophysics to Biology*, Sainte-Adèle, Quebec, Canada.

13 Birk, M., Wagner, G., and Flaud, J.-M. (1994) Experimental Line strengths of Far-Infrared Pure Rotational Transitions of Ozone, *J. Molec. Spectrosc.* **163**, 245–261.

14 Birk, M. and Wagner, G. (1997) Study on Spectroscopic Database for Millimeter and Submillimeter Wavelength, *Final Report to ESTEC*, Contract No. 11581/95/NL/CN, edited by Agnes Bauer, ESTEC, The Netherlands.

15 Birk, M. and Wagner, G. (1994) Vorrichtung zur Erzeugung eines definierten Ozon-Fremdgas-Gemisches und Verfahren zum Bereitstellen eines definierten Ozon-Fremdgas-Gemisches in einem geschlossenen Behälter, *Deutsches Patent* Nr. 43 42 624.7-52.

16 Birk, M., Wagner, G. (1997) Experimental line strengths of the ClO fundamental, *J. Geophys. Res.* **102**, 19199-19206.

17 Wagner, G. and Birk, M. (2000) Erweiterung der spektroskopischen Datenbasis von ClO-NO_2, ClOOCl, O_3, N_2O_5 im Hinblick auf die Nutzung durch die MIPAS Instrumente, *Ozone Research Program* KKZ 07DLR05, German Ministery for Education and Research, Germany.

18 Birk, M., Wagner, G., Flaud, J.-M., and Hausamann, D. (1994) Line strengths in the ν_3-ν_2 Hot Band of Ozone, *J. Mol. Spectrosc.* **163**, 262-275.

19 Wagner, G., Birk, M., Schreier, F., and Flaud, J.-M. (2000) Spectroscopic database of the three ozone fundamentals, subm. to JGR.

20 Oelhaf, H., Wetzel, G., Höpfner, M., Friedl-Vallon, F., Glatthor, N., Maucher, G., Stiller, G., Trieschmann, O., von Clarmann, Th., Birk, M., andWagner, G. (2000) Interconsistency checks of $ClONO_2$ retrievals from MIPAS-B spectra by using different bands and spectroscopic parameter sources, *Proceedings IRS 2000*, Petersburg, Russia.

21 Birk, M. and Wagner, G. (2000) A new spectroscopic database for chlorine nitrate, Poster, *6th HITRAN Conference*, Cambridge, USA.

22 von Bargen, A., Birk, M., and Wagner, G. (1997) The HO_2 Radical: The pure rotational FIR Spectrum Measured by FTS and the Molecular Constants, Poster Q1, *15th Colloquium on High Resolution Molecular Spectroscopy*, Glasgow, UK, and Hochauflösende Fourier-Transform-Spektroskopie an den Radikalen Hydroperoxyl und Hydroxyl, PhD Thesis, University of Gießen, Germany.

23 Wagner, G. and Birk, M. (1997) Line Positions and Air Broadening of Pure Rotational Transitions of BrO From FTS Measurements, Poster B6, *15th Colloquium on High Resolution Molecular Spectroscopy*, Glasgow, UK.

24 Flaud, J.-M., Birk, M., Wagner, G., Orphal, J., Klee, S., and Lafferty, W.J. (1998) The Far Infrared Spectrum of HOCl: Line Positions and Intensities, *J. Mol. Spectrosc.* **191**, 362-367.

25 Schäfer, K. (ed.) (1998) Non-Intrusive Measurement of Aircraft Engine Exhaust Emissions (AEROJET), *Final Technical Report*, Commission of the European Communities Contract No. BRPR-CT96-0142.

26 Birk, M., Colmont, J.-M., Priem, D., Wagner, G., and Wlodarczak, G. (1997) N_2-, O_2-, and Air-Broadening Coefficients of the J=3$\leftarrow$2 Line of CO and the J=342,32 $\leftarrow$341,33 Line of O_3, Measured with Two Techniques: Tunable Microwave Source and Fourier-Transform Spectroscopy, Poster F4, *15th Colloquium on High Resolution Molecular Spectroscopy*, Glasgow, UK.

REVIEW ON THE EXISTING SPECTROSCOPIC DATABASES FOR ATMOSPHERIC APPLICATIONS

Agnès PERRIN
Laboratoire de Photophysique Moléculaire, CNRS,
Université Paris-Sud, bât 350, 91405 Orsay cedex, France
email: Agnes.Perrin@ppm.u-psud.fr

1. Introduction

The ever increasing need for improvements in the accuracy of remote sensing measurements, in particular for the Earth's atmosphere, has led to numerous recent efforts to obtain improved spectroscopic parameters for the molecules of atmospheric or planetological interest. These efforts have led to the generation of numerous molecular spectroscopic databases [1-5]. As input, the radiative transfert codes require in addition to an atmospheric profile, a spectroscopic dataset which contains for an increasing number of molecules either line by line parameters (line positions, intensities and line shape data) for discrete molecular transitions or cross -sections data when a line by line description is not available or not possible. The parameters from these databases together with the absorber amount and atmospheric pressure and temperature allow the absorption/emission due to discrete molecular transitions or to quasi continua to be computed at any frequency assuming a reliable line shape function.
This presentation will be devoted to an overview of the status of the atmospheric databases. The first goal of the present paper is to define the spectroscopic parameters which are presently available in the databases. A critical analysis of the accuracy of these parameters will be performed, the deficiencies will be pointed out and recommendations for possible improvements will be given.
The second goal of this review is to perform an assessment of the scientific choices which governed the generation of these databases and to present some (possible) new directions which are under discussion for new generation databases.

2. Description of the spectroscopic parameters

2.1 DISCRETE MOLECULAR TRANSITIONS

The spectroscopic parameters which are included in the atmospheric databases involve for the discrete molecular transitions, line positions, line intensities and line shapes data.

J. Demaison et al. (eds.), Spectroscopy from Space, 235–258.

2.1.1 Line positions

Depending on the databases, the position $\tilde{\nu}$ of a line is expressed either in cm^{-1} (for HITRAN, GEISA and SAO) or in MHz (JPL) with:

$$\tilde{\nu}(in\ cm^{-1}) = \tilde{\nu}(in\ MHz)/29979.2458 \tag{1}$$

2.1.2 Line intensities

According to the Beer-Lambert law, the transmission of a monochromatic radiation through a homogeneous sample is described by:

$$I = I_0 \exp-\left(\alpha(\tilde{\nu})l\right) \tag{2}$$

where $\alpha(\tilde{\nu})$ is the absorption coefficient and l is the optical path length. The integrated absorption coefficient $k_{\tilde{\nu}}$ (in cm^{-2}) of an individual line can be written in the form:

$$k_{\tilde{\nu}} = \int_{line} \alpha(\tilde{\nu}')d\tilde{\nu}' \quad \Leftrightarrow \quad \alpha(\tilde{\nu}) = k_{\tilde{\nu}}.\Phi(\tilde{\nu}) \tag{3}$$

where $\Phi(\tilde{\nu})$ is the normalized line profile:

$$\int_{line} \Phi(\tilde{\nu}')d\tilde{\nu}' = 1 \tag{4}$$

Because $k_{\tilde{\nu}}$ depends on the concentration of the absorbing molecules, it is useful to define the line intensity $k_{\tilde{\nu}}^{N}$ (in $cm^{-1}/(molecule.cm^{-2})$ which is commonly listed in most of the databases for a given temperature T:

$$k_{\tilde{\nu}}^{N} = k_{\tilde{\nu}} / N \tag{5}$$

where N is the density of absorbing molecules (in $molecule.cm^{-3}$). In the absence of external fields, $k_{\tilde{\nu}}^{N}$ may be calculated using the formula:

$$k_{\tilde{\nu}}^{N} = \frac{8\pi^3\tilde{\nu}}{3hc} \exp\left(-\frac{E_L}{kT}\right)\left(1-\exp\left(-\frac{\tilde{\nu}}{kT}\right)\right)\frac{g_L}{Z(T)} R_L^U \tag{6}$$

where L and U refer to the lower and upper levels of the transition, $\tilde{\nu} = \left(E_U - E_L\right)$ is the wavenumber of the transition (in cm^{-1}), g_L is the nuclear spin degeneracy of the lower level of the transition, *Z(T)* is the partition function at T, and k is the Boltzman constant. $R_L^U = \left|\left\langle \varphi_U \middle| \mu_Z \middle| \varphi_L \right\rangle\right|^2$ [in Debye^2, 1Debye≈3.33564x10^{-30}C.m] is the square of the matrix element of μ_Z between the upper and lower levels of the transition, and μ_Z is the component of the electric dipole moment on the space – fixed axis OZ for a homogeneous laboratory space. Therefore, the line intensity $k_{\tilde{\nu}}^N(T)$ at a different temperature T is obtained from the line intensity at T' from the relation:

$$k_{\tilde{\nu}}^N(T) = \frac{Z(T')}{Z(T)} \frac{exp\left(-\frac{E_L}{k.T}\right)\left(1 - exp\left(-\frac{\tilde{\nu}}{k.T}\right)\right)}{exp\left(-\frac{E_L}{k.T'}\right)\left(1 - exp\left(-\frac{\tilde{\nu}}{k.T'}\right)\right)} k_{\tilde{\nu}}^N(T') \tag{7}$$

where $Z(T)$ and $Z(T')$ are the partition functions at the temperatures T and T', which are tabulated for different temperatures in the databases [3,4]. The R_L^U quantity is related to the Einstein's spontaneous emission coefficient A_L^U [s^{-1}] by:

$$A_L^U = \frac{64\pi^4}{3h} \tilde{\nu}^3 \frac{g_L}{g_U} R_L^U \, 10^{-36} \tag{8}$$

Finally, as it will be discussed latter in the text, one should underline that the conventions adopted for the line intensities quoted in the linelists differ depending on the databases (HITRAN, GEISA, SAO versus JPL for example).

2.1.3 Line profiles

Different line profiles are to be used, depending on the experimental conditions and of the considered molecules, they are:

The Gauss profile: The Gauss profile can be used for lines for which the broadening is only due to the Doppler effect:

$$\Phi_D(\tilde{\nu}') = \frac{1}{\gamma_D}\left(\frac{\ln 2}{\pi}\right)^{1/2} \exp\left(-\ln 2\left(\frac{\tilde{\nu} - \tilde{\nu}'}{\gamma_D}\right)^2\right) \tag{9}$$

where $\tilde{\nu}$ is the line center (in cm^{-1}) and γ_D is the Doppler halfwidth at half maximum (HWHM, in cm^{-1}) which is given by:

$$\gamma_D = \sqrt{2\ln 2 \frac{kT}{mc^2}}\, \tilde{\nu} \approx 3.58x10^{-7} (T/M)^{1/2} \tilde{\nu} \tag{10}$$

where k is the Boltzman constant, *c* the speed of the light, *T* the temperature (in K) and *M* the molecular weight of the absorbing gas (in atomic mass unit). This means that this broadening is, in principle, easily computable and does not need to be quoted in the databases.

The Van-Vleck-Weisskopt and the Lorentz profiles: When the broadening results from the collisions, the conventional linewidth profile in the microwave region and in the infrared or visible regions are the Van-Vleck-Weisskopf and Lorentzian lineshapes respectively:

$$\Phi_{VW}(\tilde{\nu}')=\frac{\tilde{\nu}'}{\pi\tilde{\nu}}\left(\frac{\gamma_L}{(\tilde{\nu}'-\tilde{\nu})^2+\gamma_L^2}+\frac{\gamma_L}{(\tilde{\nu}'+\tilde{\nu})^2+\gamma_L^2}\right) \tag{11}$$

$$\Phi_L(\tilde{\nu}')=\frac{1}{\pi}\left(\frac{\gamma_L}{(\tilde{\nu}'-\tilde{\nu})^2+\gamma_L^2}\right) \tag{12}$$

In both expressions γ_L is the half width at half maximum.
For the normal range of pressures encountered in the terrestrial atmosphere, one can consider that the collision broadened halfwidth for a mixture of gases is linearly dependent on the gas partial pressures of the different perturbing gases:

$$\gamma_L=\sum_i \gamma_{L,i}^0 P_i \tag{13}$$

In this expression, $\gamma_{L,i}^0$ (expressed usually in cm^{-1}/atm) and P_i (in atm, 1 atm=1013hPa) are the broadening coefficients and the partial pressures respectively for the foreign gas i. For atmospheric purposes, only air broadening (79% of N_2 and 21% of O_2) are usually to be considered. One of the exception is water vapor: for this molecule, and in addition with other problems which will be discussed latter in the text, the self broadening is not negligible at low atmospheric altitudes and has to be considered simultaneously with the air broadening. This is because the self broadening coefficients are about 5 times larger than the air broadening coefficients.
Finally one should mention that the databases devoted to the outer planets (NIMS-Galileo for example) [6] include the hydrogen and helium broadening parameters instead of the air broadening.
The temperature dependence of $\gamma_{L,i}^0(T)$ is written as:

$$\gamma_{L,i}^0(T)=\gamma_{L,i}^0(T')\left(\frac{T'}{T}\right)^n \tag{14}$$

One has to point out that contrary to the $\gamma_D(T)$ Doppler halfwidth, both $\gamma^0_{L,i}(T)$ and its n temperature dependence are uneasy to obtain and depend actually not only on the molecule but also on the rotational quantum numbers. This means that these parameters which vary from one line to the other one must be determined for each individual transition. The theoretical models used to calculate the $\gamma^0_{L,i}(T)$ Lorentzian half widths and its n temperature dependence are (depending on the molecule) not always accurate enough to provide reliable values for these parameters. Therefore, $\gamma^0_{L,i}(T)$ and n values have to be by achieved through long and delicate laboratory measurements. As an example, depending on the molecule, the n temperature exponent often differs from the value 0.5 which is predicted by the "hard spheres" classical theory.

The Voigt profile: The Voigt profile is a convolution (noted by $\otimes$) of independent Doppler and Lorentz (or Van Vleck-Weisskopf) profiles and is given by:

$$\Phi_V(\tilde{\nu}') = \Phi_D(\tilde{\nu}') \otimes \Phi_L(\tilde{\nu}') \tag{15}$$

This profile is to be used when one of the broadening (Doppler or Lorentz) is not negligible as compared to the other one. It is easily computable provided the $\gamma^0_{L,i}(T)$ broadening parameters are known.

Other profiles: The three forms of line shape discussed previously in this section (Doppler, Van Vleck-Weisskopf or Lorentz and Voigt) are the ones commonly encountered in laboratory or atmospheric spectra. However, for some of the atmospheric species which are discussed here the line shapes may differ from the usual three forms already discussed and it is necessary to use more complicated descriptions. As an example:

- For water vapor, the Voigt line shape is restricted to a narrow interval around the line center and the line shape in the far wings may differ substantially from this profile [7].
- The profiles of lines of molecules with a strong permanent dipole moment (like HCl) may be affected by the so-called Dicke collisional narrowing [8]. In this case, the lines exhibit a somehow modified Voigt profile, with a weaker Doppler linewidth.
- For rather dense spectra (carbon dioxide, methane, ...), interferences between overlapping lines may occur [9].

2.1.4 Line shifts

The pressure shift δ (in cm^{-1}/atm at 296K) is defined by:

$$\tilde{\nu}(P) = \tilde{\nu}(P_{Ref} = 0) + \delta * P \tag{16}$$

where P is the air pressure. For molecules with rather large permanent dipole moments (like HF, HCl or H_2O), the pressure line shifts may be important and therefore must be considered. These pressure shifts can change strongly (in value and in sign) with the rotational quantum numbers and with the considered band.

2.2 CROSS SECTION DATA

For numerous chemical species of atmospheric interest a line by line compilation is not available in the standard atmospheric databases. Actually this line by line description of the spectrum is not generally possible for heavy molecules (because of the strong density of lines) or becomes impossible for most of the species when going up to the visible and ultra violet spectral ranges.
Therefore only the spectral absorption cross-sections measured at atmospheric conditions can be quoted in the databases.

These absorption cross sections $\sigma_{\tilde{\nu}}$ in ($cm^2 molecule^{-1}$) are defined as:

$$\sigma_{\tilde{\nu}} = {-\ln \tau_{\tilde{\nu}}} \Big/ {N\, l} \tag{17}$$

where N is the molecular concentration (in molecule/cm^3), l (cm) is the optical path length and $\tau_{\tilde{\nu}}$ the measured spectral transmittance at the wavenumber $\tilde{\nu}$ and temperature T.

3. Available spectroscopic Catalogs. Contents of the databases

Four databases namely JPL, HITRAN, SAO and GEISA, will be described in the present paper. HITRAN, SAO and GEISA were generated mainly for the atmospheric applications. The JPL database is also devoted to the astrophysical studies. These bases include a list of discrete (line by line) molecular transitions. These linelists were designed mainly for the atmospheric applications and therefore were generated for the conditions of temperature and pressure which are encountered normally in the stratosphere or in the troposphere. More explicitly the lines present in these databases fulfill a given intensity cut off condition at the reference temperature of the database (T_{Ref}=296K in HITRAN).
Therefore, these linelists cannot be used for very high temperature conditions, since lines from a hot band, or involving high rotational quantum numbers, which are significantly strong at high temperature, may be absent from the database or poorly defined because extrapolated far away from the "reference" temperature conditions.
The HITRAN and GEISA databases include also infrared and UV/VIS cross sections. and various indices of refraction: these indices are used for the analysis of the aerosols contribution to the atmospheric opacity.

3.1 JPL

The submillimeter, millimeter, and microwave spectral line catalog published by the Jet Propulsion Laboratory (JPL) covers the frequency range from 0 to 340 cm^{-1} (0-10THz).

This catalog includes a total number of 1,845,866 lines and supplies informations on 331 atomic and molecular species [4]. The database is designed for astronomers and atmospheric scientists and contains therefore a lot of species not relevant to atmospheric observation. The JPL catalog is available on line via anonymous FTP at spec.jpl.nasa.gov and on the World Wide Web at http://spec.jpl.nasa.gov. Each molecule is treated in a separate file (labeled c_TAG.cat). There are also files (labeled d_TAG.cat) with informations about the source of the molecular data, and the values of the partition function. The JPL catalog is constantly updated and extended.
The JPL catalog includes only line positions and line intensities parameters. This means that as a matter of policy, *the line broadening and line shift parameters are completely absent from the JPL catalog.*
The catalog data tape is composed of 80-character card images, with one card image per spectral line. The format of each card image is:

TABLE 1. Format of the JPL catalog

FREQ	ERR	LGINT	DR	ELO	GUP	TAG	QNFMT	QN'	QN"
F13.4	F8.4	F8.4	I2	F10.4	I3	I7	I4	6I2	6I2

FREQ: Frequency of the line in MHz.
ERR: Estimated or experimental error of FREQ in MHz.
LGINT: Base 10 logarithm of the integrated intensity at 300 K.
DR: Degrees of freedom in the rotational partition function (0 for atoms, 2 for linear molecules, and 3 for nonlinear molecules).
ELO: Energy in cm^{-1} relative to the lower state of the transition.
GUP: Upper state degeneracy.
TAG: Species tag or molecular identifier.
QNFMT: Identifies the format of the quantum numbers given in the field QN.
QN': Quantum numbers for the upper state coded according to QNFMT.
QN": Quantum numbers for the lower state.

- Identification of the species (see the directory catdir.cat): The atoms/molecules are identified by a TAG number: This tag is a six-digit number in which the first three digits represent the mass number of the molecule or atom and the last three digits are an ordering number for the given mass. Usually there is a separate tag for each vibration-electronic state of a particular molecule.
- The line positions and intensities are computed at JPL using the parameters available in the literature. However, when this data is existing, the experimental value of the line position is quoted in the database (and indicated by a negative flag in the TAG number).
- From LGINT, the intensity $k_{\bar{\nu}}^{N}(JPL@300K)$ (in cm^{-1}/molecule.cm^{-2})) can be calculated as:

$$k_{\bar{\nu}}^{N}(JPL@300K) = 10^{LGINT}/cc \tag{18}$$

with cc=2.99792458x10[18]. It is important to underline that this intensity is quoted *for a pure isotopic sample and a temperature of T_{Ref}=300K.*
The error on the quoted intensities is estimated to be less than 1% except for few exceptions corresponding to unanticipated resonances or poorly defined dipole moments. This accuracy on line positions and line intensities is also documented explicitly by checking the source of the line positions and intensities parameters as stated in the JPL catalog description. Also the partition functions are given.
The JPL catalog has more lines included within the 0 to 340 cm^{-1} (0-10THz) frequency range than the other catalogs and the center frequency is given with a higher accuracy. It is easy to access and constantly updated. Major disadvantage is the absence of the air- and self- pressure-broadening coefficients and of pressure shift parameters.

3.2 HITRAN

The high-resolution transmission molecular absorption database of the Air Force Geophysics Laboratory (HITRAN) is available on CD-ROM, the latest edition is that of 1996 [3], and the contact is Dr. Laurence Rothman (email: lrothman@cfa.harvard.edu) It includes 37 species with 91 isotopes [3]. The HITRAN catalog is available as a CD ROM and is updated every few years. Corrections can be obtained from internet at http://www.HITRAN.com. The HITRAN CD-ROM includes a line by line compilation, two sets of cross sections parameters (for the infrared and UV spectral ranges respectively), one set of line by line parameters in the UV (the O_2 Schumann-Runge system), some indices of refraction, and various documentations and softwares.

3.2.1 The line by line catalog:

The line by line catalog involves almost 1,000,000 atmospheric lines for 35 molecules in the wavenumber region from 0 to 23,000 cm^{-1} (0-700THz) . In addition, separate files give parameters for NO^+, HOBr, the oxygen lines in the Schumann- Runge system, and Non Local Thermodynamic Equilibrium (NLTE) ozone lines which are stored in separate directories. The present version of the HITRAN catalog uses the following 100 character format for the line information. However, it is important to underline that this format will be modified during the next year.

TABLE II: Format of the HITRAN line by line catalog

MO, ISO	$\tilde{\nu}$	S	R	AGAM SGAM	E_L	n	δ	V1 V2	Q1 Q2	IERF IERS IERH	IREFF IREFS IREFH
I2 , I1	F12.6	E10.3	E10.3	2F5.4	F10.4	F4.2	F8.6	2I3	2A9	3I1	3I2

Each item is defined as below:

- *MO, ISO*: Identification of the species. The molecule is identified by a specific number MO which is given in Ref.[3]. ISO is the isotopic number (1=most abundant, 2 = second, etc…) according to the natural terrestrial isotopic abundance a_{iso}. Both MO and a_{iso} are quoted in the attached MOLPARAM.TXT document file.
- $\tilde{\nu}$: The line position ($\tilde{\nu}$) is given in cm^{-1}

- *S:* In HITRAN, *the line intensity S_{HITRAN}* (in $cm^{-1}/(molecule/cm^{-2})$) *is given for a reference temperature T_{Ref}=296K and is weighted according to the natural terrestrial isotopic abundance a_{iso}* i.e:

$$S_{HITRAN} = k_{\sigma}^{N}(296K).a_{iso} \tag{19}$$

- *R :* the transition probability squared (in $Debyes^2$). R is appearing as R_L^U in the right hand side of Eq.(6). This quantity will be replaced in the next coming version of HITRAN by the Einstein coefficient (see Eq. (8)) which is more useful for calculations performed for Non Local Thermodynamic Equilibrium (NLTE) purposes.
- *AGAM, SGAM:* The air -broadening AGAM= $\gamma_{air}(T_{Ref}=296K)$, the self broadening SGAM= $\gamma_{self}(T_{Ref}=296K)$ halfwidths (HWHM) are quoted for a T_{Ref}=296K reference temperature in (cm^{-1}/atm) unit.
- E_L : lower state energy (in cm^{-1}).
- *n:* n-temperature dependence (see Eq.(14)) of the air- broadening parameter.
- *δ:* air-pressure shift $\delta=\delta_{air}(T_{Ref}=296K)$ parameters quoted for a T_{Ref}=296K reference temperature in (cm^{-1}/atm) unit.
- *V1, V2*: upper and lower state global quanta indices
- *Q1, Q2:* upper and lower state local quanta indices
- *Accuracy and references:* The accuracy of the parameters is, in principle, stated by the IERF, IERS and IERH indexes for the frequencies, intensities and pressure halfwidths accuracy respectively.
- *IREFF, IREFS and IREFH* give the references for the sources of the data.

3.2.2 Cross sections

The infrared cross sections directory involves numerous heavy molecules (mainly chlorofluorocarbons (CFC's) hydrochlorofluorocarbons (HCFC's), SF_6 and heavy oxides of nitrogen (like N_2O_5)). For the UV/VIS region, only nitrous oxide and sulfur dioxide are presently considered.
The file structure includes, in front of the data list, a header which gives the identification of the molecule, the frequency range, the number of points, the temperature, the pressure, references and additional informations.

3.2.3 Indices of refraction

In addition to the line by line parameters and to the cross section data, the HITRAN database contains some indices of refraction (real and imaginary parts) for various aerosols particles. These aerosols particles which are present in clouds (water liquid and solid), in stratospheric clouds (nitric acid trihydrate etc...) or in volcanic dusts (aqueous sulfuric acid) contribute to the opacity of the atmosphere.

3.3 GEISA

The Gestion et Etude des Informations Spectroscopiques Atmosphériques (GEISA) database of the Laboratoire de Météorologie Dynamique is available on request [5].

The person to contact is Dr. Nicole Jacquinet-Husson (email: husson@ara01.polytechnique.fr). Updates are made every 1 to 2 years. This database contains a line by line catalog involving positions, intensities, pressure broadening and pressure shift. GEISA includes 42 molecules with 96 isotopic species between 0 and 23,000 cm^{-1} (0-700THz) with more than 1,346,266 lines. In addition, GEISA includes two sets of cross sections data (infrared and UV/VIS). and indices of refraction for the aerosols. An extensive description of the GEISA catalog (content of the database, format of the data, software etc..) is given in Ref.[5]. This database was designed not only for the molecules of the Earth's atmosphere, but also for planetary atmospheres. Consequently, GEISA differs from HITRAN mainly by the inclusion of line parameters involving some molecules (like C_3H_8, GeH_4, etc...) relevant to Giant Planets. On the other hand, except for some updates performed for H_2O, CO_2, O_3, N_2O, CO, CH_4, NO, PH_3, HNO_3, OH, OCS, C_2H_6, CH_3D, HCN, N_2, H_2O_2, and H_2S most of the data relevant to the Earth's atmosphere present in GEISA are the same as in the HITRAN catalog.

3.4 SAO

The molecular line by line database of the Smithsonian Astrophysical Observatory (SAO) is available upon request to Dr. Kelly Chance (kelly@cfa.harward.edu). A description of this database is given at http://pluto.harvard.edu/firs/sao92.html. It is constantly updated. The files are split up by molecules and wavenumber range, which makes access easy. The covered wavenumber region is 10-800cm^{-1} (300GHz -24THz) with 39 molecules considered [2]. Except for its restricted frequency range, this database is very similar in its form to HITRAN as it does provide the same type of line parameters (involving positions, intensities, pressure broadening and pressure shift). The main goal of this database is to provide refined parameters for some "*target molecules*" which may be retrieved at given frequencies by the SAO instruments: the quoted data combines the "best parameters" from HITRAN catalog, JPL catalog, and other measurements and calculations available in the literature.

The data are quoted in the same format than in HITRAN (see Table II). Also, SAO uses the error and reference code of the HITRAN catalog. The molecule identification follows mostly the HITRAN pattern, but there are some molecules that are treated within one number in HITRAN, which are split up in SAO. For example the hot bands of O_3 have a separate number, to account for possible effects from non local thermodynamic equilibrium [2].

3.5 OTHER DATABASES

3.5.1 ATMOS

The line list catalog of the Atmospheric Trace MOlecule Spectroscopy instrument (ATMOS) was generated focusing on the particular requirements of the ATMOS instrument which was an FTS interferometer observing in the occultation mode and performing measurements in the infrared spectrum (580 - 4500 cm^{-1}). The ATMOS catalog involves 46 minor and trace species documented in the 1 to 10,000 cm^{-1}

wavenumber region [1]. However this catalog will not be discussed in this paper because it was not updated recently.

3.5.2 HITEMP

It was already underlined that the HITRAN , SAO, GEISA and JPL databases are generated mainly for atmospheric applications. For high temperature conditions it may be preferable to use specific databases. The HITEMP database (contact Dr. Laurence Rothman, email: lrothman@cfa.harvard.edu) involves line by line datasets for water, carbon dioxide, carbon monoxide with an extensive contribution of lines belonging to hot bands and/or involving high rotational quantum numbers. Therefore, HITEMP is designated for simulations and analysis of biomass fires, plumes from industrial chimneys, combustions etc...) .

3.5.3 PARTRIDGE/SCHWENKE

This database (positions and intensities) was generated recently for water vapor for very high temperature conditions (access Dr. D.W.Schwenke schwenke@pegasus.arc.nasa.gov). The computation was performed by scaling extensive ab initio calculations and experimental data [10]. The accuracy of the line positions and intensities (between $\approx 0.1 cm^{-1}$ and a few cm^{-1} for the line positions), which depends on the spectral region, is significantly weaker than in "classical databases" like HITRAN or GEISA. On the other hand, this very extensive water linelist can be used fruitfully in rather extreme temperature conditions where the "classical databases" are anyway of little use. For example this linelist is presently used to identify water lines in the sunspots spectra [11].

3.5.4 TDS/STDS

The TDS/STDS (Spherical Top Data System) was developed for the simulation (line positions and intensities) of the spectra of spherical top molecules (CH_4, SiH_4 , GeH_4 , SnH_4 and CF_4 with their related isotopomers). An extensive description is given in Ref. [12]. STDS is freely accessible by anonymous ftp at jupiter.u-bourgogne.fr. For specific problems involving spherical top molecules, contact Dr. Jean-Paul Champion (Jean-Paul.Champion@u-bourgogne.fr).

3.5.5 NIMS/GALILEO

It is important to mention the existence of planetologic databases. For example, the NIMS/Galileo database was generated for the interpretation of the data collected by the Near Infrared Mapping Spectrometer (NIMS) experiment on board the Galileo Orbiter spacecraft [6], and will be extended for the analysis of the Visual and Infrared Mapping Spectrometer (VIMS) on board the Cassini spacecraft. These parameters were generated for the analysis of the atmosphere of Giant Planets (Jupiter, Venus, Saturn for example), or of their satellites (Titan...). As compared to the atmospheric databases (HITRAN or GEISA) this catalog involves specific paramaters for molecules selected because of their planetologic implications. More explicitly the NIMS/Galileo database includes line positions, line intensities, and H_2/He pressure broadening parameters (instead of air-pressure line widths in HITRAN) for methane, water, germane, hydrogen sulfur, ammoniac, phosphine and arsine. For further details, please contact Ðr. Pierre Ðrossart (Pierre.Ðrossart@obspm.fr) or Ðr. Linda R.Brown (linda@caesar.jpl.nasa.gov).

3.6 CONCLUSION

Let us summarize now the main features of the various databases. The JPL database includes an extensive linelist of positions and intensities for numerous molecules in the microwave - submillimeter region. On the other hand, this database does not give any information concerning the line broadening parameters. The HITRAN database includes an extended linelist of individual line by line parameters for the positions, intensities and pressure line widths. Also cross sections parameters and indices of refraction are available for some molecules in the infrared and UV/VIS regions. The GEISA database was developed in close cooperation with HITRAN. The line by line parameters gathered in SAO are, for some target molecules, more accurate and more complete than those available in other databases, but this list is limited to the 10-800cm^{-1} frequency range. Finally, some other databases were described which may be useful for specific purposes

4. Status in the current atmospheric databases.

Many efforts are currently done to improve or to validate the existing spectroscopic parameters which are needed as input for atmospheric remote sensing investigations. As pointed out in the text, the spectroscopic databases are currently updated using the best existing data. However, these databases have still their limitations and flaws. These limitations may occur because:

- the data available in the literature are not yet included in the database, or were implemented incorrectly.
- the parameters existing in the literature are not existing at the required accuracy
- a restricted scientific policy was adopted for the generation of the database. For example, spectroscopic effects such as line mixing, continua etc.. which affect significantly the atmospheric spectra are not accounted for in the databases.

Some recent and extended review papers [3,4,5,13-17] pointed out some of these deficiencies in the JPL, HITRAN or GEISA databases and the reader is reported to these very dense documents for a detailed information. The next sections paragraph will present some examples. Some of the existing deficiencies in the databases will be described, and propositions for updates or corrections will be given.
Finally, it is under discussion to extend to new types of parameters the scientific policy which was adopted for the generation of these databases. These propositions will be presented and discussed.

4.1 LINE BY LINE PARAMETERS:

4.1.1 Examples of line positions or line intensities deficiencies:
Numerous problems were evidenced in the existing line by line databases.

For example, the line by line parameters need to be validated systematically for the stronger absorbers (water, carbon dioxide, methane, ozone,...) because absorption from these molecules is ubiquitous in the whole infrared and near infrared region.

For water vapor the line positions and intensities parameters quoted in the databases are not always known to the required accuracy in the infrared region and this comes from existing deficiencies in the literature.

The so-called "weak " water lines are often incorrect or absent. These lines correspond to transitions involving high rotational quantum numbers or large differences in these quantum numbers [18]. Therefore these water lines are difficult to predict or to identify by usual models because the HÔH bending mode is a large amplitude motion [19]. In addition the position and intensities measurements are difficult for these weak lines (need of a long path absorption cell) [20] which however, contribute significantly to the observed IR activity of water in the atmospheric spectra which involve very long optical pathlength in the atmosphere.

The absolute band intensities are not always known to the required accuracy, this is the case for ozone [13,21] and nitric acid [15]. For example for the 10 μm band of ozone, it has to be underlined that the accuracy of the absolute line intensities available in the databases is probably not better than 3-5% given the fact that such a discrepancy exists between various results in the literature [13,21].

Some infrared bands which are strong in atmospheric spectra are still totally absent in the databases even for molecules already well documented in the database. This is the case at 3551 cm^{-1} for the ν_1 band of HNO_3 whose signature was unambiguously observed (with about 30% of absorption) by the ATMOS/ATLAS3 experiment [15]. Figure 1 shows the contribution of the ν_1 band of HNO_3 in the ATMOS spectra at 3551 cm^{-1}. However, parameters for this band are still completely missing in the literature and therefore also in the databases.

4.1.2 Line shape parameters:

The air- broadening parameters quoted in the databases come either from measurements or from calculations. These parameters are often not accurate enough because:

- The measurement of precise line broadening parameters is a difficult task. As an example, significant differences between the existing experimental line broadening parameters were pointed out for water in a recent review paper [22].
- Even the most sophisticated theoretical models are not always accurate enough to predict or explain the observed broadening lineshapes [23].
- The n- temperature dependencies of the air-broadening parameters are uneasy to obtain, and therefore usually set in the databases to a "default" value (n=0.5 or n=0.7) which has no real physical meaning.
- Finally for some species the line shape differs from the "classical" Lorentz or Voigt profile and therefore cannot be modeled correctly using the parameters quoted in the databases: this is the case for water in the "far wings" for example [7].

To finish with the problem of line shape parameters, one should mention that the self-broadening parameters are often missing in the databases (i.e. set to a zero default value). However, with the exception of water vapor because of the strong value of its

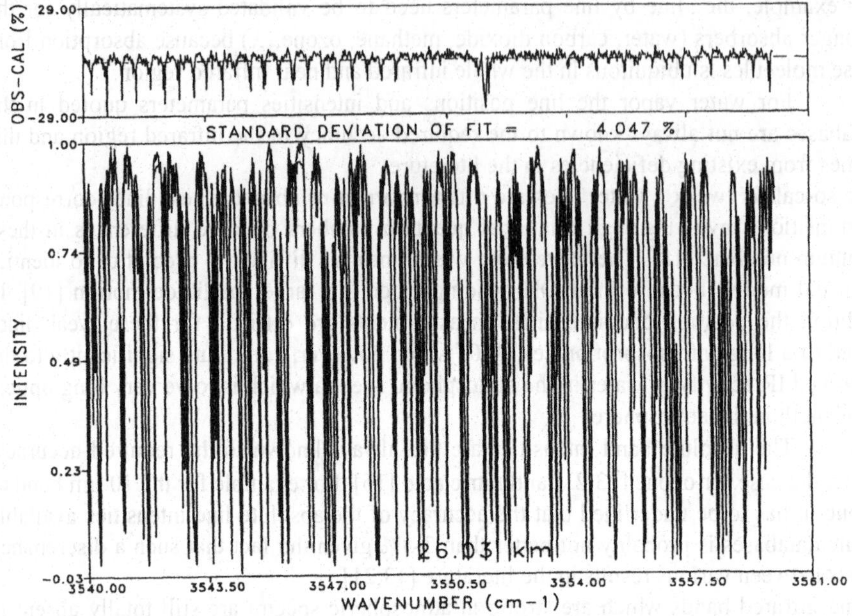

Figure 1: Spectral least squares fitting to an ATMOS/ATLAS 3 spectrum in the 3551 cm^{-1} region. The ATMOS spectrum was recorded with a tangent height of 26.03km. The spectral lines in this region are mostly due to CO_2. The line parameters database does not include yet the ν_1 band of HNO_3 lines which are evidenced in the difference spectrum. [from Goldman,A., Rinsland, C.-P., Perrin, A., and Flaud, J.-M., (1998) *J. Quant. Spectrosc. Radiat. Transf.* **60**, 851-861.]

self-broadening parameter (or of CO_2 because of its high concentration in the Venus spectra), the self-broadening effects are usually weak in atmospheric spectra because of the low concentration of the involved species.

4.1.3 Accuracy and references:

The accuracy of the parameters are "*in principle*" given by the IERF, IERS and IERH indexes for the frequency, intensity and linewidths respectively. Also, IREFF, IREFS IREFH give, in principle, the references for the sources of the data. In fact these indexes were not always updated in the current version of HITRAN and set to an undefined error IERF = 0 in the catalog version of 1996. This also holds for the SAO and GEISA lines which have been taken from HITRAN. Therefore, it is recommended that this accuracy be estimated by checking carefully the references which are given in the review papers describing the HITRAN database [3,13-17].

4.2 CROSS-SECTIONS

For heavy molecules in the infrared region, or for most of the species in the visible and ultra violet spectral ranges, a line by line compilation is not possible or not available in the standard atmospheric databases, and it is necessary to use cross sections parameters.

In the infrared a number of species (CFCs, HCFCs, SF_6, $ClONO_2$, N_2O_5, HO_2NO_2,...) exhibit a pseudo continuum. Indeed, because of the large moments of inertia of these heavy species, the number of transitions of the cold bands is about 10 times larger than for lighter molecules (water, ozone, nitrogen dioxide etc...). Furthermore, the first vibrational states are located at low energy, and hence the spectra are complicated by the existence of numerous hot bands even for the rather low temperature of the stratosphere. Finally, for some of them, the presence of various Cl isotopes should be considered. Therefore their infrared signatures must be characterized carefully using cross section parameters since a line by line description is not possible.

The ultraviolet /visible (UV/VIS) spectral region corresponds for numerous molecular species to various electronic transitions with complicated interactions within the electronic states. Because of a lack of understanding of these very dense spectra a line by line compilation is not possible either and experimental cross sections are to be used.

Because the involved individual transitions are unidentified, it is usually not possible to predict the dependances of the absorption cross sections. Therefore, these cross-sections can only be obtained through meticulous and extensive laboratory measurements performed for a large set of well-characterized pressure and temperature conditions corresponding to different stratospheric conditions. In particular, these measurements have to be performed at a high resolution (at least at a resolution about 10 times better than that used in the applications). The ideal situation being when the resulting spectral resolution of the reference data should be limited only by the molecular absorption structure and not by the instrument line shape with a very careful wavelength calibration and highly sampled datasets. In addition for each molecule a reasonable interpolation algorithm should be provided for the description of the situations corresponding to intermediate temperature and pressure values

A critical review on the existing cross section parameters for CFC's, CCl_4, N_2O_5 and $ClONO_2$ in the HITRAN or GEISA databases was performed in Refs. [3,16,24-27]. For these data one may underline the following points.

- For some species like CFC-114 ($C_2Cl_2F_4$), the cross sections are only given for different temperatures and not for different sets of (pressure - temperature) pairs which allow to perform more reliable extrapolations for different atmospheric conditions. Furthermore, the influence of the air broadening which changes significantly for example the fine structure CFC-12 absorption coefficient is not modeled.
- Concerning $ClONO_2$ tentative line by line calculations were performed recently [27]. These linelists are however still unable to account for the observed spectra in the stratospheric conditions. The discrepancies occur partly because of the difficulties to account for the hot bands contribution even at the low temperatures of the stratosphere.

In the 1998 version of HITRAN only two molecules, nitrous oxide (N_2O) and sulfur dioxide (SO_2), were considered for the (UV/VIS) cross section directory, and this is far from being sufficient. In fact the (UV/VIS) spectral region is widely used for the remote sensing of numerous molecules like O_3, O_2, NO_2, H_2O, $(O_2)_2$, SO_2, H_2CO, BrO, OClO, ClO and NO_2. For example two recent multi channel diode array spectrometers are operating (or will operate) in the UV-visible (230-790 nm) spectral range: the first one is the Global Ozone Monitoring Experiment (GOME) which is in orbit on board the

ERS-2 satellite since 1995 [28] and the second one is the Scanning Imaging Absorption Spectrometer for Atmospheric Cartography (SCIAMACHY) which is scheduled for launch onboard ENVISAT (ENVironmental SATellite) in 2001 [29]. The (UV-VIS) spectral range is also used by the Earth Probe TOMS (Total Ozone Mapping Spectrometer) on the ADEOS (Advanced Earth Observation Satellite) instrument [30]. Finally, the UV/VIS spectral region is commonly used by various balloon or satellite experiments for NO_2 measurements [31] and was in particular used for the tracking of SO_2 clouds after the Pinatubo eruption [30].
New cross sections UV/VIS parameters are now available in the literature ([32,33,34] and Refs. therein) for O_3, O_2, NO_2, H_2O, $(O_2)_2$, SO_2, H_2CO, BrO, OClO, ClO and NO_2. It is anticipated that these high-resolution cross-sections will be included in the upcoming edition of HITRAN2000 [34].

4.3 INDICES OF REFRACTION

In the presence of substantial aerosols loading, occultation spectra will have non-uniform background envelopes which deviate from 1.00. Emission experiments will have enhanced radiances, which can be misinterpreted as enhanced mixing ratio of gaseous species. A further complication is that aerosol particles both scatter and absorb radiation. Thus the equations which are needed to interpret observations in retrievals programs are complex. Therefore the aerosols opacity complicates the interpretation of remote sensing observations of gaseous species.
Because of the importance of taking into account the optical activity of aerosols, the real and complex indices for various aerosols (water liquid and ice, H_2SO_4/H_2O, nitric acid trihydrate, sulfate aerosols etc...) were tabulated in the 1996 version of HITRAN [3,35] and were also included very recently in the GEISA database.
Water liquid and ice aerosols play an important role for the description of the optical activity of clouds (for example for cirrus clouds in the tropical and sub tropical upper troposphere) [36]. Interest in the compounds formed by the nitric acid trihydrate and by $H_2SO_4/HNO_3/H_2O$ aerosol particles is motivated by their influence in altering the gas phase chemistry in the polar stratospheric clouds (PSCs) during the development of the Antarctic ozone hole during the winter [37]. Finally, sulfate aerosol particles also contribute to the infrared opacity of the atmosphere and reduce the visible light transmission through the atmosphere especially during the several years periods which follow major volcanic eruptions like the eruptions of El Chichon in 1982 and of the Mt. Pinatubo in 1991 [38].]. As an example, Figure 2 from Rinsland *et al.* [38] shows a portion of the ATMOS spectrum recorded in 1992 (then after the Pinatubo eruption): in the 750-2000 cm^{-1} spectral region, broadband structures due to sulfate aerosols are clearly evidenced.
According to the review of the existing parameters [35], ongoing laboratory studies will likely lead to indices of refraction of the cold sulfuric acid and nitric acid compounds in the near future. Also recent laboratory measurements show a strong temperature dependence for ice optical constants in the 8-12 μm infrared region [35]. Hence future aerosols investigations need to incorporate temperature dependent optical constants in their models of the radiative properties of aerosols [36]. Finally, it is also necessary to

improve the algorithms used to retrieve particle size distributions from atmospheric measurements [35,36,38].

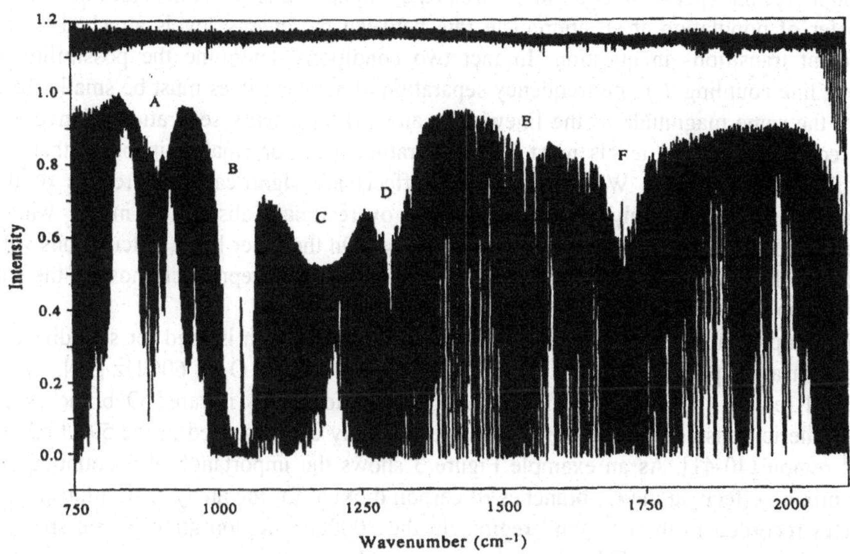

Figure 2: Portion of the solar spectra recorded on the 1 April 1992 by ATMOS (after the Pinatubo eruption). The tangent heights of the upper and lower spectra are 85.9 and 20.9 km respectively, and the spectra are normalized to the highest measured intensity. Six prominent atmospheric features are marked: (A) identifies a sulfate aerosol feature superposed on the $2\nu_9$ band of HNO_3; (B) the 1043 cm^{-1} O_3 ν_3 band; (C) a strong broad sulfate aerosol feature at 1190 cm^{-1}; (D) the ν_4 band of CH_4 AT 1307 cm^{-1}; (E) the pressure induced (1-0) vibration rotation band of O_2 at 1556 cm^{-1}; and (F) a broad aerosol feature at 1720 cm^{-1} superposed on HNO_3 and O_3. [from Rinsland, C.-P., Yue, G.K., Gunson, M.R., Zander, R., and Abrams, M.C. (1994) Mid infrared extinction by sulfate aerosols from Mt Pinatubo eruption, *J. Quant. Spectrosc. and Rad. Transf.* **52**, 241-252.).]

4.4 FUTURE PARAMETRISATION:

In some cases it appears that the type of data present in the usual databases cannot describe some characteristics of the atmospheric spectra. This is because (by definition) the databases are generated under a given well-defined and restricted scientific policy.

To account for these problems, propositions are in discussion for extending the scientific range of the databases [3]. To illustrate this point, examples will be presented which underline the importance for accounting for the line mixing effects (in carbon dioxide or methane Q- branches), for the water continuum or for the collision induced absorption by O_2 and N_2. Also the importance of generating linelists for molecules in Non Local Thermodynamic Equilibrium (nonLTE) situations in the stratosphere will be presented.

4.4.1 Line mixing:

The need to account in computations of atmospheric spectra for line mixing effects in some spectral regions is now well established [9,39-42]. This effect involves a global modification of the usual Lorentzian spectra profile in situations where (i) lines of a given molecular species overlap in the frequency domain, and (ii) collisions can effect a transfer of population at an appreciate rate between the energy levels attached to the different transitions in question. In fact two conditions determine the possibility of strong line coupling: (i) the frequency separation of adjacent lines must be smaller than or of the same magnitude as the linewidths, and (ii) the energy separation of involved consecutive rotational levels must be comparable to kT or smaller in order that the coupling be facilitated. When line mixing effects are significant, the results of the modeling accounting only for Voigt profiles overestimate absorption in the wings, whereas absorption is underestimated at the peaks. On the other hand, calculations with a reliable line mixing model lead to satisfactory results and reproduce most of the sub Lorentzian behavior of the wings.

For example, the millimeter region spectrum of oxygen which is used for soundings of the lower atmosphere is dominated by line mixing effects for O_2 at 60GHz [39]. In the infrared region, the influence of line-mixing in various CO_2 infrared Q branches on atmospheric transmission or emission has been widely demonstrated in the 597 to 2230 cm^{-1} region [40-41]. As an example Figure 3 shows the importance of accounting for line mixing effects in the Q- branches of carbon dioxide for the analysis of atmospheric spectra recorded in the 618 cm^{-1} region. In the 3000cm^{-1} region atmospheric spectral calculations using the HITRAN methane are inadequate for experiments such as the Halogen Occultation Experiment (HALOE) on the UARS satellite or the LPMA balloon borne FTS spectrometer [9] when using only Voigt parameters: on the contrary the inclusion of line mixing for the P, R and Q manifold of the ν_3 band of methane solves this problem. Also it is important to account for the line mixings in the Q branches of N_2O [42].

The inclusion of collisional line mixing has important implications for temperature soundings of the earth's atmosphere which are accomplished via the inversion of molecular infrared and millimeter spectra. Actually, CO_2 (and O_2) are good tracer species for atmospheric temperature profile retrievals obtained from satellite data because both molecules are uniformly mixed in the earth's troposphere and stratosphere. The new temperature sounders such as AIRS (Atmospheric InfraRed Sounders) and IASI (Infrared Atmospheric Sounding Interferometer) are based on improved resolution instruments. Therefore, their ultimate performance will strongly depend on the precision of the modeling of CO_2 absorption and accurate predictions of the spectra around the pressure and temperature sensitive Q lines are required.

Numerous experimental and theoretical studies were performed for an accurate description of the line mixing. The theoretical background in the so-called "impact approximation" requires, for the calculation of the absorption coefficient which accounts for the line mixings in a Q branch, the diagonalization of a rather complex and large W matrix. Fortunately, for rather low pressures of the broadeners, the calculations

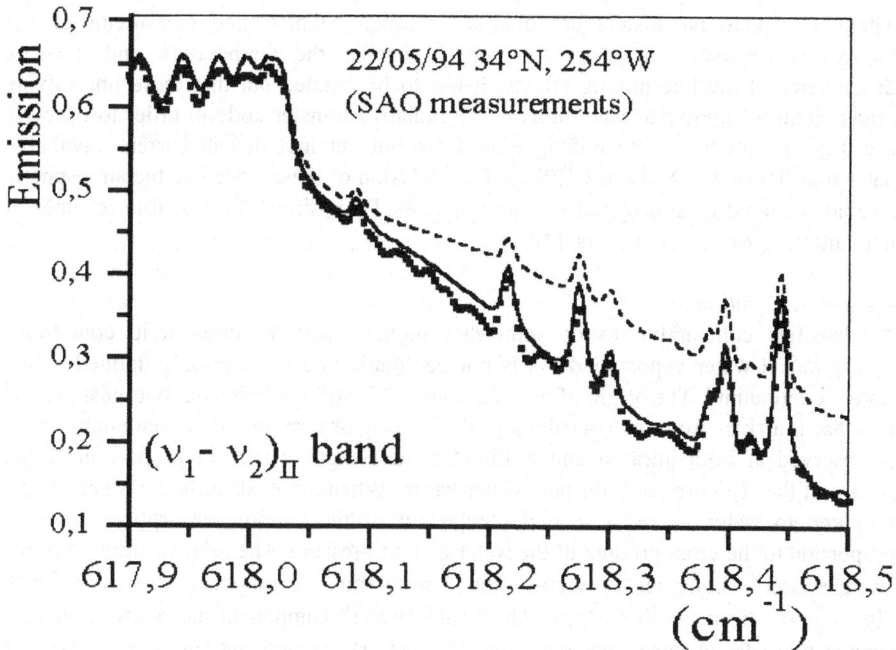

Figure 3: Measured atmospheric emission (SAO measurements) in the spectral ranges corresponding to the $(\nu_1\text{-}\nu_2)_{II}$ (at 618 cm^{-1}) band of CO_2. Observed spectra (●); calculations accounting for the line mixing (——); calculations neglecting the line mixing (.....). [from Jean-Michel Hartmann, *private communication* and from Rodrigues, R., Jucks, K.W., Lacome, N., Blanquet, Gh., Walrand, J., Traub, W.A.., Khalil, B., Le Doucen, R., Valentin, A.., Camy-Peyret, C., Bonamy, L. and Hartmann, J.-M. (1999)) Model , software and database for computation of line mixing effects in infrared Q branches of atmospheric CO_2 - I symmetric isotopomers, *J. Quant. Spectrosc. and Rad. Transf.* **61**, 153-184]

can often be performed with a reasonable accuracy in the "Rosenkranz" approximation [39]. In this condition the absorption is given by the addition of line contributions in which line mixing is taken into account through first order coupling Y_j mixing coefficients (in atm^{-1}). Therefore when Doppler broadening are negligible the line mixing contributions appear as extra terms in Y_j in the usual Lorentz line shape:

$$k_{LM}(\tilde{\nu}) = \frac{N}{\pi} \sum_j S_j \left(\frac{P\gamma_L + (\tilde{\nu} - \tilde{\nu}_j) P Y_j}{(\tilde{\nu} - \tilde{\nu}_j)^2 + (P\gamma_j)^2} \right) \tag{20}$$

where $\tilde{\nu}_j$ is the center wavenumber of line j, γ_j is the Lorentz line halfwidth ($cm^{-1}atm^{-1}$), S_j is the line strength, and N is the absorber number density.
When the Doppler broadening is not negligible, the actual absorption coefficient can be computed from the expression given in Eq.(20) by a convolution with the Doppler lineshape using standards methods.

These Y_j mixing parameters are obtained through sophisticated measurements and theoretical analyses in order to properly handle the temperature and pressure dependence of the line mixing effects. It has to be pointed out that these line mixing effects require significant alterations in the radiative transfer code in order to properly use these parameters. Line mixing effects are not yet included in current available databases (HITRAN, SAO or GEISA). The inclusion of these effects in the atmospheric databases should be considered in future updates. Propositions for possible schemes of implantation were given in Refs. [40,41].

4.4.2 Water continuum:

At densities comparable to or somewhat higher than in atmospheric conditions, absorption in water vapor windows is not negligible and is commonly defined as the water "continuum". The origin of this "continuum" is still controversial but most experts feel that it arrives from the contribution of far wing absorption. This continuum which is observed at both infrared and millimeter wavelength. varies in proportion to the square of the H_2O pressure for pure water vapor. When a non absorbing gas such as air is added to water vapor, a second component of the window absorption varies in proportion to the cross product of the water and air pressure. The relative strength of the "water*water" component versus the "water*air" component increases with displacement from the line center. The "water*water" component has a strong inverse temperature dependence, whereas the "water*air" component does not. Neither component can be correctly calculated by means of line shapes based on impact approximation.

A semi empirical approach can be used for the description of this "continuum" [43-44]. However, it is clear from these studies than an appropriate water continuum dataset should involve a precise and coherent definition of both the "water continuum" and of the so-called water "local lineshape". The "local lineshape" [7] which describes the absorption of water near each line center differs in Ref.[43] and in Ref.[44], leading therefore to a differing definition of the water continuum which is the water absorption coefficients remaining after removing this "local contribution".

4.4.3 Collision induced absorption parameters:

It is well know that tropospheric absorption spectra obtained over long paths in the atmospheric window are affected by continuum absorption by O_2 near 1600 cm^{-1} and near 7900 cm^{-1} [45-46] and by N_2 and CO_2 near 2400 cm^{-1} [47-48]. Therefore, numerous experimental studies [45,47] were performed in order to provide temperature dependent coefficients for semi- empirical modeling of atmospheric spectra. Actually, this modeling could rather easily be implemented in new versions of the databases.

4.4.4 Non Local Thermodynamic Equilibrium effects (nonLTE):

Under a Local Thermodynamical Equilibrium (LTE) the relative populations of the lower and upper states of a vibrational rotation transition are determined by the Botzmann equation evaluated at a given local kinetic temperature. With increasing altitude, the decrease of collisional energy exchange combined with higher radiative or chemical pumping lead to vibrational populations differing from the LTE conditions.

For some stratospheric species (li e carbon dioxide, ozone, water, methane and nitrogen dioxide) some nonLTE situations were noticed from the results of the Limb Infrared Monitor of the Stratosphere (LIMS) that flew on NIMBUS-7 [49]. Therefore, it was necessary to generate specific linelists for these molecules in order to evaluate the possible effects of nonLTE effects on the limb radiances which will be measured by the future Michelson Infrared for Passive Atmospheric Sounding (MIPAS) instrument which is scheduled to fly on ENVISAT-1 platform [50]. As compared to "classical databases" these linelists include numerous vibrational hot bands which are not to be considered for LTE conditions.

5 Conclusion

We presented an overview of the existing atmospheric databases. These databases were generated within a given scientific policy for atmospheric applications to describe the absorption or emission of various atmospheric species in the microwave, infrared or UV/VIS spectral regions. These catalogs are constantly updated and involve, for an increasing number of molecules either line by line parameters (line positions, intensities and line shape data) for discrete molecular transitions or cross sections data when a line by line description is not available or not possible. The quality of these parameters was evaluated and some recommendations were proposed.

From the analysis of the literature devoted to the status of the individual linelist, it seems that some of the main deficiencies come from absolute line intensities and from the linewidths parameters. Also, because these molecules contribute ubiquitously in the whole infrared spectral region, it is important to validate systematically the line by line parameters for the stronger absorbers (water, carbon dioxide, methane, and ozone). For the heavy molecules in the infrared region, or for most of the molecules in the UV/VIS spectral region, it is necessary to have very accurate cross sections parameters. These data should be obtained in well characterized laboratory conditions which reproduce as much as possible the atmospheric situations. Also, these databases should have well documented literature references for the source of the data (frequency, intensity, and halfwidth or cross sections) and include precise error codes. Finally, we present some possible extension of the existing scientific policy which are under discussion for the generation of future versions of the databases.

Acknowledgments:

The author is deeply indebted to Dr. Jean-Marie Flaud and Dr. Jean Demaison for numerous encouragements during the course of this wor and for a careful reading of the manuscript. Also fruitful discussions with Dr. Jean-Michel Hartmann and Dr. Johannes Orphal are gratefully ac nowledged. Finally, the author would li e to than Dr. Aaron Goldman, Dr. Curt Rinsland and Dr. Jean-Michel Hartmann for giving her the permission to use figures 1, 2 and 3.

6. References

1. Brown, L.R., Farmer, C.B., Rinsland, C.P. and Toth, R.A. (1987) Molecular line parameters for the atmospheric molecule trace spectroscopy experiment, *Appl. Opt.* **26**, 5154-5182.

2. Chance, K., Jucks, K.W., Johnson, D.G., and Traub, W.A. (1994) The Smithsonian astrophyical Observatory Database SAO92, *J. Quant. Spectrosc. Radiat. Transf.* **52,** 447-457.
3. Rothman, L.S., Rinsland,C.-P., Rinsland, C.-P., S.T.Massie, D.P.Edwards, Flaud, J.-M., Perrin, A., Camy-Peyret, C., Dana, V., Mandin, J.-Y., Schroeder, J., McCann, A., Gamache, R.R., Wattson, R.B., Yoshino, K., Chance, K.V., Jucks, K.W., Brown, L.R., Nemtchinov , V., and Varanasi, P. (1998) The HITRAN molecular spectroscopic database and HAWKS (HITRAN atmospheric workstation) 1996 edition, *J. Quant. Spectrosc. Radiat. Transf.* **60**, 665-710.
4. Pickett, H.M., R.I.Poynter, R.L., Cohen, E.A., Delitsky, M.L., Pearson, J.C., Muller, H.S.P. (1998) Submillimeter, millimeter and microwave spectral line catalog, *J. Quant. Spectrosc. Radiat. Transf.* **60,** 883-890.
5. Jacquinet-Husson, N., Arié, E., Ballard, J., Barbe, A., Bjoraker, G., Bonnet, B., Brown, L.R., Camy-Peyret, C., Champion, J.P., Chédin, A., Chursin, A., Clerbaux, C., Duxbury, G., Flaud, J.-M., Fourrié, N., Fayt, A., Graner, G., Gamache, R., Goldman, A., Golovko, VL., Guelachvili, G., Hartmann, J.-M., Hilico, J.C., Hillman, J., Lefèvre, J.G., Lellouch, E., Mikhaïlenko, S.N., Naumenko, O.V., Nemtchinov, V.D., Newnham, A., Nikitin, A., Orphal, J., Perrin, A., Reuter, D.C., Rinsland, C.-P., Rosenmann, L., Rothman, L.S., Scott, N.A., Selby, J., Sinitsa, L.N., Sirota, J.M., Smith, M.A., Smith, K.M., Tyuterev, Vl.G., Tipping, R.H., Urban, S., Varanasi, P., Weber, M. (1999) The 1997 GEISA databank, *J. Quant. Spectrosc. Radiat. Transf.* **62**, 205-254.
6. Carlson, R., Smythe, W., Baines, K., Barbinis, E., Becker, K, Burns, R., Calcutt, S., Calvin, W., Clark, R., Danielson, G., Davies, A., Drossart, P., Encrenaz, T., Fanale, F., Granahan, I.J., Hansen, G., Herrera, P., Hibbitts, C., Hui, J., Irwin, P., Johnson, T., Kamp, L., Kieffer, H., Leader, F., Lellouch, E., Lopes, Gautier, R., Matson, D., McCord, T., Mehlman, T.R., Ocampo, A., Orton, G., Roos, Serote, M., Segura, M., Shirley, J., Soderblom, L., Stevenson, A., Taylor, F., Torson, J., Weir, A., Weissman, P. (1996) Near-infrared spectroscopy and spectral mapping of Jupiter and the Galilean satellites: results from Galileo's initial orbit, *Science* **274**, 385-388.
7. Ma, Q. and Tipping, R.H. (1999) The averaged density matrix in the coordinate representation: Application to the calculation of the far wing lineshapes for H_2O, *J. Chem. Phys.* **111**, 5909-5921.
8. Pine, A.S., Looney, J.P. (1987) N_2 and air broadening in the fundamental bands of HF and HCl, *J. Mol. Spectrosc.* **122**, 41-55.
9. Pieroni, D., Hartmann, J.-M. , Camy-Peyret, C., Jeseck, P., and Payan, S. (2000) Influence of line mixing on absorption by CH_4 in atmospheric balloon borne spectra near 3.3 μm, *J. Quant. Spectrosc. and Rad. Transf.* **68,** 117-133.
10. Partridge, H., Schwenke, D.W. (1997) The determination of an accurate isotope dependent potential energy surface for water from extensive ab initio calculations and experimental data, *J. Chem. Phys.* **106**, 4618-4639.
11. Polyansky, O.L., Zobov, N.F., Viti, S., Tennyson, J., Bernath, P.F., Wallace, L. (1997) High, temperature rotational transitions of water in sunspot and laboratory spectra, *J. Mol. Spectrosc.* **186**, 422-447.
12. Wenger, Ch., Champion, J.P. (1998) Spherical Top Data System (STDS) software for the simulation of spherical top spectra, *J. Quant. Spectrosc. Radiat. Transf.* **59**, 471-480.
13. Rinsland, C.-P., Flaud, J.-M., Goldman, A., Perrin, A., Camy-Peyret, C., M.A.Smith, Malathy Devi, V., D.Ch.Benner, Barbe, A., Stephen, T.M., and Murcray, F.J. (1998) Spectroscopic parameters for ozone and its isotopes: current status, prospect for improvement, and the identification of $^{16}O^{16}O^{17}O$ and of $^{16}O^{17}O^{16}O$ lines in infrared ground -based and Stratospheric Solar absorption spectra, *J. Quant. Spectrosc. Radiat. Transf.* **60**, 803-814.
14. Perrin, A., Flaud, J.-M., Goldman, A., Camy-Peyret, C., Lafferty, W.J., Arcas, Ph., and Rinsland, C.-P. (1998) NO_2 and SO_2 line parameters: 1996 HITRAN update and new results, *J. Quant. Spectrosc. Radiat. Transf.* **60,** 839-850.
15. Goldman, A., Rinsland, C.-P., Perrin, A., and Flaud, J.-M. (1998) HNO_3 line parameters ; 1996 HITRAN update and new results *J. Quant. Spectrosc. Radiat. Transf.* **60**, 851-861.
16. Goldman, A., Rinsland, C.-P., Flaud, J.-M., and Orphal, J. (1998) $ClONO_2$: spectroscopic line parameters and cross sections in 1996 HITRAN, *J. Quant. Spectrosc. and Rad. Transf.* **60,** 875-882 .
17. Giver, L.P., Chackerian, Ch., Varanasi, P. (2000) Visible and near infrared $H_2{}^{16}O$ line intensity corrections for HITRAN-96, *J. Quant. Spectrosc. Radiat. Transf.* **66**, 101-105.
18. Flaud, J.-M., Camy-Peyret, C., and Toth, R.A. (1981) *Water Vapour Line Parameters from Microwave to Medium Infrared*, Pergamon press, Oxford.

19. Coudert, L.H. (1999) Line frequency and line intensity analyses of water vapor, *Mol Phys*. **96**, 941-954.
20. Toth, R.A. (1998) water vapor measurements between 590 and 2582 cm^{-1}: line positions and strenghts, *J. Mol Spectrosc*. **190**, 379-396.
21. Flaud, J.-M., Camy-Peyret, C., Rinsland, C.-P., Smith, M.A.H., and Malathy Devi, V. (1990) *Atlas of ozone spectral parameters from microwave to medium infrared*, Academic Press, San Diego.
22. Gamache, R.R., Hartmann, J.-M. and.Rosenmann, L. (1994) Collisional broadening of water vapor lines: a survey of experimental results *J. Quant. Spectrosc. Radiat. Transf.* **52**, 481-499.
23. Gamache, R.R., Lynch, R. and Neshyba, S. (1998) New development in the pressure broadening and pressure shift of spectral lines of H_2O ; the complex Robert Bonamy formalism, *J. Quant. Spectrosc. Radiat. Transf.* **59**, 319-336.
24. Massie, S.T. and Goldman, A. (1992) Absorption parameters of very dense molecular spectra for the HITRAN compilation, *J. Quant. Spectrosc. and Rad. Transf.* **48,** 713-719.
25. Goldman, A., Rinsland, C.-P., Murcray, F.J., Blatherwick, R.D., and Murcray, D.G. (1994) High resolution studies of heavy NO_y molecules in atmospheric spectra, *J. Quant. Spectrosc. and Rad. Transf.* **52**, 367-377.
26. Varanasi, P., Li, Z., Nemtchinov, V., and Cherukuri, A. (1994) Spectral absorption coefficient data of HCFC-22 and SF_6 for remote sensing applications, *J. Quant. Spectrosc. and Rad. Transf.* **52**, 323-332.
27. Domenech , J.L., Flaud , J-M., Fraser ,G.T., Andrews, A.M., Lafferty ,W.J., and Watson, P.L. (1997) Infrared diode-laser molecular-beam spectrum of the ν_2 of chlorine nitrate at 1293 cm^{-1}, *J. Mol. Spectrosc.* **183**, 228-233.
28. Burrows, J.P., Weber, M., Buchwitz, M., Rozanov, V., Ladstatter-Weissenmayer, A., Richter, A., DeBeek, R., Hoogen, R., Bramstedt, K., Eichmann, K.-U., Eisinger, M., and Perner, D. (1999) The Global Ozone Monitoring Experiment (GOME) : Mission concept and first scientific results, *J. Atm. Sci.* **56**, 151-175.
29. Bovensmann, H., Burrows, J.P.., Buchwitz, M., Frerick, J., Noël, S., Rozanov, V.V., Chance, K.V. and Goede, A.P.H. (1999) SCIAMACHY - Mission objectives and measurements modes, *J. Atm. Sci.* **56**, 127-150.
30. Krueger, A.J., Walter, L.S., Bhartia, P.K., Schneltzler, C.C., Krotkov, N.A., Sprod, I., Bluth, G.J.S. (1995) Volcanic sulfur dioxide measurements from the total ozone mapping spectrometer (TOMS) instruments, *J. Geophys. Res.* **D100**, 14057-14076
31. Hofmann, D., Bonasoni, P., De Maziere, M., Evangelisti, F., Giovanelli, G., Goldman, A., Goutail, F., Harter, J.W., Jakoubek, R., Johnston, P., Kerr, J., Kerr, W.,Matthews, W.A., McElroy, T., McKenzie, R., Mount, G., Platt, , H., Pommereau, J.P., Sarkissian, A., Simon, P., Solomon, S., Stutz, J., Thomas, A., Van Roozendael, M., and Wu, E. (1995) Intercomparison of UV/visible spectrometers for measurements of stratospheric NO_2 for the Network for the Detection of Stratospheric Change, *J. Geophys. Res.* **D100**, 16765-16791.
32. Vandaele, A.C., Hermans, C., Simon, P.C., Carleer, M., Colin, R., Fally, S., Merienne, M.F., Jenouvrier, A., and Coquart, B. (1998), Measurements of the NO_2 absorption cross-section from 42 000 cm^{-1} to 10 000 cm $^{-1}$ (238-1000 nm) at 220 K and 294 K, *J. Quant. Spectrosc. and Rad. Transf.* **59**, 171-184.
33. Burrows, J.P.., Rischter, A., Dehn, A., Deters, B., Himmelmann, S., Voigt S., and Orphal, J. (1999) Atmospheric remote sensing reference data from GOME-2: Part 2. Temperature dependent absorption cross sections of O_3 in the 231-794 nm range, *J. Quant. Spectrosc. and Rad. Transf.* **61**, 509-517.
34. Chance K., Krosu T.P., Yoshimo K., Parkinson W., Rothman L. , Goldman A., and Orphal J. (1999) UV Spectral data for HITRAN2000, *Atmospheric Spectrosc. Appl. ASA-1999, Prodeedings* 237-242.
35. Massie, S.T. (1994) Indice of refraction for the HITRAN compilation, *J. Quant. Spectrosc. and Rad. Transf.* **52,** 501-513.
36. Rinsland, C.-P., Gunson, M.R., Wang, P.H., Arduini, R.F., Baum, B.A., Minnis, P., Goldman, A., Abrams, M.C., Zander, R., Mahieu, E., Slawitch, R.J., Michelsen, H.A., Irion, F.W. and Newchurch, M.J. (1998) ATMOS/ATLAS 3 infrared profile measurements of clouds in the troposphere and subtroposphere, *J. Quant. Spectrosc. and Rad. Transf.* **60**, 903-919.
37. WMO (World Meteorological Organisation), Global Ozone Research and Monitoring Project / United Nations Environment Programme (WMO/UNEP) (1999) *Scientific Assessment of Ozone Depletion, 1998*, Rep n° **44** , Geneva.

38. Rinsland, C.-P., Yue, G.K., Gunson, M.R., Zander, R., and Abrams, M.C. (1994) Mid infrared extinction by sulfate aerosols from Mt Pinatubo eruption, *J. Quant. Spectrosc. and Rad. Transf.* **52**, 241-252.
39. Liebe, H.J., Rosenkranz, P.W., and Hufford, G.A. (1992) Atmospheric 60-GHz oxygen spectrum: new laboratory measurements and line parameters, *J. Quant. Spectrosc. and Rad. Transf.* **48**, 629-643.
40. Larrabee Strow, L., Tobin, D.C., and Hannon, S.E. (1994) A compilation of first order line mixing coefficients for CO_2 Q branches, *J. Quant. Spectrosc. Radiat. Transf.* **52**, 281-294.
41. Rodrigues, R., Jucks, K.W., Lacome, N., Blanquet, Gh., Walrand, J., Traub, W.A., Khalil, B., Le Doucen, R., Valentin, A., Camy_Peyret, C., Bonamy, L., and Hartmann, J.-M. (1999) Model , software and database for computation of line mixing effects in infrared Q branches of atmospheric CO_2 - I symmetric isotopomers, *J. Quant. Spectrosc. Radiat. Transf.* **61**, 153-184.
42. Hartmann, J.-M. , Bouanich, J.P., Jucks, K.W., Blanquet, Gh., Walrand, J., Bermejo, D., Domenech, J.-L., and Lacome, N. (1999) Line mixing effects in N_2O Q branches: model, laboratory and atmospheric spectra, *J. Chem. Phys.* **110**, 1959-1969.
43. Clough, S.A., Kneizys, F.X., and Davies, R.W. (1989) Line shape and the water vapor continuum , *Atmospheric Research*, **23**, 229-241.
44. Larrabee Strow, L , Robin, D.C., McMillan, W.W., Hannon, S.E., Smith, W.L., Revercomb H.E., and Knuteson, R.O. (1998) Impact of a new water vapor continuum and line shape model on observed high resolution infrared radiance, *J. Quant. Spectrosc. Radiat.Transf.* **59**, 307-317.
45. Thibault, F., Menoux, V., Le Doucen, R., Rosenmann, L., Hartmann, J.-H., and Boulet, Ch. (1997) Infrared collision induced absorption by O_2 near 6.4 μm for atmospheric application: measurements and empirical modeling, *Appl. Opt.* **36**, 563-567.
46. Mlawer, E.J., Clough, S.A., Brown, P.D., Stephen T.M., Landry, J.C., Goldman, A., and Murcray, F.J. (1998) Observed atmospheric collision induced absorption in near infrared oxygen band, *J. Geophys. Res.* **D103**, 3859-3863
47. Lafferty, W.J., Solodov, A..M., Weber, A.., Olson, Wm.B. and Hartmann, J.-M. (1996) Infrared collision induced absorption by N_2 near 4.3 μm for atmospheric applications: measurements and empirical modeling, *Appl. Opt.* **35**, 5911-5917
48. Rinsland, C.P., Smith, M.A.H., Russell III, J.M., Park, J.H., and Framer, C.B. (1981) Stratospheric measurements of continuous absorption near 2400 cm^{-1}, *Appl. Opt.* **20**, 4167-4171
49. Kerridge, B.J., Remsberg, E.E. (1989) Evidence from the Limb Infrared Monitor of the Stratosphere for non local thermodynamic equilibrium in the ν_2 mode of mesospheric water vapor and the ν_3 mode of stratospheric nitrogen dioxide, *J. Geophys. Res.* **94**, 16323-16342.
50. Lopez-Puertas, M., Zaragoza, G., Lopez-Valverde, M.A.., Martin- Torres, F.J., Shved, G.M., Manuilova, R.O., Kutepov, A.A., Gusev, O., von Clarmann, T., Linden, A., Stiller, G., Wegner, A.., Oelhaf, H., Edwards, D.P., and Flaud, J.-M. (1998) Non local thermodynamical equilibrium limb radiances for the MIPAS instrument on Envisat-1, *J. Quant. Spectrosc. Radiat.Transf.* **59**, 377-403.

SPECTROSCOPIC MEASUREMENTS FROM SPACE WITH THE FOCUS SENSOR SYSTEM TO ANALYSE GAS AND SMOKE PROPERTIES OF HIGH TEMPERATURE EVENTS

V. TANK †, P. HASCHBERGER †, K. BOCHTER †,
D. OERTEL ‡, K. BEIER †, F. SCHREIER †, M. BIRK †,
E. LINDERMEIR †, G. WAGNER †

† *DLR Remote Sensing Technology Institute*
Oberpfaffenhofen, D-82234 Wessling, Germany

‡ *DLR Institute of Space Sensor Technology and Planetary Exploration*
Rutherfordstrasse 2, D-12489 Berlin, Germany

Abstract

In 1998 the European Space Agency (ESA) selected the proposal FOCUS as the first and only Earth observation experiment for the Early Utilization Phase of the International Space Station (ISS). FOCUS consists of imaging sensors in the visible and infrared spectral region as well as an infrared Fourier transform (FTIR) spectrometer. It aims at autonomous detection and geolocation of high temperature events (HTE) on Earth like forest and bush fires, volcano eruptions or coal seam fires. Furthermore HTEs and their emissions are to be analyzed with respect to their spatial extension, temperatures, gas and smoke composition and distribution. This will be achieved by a combination of spectral and imaging data of the FOCUS sensor system and analyses based on unique fusion of these data. FOCUS is scheduled to be launched with ISS utilization flight UF-3 in September 2004.

This paper introduces to the FOCUS experiment and briefly describes its imaging part. It then concentrates on the spectrometry and data analysis. The FTIR spectrometer based on the MIROR concept (developed by DLR) is presented as well as are the optical modifications necessary for the FOCUS experiment requirements. High accuracy radiometric calibration demands are investigated. Modeling of radiation transfer, including gases of relevance like CO_2. CO, CH_4, and SO_2 as well as smoke is part of the development of data analysis procedures. Retrieval algorithms for the determination of combustion trace gas columns and aerosol particle size distribution are under development. The retrieval approach based on data fusion is discussed.

1. FOCUS Multisensor System Overview

Existing spaceborne sensors are able to provide information on fire risk as well as to monitor burnt areas. However, if used for active fire recognition, they suffer from serious limitations, such as saturation effects in the mid infrared channel. Furthermore there is no spaceborne sensor available which might provide quantitative data on the emission products from fires and volcanic activities.

J. Demaison et al. (eds.), Spectroscopy from Space, 259–273.

The concern of scientific communities and international politicians about climate change and environmental degradation has highlighted the requirement for better information on the dynamics of biomass burning. Only spaceborne observation can provide the required information on High Temperature Events (HTE), such as wildfires and volcanic activities, on a global scale [1].

FOCUS is an intelligent infrared sensor system for autonomous on-board detection, for classification of HTE and for the IR-spectrometry of the HTE emissions. Its main functional requirements are (1) autonomous on-board hot spot detection by the Fore-Field Sensor (FFS) and processor and (2) record of detailed HTE data by the FOCUS Main Sensor (MS) for on-ground parameter determination of the HTE and their emissions. The whole instrument and its main parameters are shown in *Figure 1.*

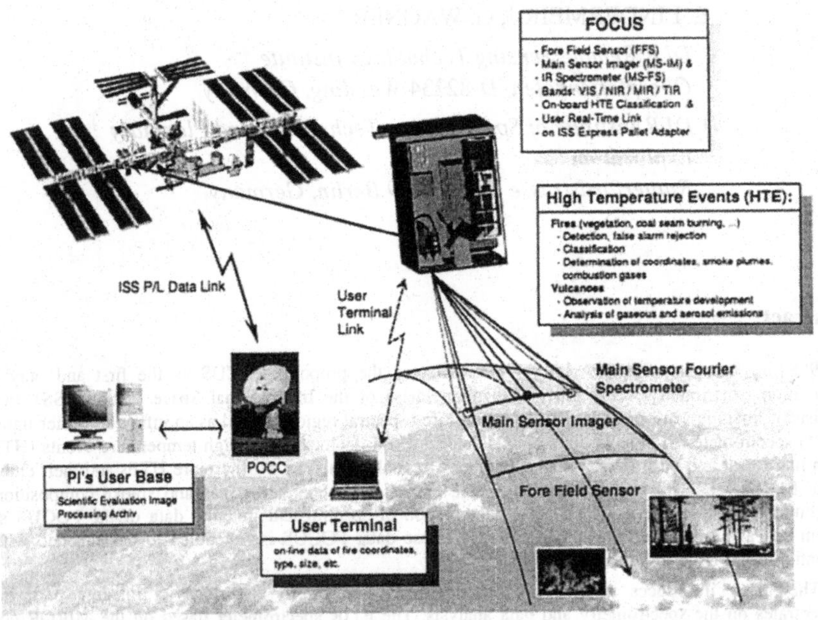

Figure 1. FOCUS System Overview

The Fore-Field Sensor – a forward looking multi-spectral imaging sensor – provides hot spot detection capability by two IR channels in the atmospheric windows at 3 – 4 μm (MIR) and 8 – 10 μm (TIR) and a VIS channel. It has a sufficient viewing angle for the coverage of a reasonable image swath of 355 km width, spatial resolution is about 700 m (IR) resp. 180 m (VIS).

The Main Sensor (MS) consists of a combination of spatially highly resolving imaging sensors (MS-IM) in different spectral bands and a profiling FTIR spectrometer with high spectral resolution (MS-FS). It provides nadir and off-nadir sounding geometry in a narrow Field of View (FOV), which is tilted across track within a field of regard of ± 30° by FFS positioning commands. MS-IM consists of two IR channels, operating in the MIR and TIR atmospheric windows, and two VIS/NIR channels in the 0.5 – 0.9 μm wavelength region. Swath width is 70 km and spatial resolution is 136 m resp. 34 m in IR and VIS. MS-FS provides high spectral resolution sounding in the wavelength region from 3 – 15 μm. Its approx. 8 km footprint is swiveled by a micro scan mirror across the 70 km MS-IM swath.

2. FOCUS Imaging Sensors

The MIR and TIR detectors and front end electronic solutions of the DLR small satellite Bi-spectral IR Detection (BIRD) – foreseen to be launched in 2001 – have been selected as the base line for the MIR/TIR channels of the FOCUS FFS and MS-IM.

The high dynamic MIR and TIR channels are built on staggered IR line arrays with an identical layout. An integrated Mercury Cadmium Telluride (HgCdTe) 2x512 elements line array detector/Stirling cooler assembly is the core of each of these channels [2]. The detectors require a special electronic configuration with a low noise design for detector, temperature sensor, preamplifier, clock driver, and digital controlled bias voltages and current. These components are arranged near the focal plane. Both of the detector arrays have four outputs working simultaneously with a pixel rate of 1 MSamples/s and 14 bit digitising [3]. The MIR/TIR sensors have been tested onboard DLR research aircraft [4].

The VIS/NIR channels are equipped with 2048 elements CCDs. A dynamic range of 12 bit is sufficient for HTE observation in this spectral region. The electronic layout is derived from the DLR WAOSS camera [5].

3. FOCUS Main Sensor Fourier Spectrometer (MS-FS)

3.1. FTIR SPECTROMETER MIROR

As basic design concept for the MS-FS instrument the MIROR (Michelson Interferometer with ROtating Retroreflector) setup is selected (*Figure* 2). The MIROR design is based on the classical Michelson interferometer, however significant modifications are incorporated.

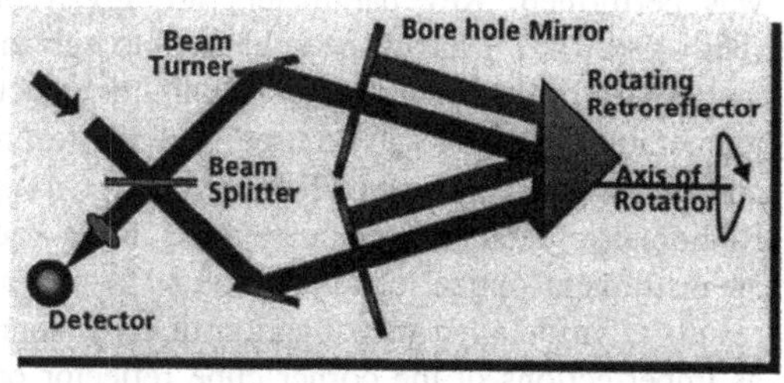

Figure 2. MIROR Interferometer Concept

Instead of the two plane mirrors of the classical Michelson interferometer a retroreflector (hollow corner cube mirror) is used in the MIROR design to generate optical path differences. This is achieved by directing the beam of each interferometer arm into this device. As the retroreflector rotates eccentrically with reference to the driving axle the optical path in one arm is shortened during rotation while it becomes longer in the other arm, and vice versa. The fixed bore hole mirrors (actually resembling the plane Michelson mirrors) reflect the radiation of each arm through the retroreflector back to the beam splitter. The maximum optical path difference s_{max} generated by this arrangement is determined by the angle α between the radiation entering the reflector and the reflector's axis of rotation, and the eccentricity E of the reflector's center with respect to the axis of rotation:

$$s_{max} = 16 \cdot E \cdot \sin(\alpha) \quad (1)$$

Two interferograms can be measured per reflector revolution. Further details of the MIROR design and characteristic features are given elsewhere [6, 7, 8]. Compared to the translatory mirror movement of the conventional Michelson interferometer the MIROR concept presents several advantages and is therefore especially suited for space applications: (1) Due to the continuous rotation of the retroreflector the stop-and-go operation of conventional interferometers is overcome, linear accelerations are exhibited. (2) Reduced effort in reflector bearing: No linear slidings are necessary. The retroreflector is simple to adjust since it reflects an incoming beam of radiation parallel to itself independently of its orientation (tilt and shear compensation). (3) Insensitivity against mechanical disturbances in axial direction. (4) Low, constant power consumption. (5) High independence between optical path difference and measurement rate: For the classical set-up trade-offs have to be made between the optical path difference (determining the spectral resolution) and the measurement rate. Increasing both parameters leads to strong linear accelerations and an unstable mechanical performance (vibrations). In the MIROR design both parameters are largely independent. According to eq. (1) the optical path difference is characterised by two geometrical settings only, while the measurement rate is defined by the reflector's speed of rotation. Thus, from the mechanical point of view both parameters can be set independently. Limitations arise only from the electrical bandwidth of the detectors and the data acquisition electronics capability.

3.2. MS-FS POINT DESIGN

Based on the FOCUS experiment requirements a first approach MS-FS point design was performed. Key parameters are given in TABLE 1. The wide spectral range (650 – 3000 cm^{-1}) has to be split up into sub-ranges covered by different detectors for increased spectral sensitivity. Quantum detectors operated close to liquid nitrogen temperature (≈80 K) will be used: Indium-Antimonide (InSb) for the MIR (3 – 5.5 μm) and Mercury Cadmium Telluride (HgCdTe) for the TIR range (5.5 – 16 μm). Radiometric data were calculated assuming background limited detector performance, an instrument optical efficiency of 10 % and a modulation factor of 80 %. The latter results from detailed investigation of error sources in the opto-mechanical set-up such as imperfections of the corner cube reflector or misalignment of the flat surface fixed mirrors.

As the MS will not be equipped with a device compensating the ISS alongtrack motion, acquisition time for a single interferogram must be as short as possible to avoid smearing effects. On the other hand, acquisition time mainly drives the noise equivalent spectral radiance (NESR). From the radiometric point of view both contradictory requirements can be met by increasing the ground spot diameter, i.e. the instrument's FOV: A larger FOV allows to reduce the acquisition time, thus reducing the smearing effects while deteriorating the spatial resolution. A ground spot diameter in the range of several kilometres represents an adequate compromise.

TABLE 1. Key parameters of MS-FS Point Design

Spectral ranges: HgCdTe InSb	 650-1800 cm^{-1} 1800-3000 cm^{-1}
Detector data: diameter quantum efficiency· D^* (calculated)	1 mm 60% HgCdTe: 5.5 10^{11} cm√Hz/W @1000 cm^{-1} InSb: 2.3 10^{12} cm√Hz/W @2000 cm^{-1}
max. optical path difference	6.7 cm
used optical path difference	6 cm (± 3 cm symmetrically)
Spectral resolution	0.17 cm^{-1}
Beam divergency (half angle)	8 mrad
FOV	16 mrad ≈ 1°
Ground spot diameter (for ISS altitude 400 km)	6400 m
Beam diameter	2 cm
Optical throughput	6.3 10^{-4} cm^2 sr
NESR (for 300 K background)	HgCdTe: 0.1....0.25 μW/cm^2 sr cm^{-1} InSb: 0.03...0.05 μW/cm^2 sr cm^{-1}

3.3. MS-FS DETECTOR FOREOPTICS

Investigating HTE scenes with an FTIR spectrometer puts higher demands on the imaging optics than current laboratory spectroscopy because of the inhomogeneous target radiance distribution. For FOCUS it is planned to have no optical elements in front of the interferometer entrance. Therefore a thorough design of the detector foreoptics is necessary to get a well defined ground spot point spread function (PSF) and instrumental line shape (ILS). In case of the long wavelength thermal infrared a constraint on the optics is given by background radiation (instrument´s internal radiation) causing photon noise. Thus, the foreoptics must restrict the optical throughput of the background as far as possible. This is done by cold stops. To decrease thermal excitation noise the photon detectors will be operated at about 80 K. For further reduction the optical design process aims at minimizing the detector areas, which puts an additional demand on the foreoptics.

Four different optical arrangements outlined in *Figure 3* were investigated. The figure also shows the positions of the different field and aperture stops for target and background radiation. A parametric model linking the detector surface area to thermal excitation noise and the background photon noise was developed. Raytracing was applied to design the optics for selected case studies within the different options.

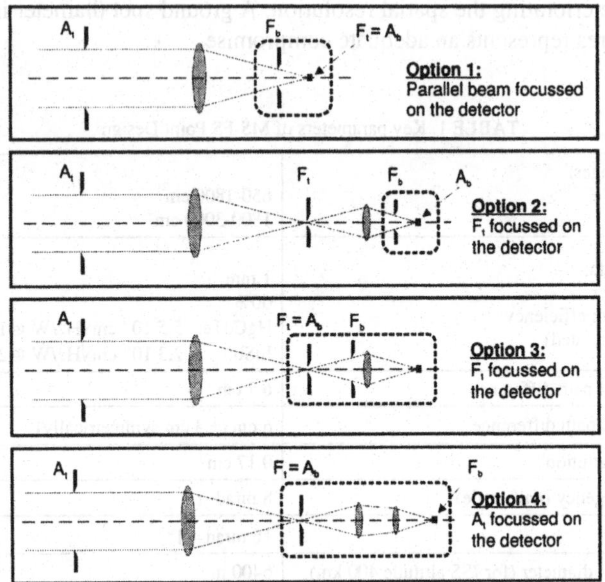

Figure 3. Schematic representation of different detector foreoptics options. For simplicity imaging optical elements are shown as lenses. The area within the dashed line represents the cooled volume, the black square the detector. F = field stop, A = aperture stop, t = target radiation, b = background radiation

Option 1 has the simplest layout, but the disadvantages of astigmatism due to the fast optics (allowing for rays with large off-axis angles) associated with the small detector size and high sensitivity of ILS, PSF, and viewing direction on detector misalignment as well as large background radiation throughput. In option 2 the ILS, PSF, and viewing direction are influenced by the additional stop F_t instead of the detector. Because F_t is of larger diameter fast optics and thus astigmatism is avoided and adjustment sensitivity reduced. In option 3 F_t is placed within the cooled volume and F_b is positioned at a distinct image of A_t. This allows to reduce background radiation.

In options 1 to 3 the inhomogeneous HTE scene is imaged on the detector causing non uniform radiance distribution on its surface. In contrast the radiometric calibration of the spectrometer is done by use of black bodies with homogeneous radiance distribution. Because of the nonlinear behavior of the photoconductive HgCgTe detector this may lead to scene depending signal deviations. This drawback is removed by option 4 where the homogeneously illuminated stop A_t, e.g. the beamsplitter, is imaged onto the detector.

Option 4 was found to be best suited. Its specific features and advantages are given in TABLE 2. It shows the key output parameters of the parametric study based on the FOCUS Phase A point design of the spectrometer.

TABLE 2. Specific features of Option 4.

Advantages of option 4
• Nearly ideal groundspot PSF
• Nearly ideal ILS
• Low sensitivity of MS-FS viewing direction on detector misalignment
• Low sensitivity of PSF and ILS to detector adjustment
• Lowest NESR
• No calibration errors by inhomogeneous scene illumination

Key output parameters of option 4	
Sensitivity loss factor*	1.6 **
Detector size	0.6 x 0.6 mm^2
Size of cold module	65 x 35 x 20 mm^3
Number of imaging mirrors	3 (2 cooled)
Number of cooled stops	1
Cooling power estimate	0.16 W

* relative to min. achievable noise
** option 1+2: 2.3, option 3: 1.9

3.4. MS-FS CALIBRATION

The MS-FS will perform spectrometric gas remote sensing and deliver quantitative infrared combustion gas emission, atmospheric gas and background (earth surface) spectra for the data fusion analysis of the HTEs. This requires infrared radiometric calibration and as such is subject to systematic calibration errors which will propagate throughout the analysis procedure, and influence the accuracy of the results.

Error propagation has to been taken into account when investigating the calibration requirements. A calibration and analysis model was used to determine the impact of calibration parameter uncertainties like those of calibration source temperature and emissivity. The derived requirements have been compared to technical feasibility.

3.4.1. *Calibration Objectives*

Spectrometers by all means deliver raw spectra $S(\sigma)$ as direct output, where σ is the wavenumber. These spectra are composed of the investigated object's infrared spectral radiance $L_{HTE}(\sigma)$ and the radiance $L_{inst}(\sigma)$ of the instrument itself (its inner surfaces, which contribute to the overall radiation detected), see *Figure 4*. Furthermore the optical and electronic properties of the spectrometer summarized as its spectral responsivity $R(\sigma)$ influence the raw spectrum.

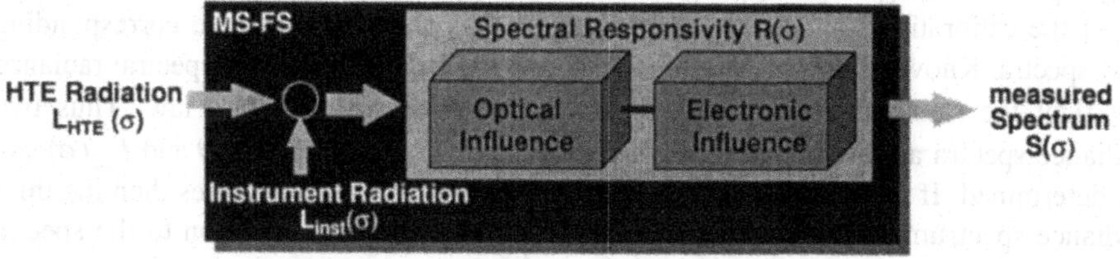

Figure 4. Radiometric Calibration Model

The raw spectrum is thus given by:

$$S(\sigma) = R(\sigma) \cdot [L_{HTE}(\sigma) + L_{inst}(\sigma)] \tag{2}$$

where $S(\sigma)$ is the raw spectrum, $R(\sigma)$ the responsivity of spectrometer, $L_{HTE}(\sigma)$ the HTE spectral radiance, $L_{inst}(\sigma)$ the spectral radiance of the spectrometer.

$L_{HTE}(\sigma)$ has to be determined from the raw spectrum $S(\sigma)$. Hence $R(\sigma)$ and $L_{inst}(\sigma)$ need to be known. This is achieved by calibration. $L_{HTE}(\sigma)$ can then be determined. $R(\sigma)$ and $L_{inst}(\sigma)$ are not necessarily constant over time since they depend on the optical alignment and the temperature of the instrument which may undergo alterations depending on ambient conditions. Therefore, in general, recurrent calibrations have proven useful. Hence the objective of calibration is twofold: It yields absolute radiance spectra and also serves to monitor the properties and quality of the spectrometer itself.

3.4.2. *Calibration Principle*

As shown the calibration needs to yield $R(\sigma)$, the spectral responsivity and $L_{inst}(\sigma)$, the spectral radiance of the spectrometer. Two different calibration procedures have been investigated in detail: With the two-temperature calibration method the calibration source is used at two different temperatures. $T_c(1)$ and $T_c(2)$ are measured by contact sensing and $S(\sigma)$ and $L_{inst}(\sigma)$ are determined. The three temperature calibration method [9] allows to determine $S(\sigma)$, $L_{inst}(\sigma)$ and $T_c(1, 2, 3,...)$ by a nonlinear least squares fit.

The first procedure is the widely used standard method. In practice two calibration standards kept at the two different temperatures are utilized to avoid heat up and cool down times that would be associated with the use of a single standard. The added effort of three calibration standards is compensated by the fact that the source's temperatures result from the spectra via least squares fit calculations and no sensors are needed for temperature measurement.

As calibration standards black body sources are used which are defined by their spectral radiance determined by their temperature according to Planck's law. The influence of atmospheric absorption and emission is neglected which is appropriate for space applications. Ideally for black body calibration standards holds $\varepsilon_c = 1$, and then a spectrum measured of the black body standard is given by:

$$S(\sigma) = R(\sigma) \cdot [L_C(\sigma) + L_{inst}(\sigma)] \tag{3}$$

with $L_c(\sigma) = f(\sigma, T_c)$. The calibration can then be performed by setting the temperature T_c of the calibration standard to two different levels and measuring the corresponding raw spectra. Knowing the temperatures T_c allows the calculation of the spectral radiance $L_c(\sigma)$ of the standard at the two temperatures according to Planck's law. Thus two radiance spectra are available from which the unknown spectra of $R(\sigma)$ and $L_{inst}(\sigma)$ can be determined. If measurement takes place at three different temperatures then the third radiance spectrum serves to determine all three temperatures in addition to the spectra of $R(\sigma)$ and $L_{inst}(\sigma)$.

In reality, however, the emissivity ε_C of a black body is always lower than 1 because of its non-zero reflexivity ρ_C. Hence the raw spectrum is determined by a less simple equation:

$$S(\sigma) = R(\sigma) \cdot [\varepsilon_C L_C(\sigma) + \rho_C L_F(\sigma) + L_{inst}(\sigma)] \quad (4)$$

with $L_C(\sigma) = f(\sigma, T_C, \varepsilon_C)$. In this equation $\varepsilon_C L_C(\sigma)$ describes the radiance of one of the calibration standards and $\rho_C L_F(\sigma)$ the radiance of foreground reflected by the standards.

With the two-temperature calibration method the temperatures T_C have to be measured by contact sensors which can not be done free of errors:

$$T_C = T_{Cmeas} + \Delta T_{Cerr} \quad (5)$$

T_{Cmeas} describes the measured calibration standard temperature and ΔT_{Cerr} the temperature error. Since neither ΔT_{Cerr} is known, nor equation (4) suites practical application, common calibration procedures assume $\varepsilon_C = 1$ and $\Delta T_{Cerr} = 0$, which leads to the mentioned errors that propagate throughout the data analysis process. The maximum allowable deviations from the ideal $\varepsilon_C = 1$ and $\Delta T_{Cerr} = 0$ have been defined with reference to the measurement task and the calibration procedure and also to the spectrometer's NESR. In this way it is assured that NESR is the dominating parameter determining the accuracy of the retrieval results and systematic calibration errors are of minor influence. The actual MS-FS calibration requirements assuming a two or three temperature calibration are listed in TABLE 3.

TABLE 3. Black Body Requirements

Spectral emissivity	$\varepsilon(\sigma) > 0.995$
Temperature inhomogenity	$\Delta T < 0.1$ K
Absolute temperature uncertainty	$\Delta T < 0.1$ K
Temperature instability during calibration time	$\Delta T < 0.1$ K
Desired Temperatures	$T_1 = 300$ K, $T_2 = 400$ K, ($T_3 = 500$ K)

4. Simulation of IR Spectra

Radiative transfer simulations utilizing appropriate low and high spectral resolution models, e.g. MODTRAN [10] or FASCODE [11], are required for a variety of purposes: First, simulations are used to estimate images and spectra expected to be seen by FOCUS MS-IM and MS-FS, respectively, e.g., in order to get an idea about the order of magnitude change of fire spectra compared to cold atmosphere spectra. Furthermore, generation of synthetic "measured" spectra is necessary as these are used as input for the retrieval feasibility study.

A complete description of the state of the hot atmosphere above fires, i.e., pressure, temperature, gas, smoke, and aerosol profiles as a function of altitude, is required as input for realistic radiative transfer simulations.

4.1. TEMPERATURE AND GAS CONCENTRATIONS

To our knowledge the only airborne IR spectroscopic observation of wild fires has been performed with the Airborne Emission Spectrometer (AES, an airborne simulator of the TES instrument to be flown with EOS-Chem). Worden *et al.* [12] concluded that for the analysis of fire spectra it is reasonable to distinguish between a few generic types of fires contributing to the observed spectrum, that – due to the large footprint of the spectrometer – is a mixture of these generic contributions. This approach has been adopted for the FOCUS retrieval study, and scenarios have been defined for forest and savannah fires.

In the *flaming zone* of forest fires gas temperatures reach some 1000 – 1800 K at about 30 m above ground. Due to the reduced amount of burning material, savannah fires are colder with typical peak temperatures of about 800 K. Surface temperatures vary significantly, and an average of 1000 K has been assumed. Average gas temperature of the *smouldering zone* near ground is between 400 and 500 K, and the average surface temperature is about 500 K. The *zone of smoke plume over unburned ground* is characterized by average surface and gas temperatures of 350 K. For the *unperturbed zone* standard atmospheric models are applicable.

4.2. OPTICAL PROPERTIES OF SMOKE PARTICLES

Required input data are the optical properties of particles and the vertical particle density and temperature profiles in the smoke plumes. The optical properties of the smoke particles are derived from measured microphysical parameters by means of Mie theory using the computational aerosol model DBASE [13].

The microphysical parameters and vertical profiles of smoke are adopted and assembled from different sources, mainly international experiments such as TRACE-A and SAFARI [14], and SCAR-B [15]. For savannah or forest fires the total vertical optical depths of smoke plumes, including natural background aerosols, are within 1 to 6 in the visible [16]. The extinction, absorption, and scattering coefficients, the single scattering albedo, and the asymmetry parameter are wavelength dependent and are calculated over the relevant wavelength range. Typical maximum smoke particle mass concentrations range from 300 – 450 μg/m^3 which results in extinction coefficients from 1 to 1.5 km^{-1} at 0.55 μm.

The size distribution is bimodal and can be represented by two lognormals. Parameters for the lognormal distribution are mean modal radius, standard deviation of the natural logarithm of radius, and number or volume of particles per cross section of the atmospheric column. The two modes are the accumulation mode (small particles) and the coarse mode (large particles). For biomass burning the optical properties are dominated by the smaller particles. For simplicity we assumed that smoke particles are spherical.

The smoke particle size distribution consists of an external mixture of a large number of small particles with a median radius from 0.02 to 0.1 μm, and a small number of large particles with diameters from 1 to 30 μm. The small particles themselves consist of an internal mixture of organic liquids surrounding a solid spherical core of black carbon. The coarse mode particles are of crustal origin. The smoke aerosol is assumed to have an average index of refraction of 1.43 – 0.0035i.

4.3. SIMULATION RESULTS

For an assessment of radiative properties of smoke and gases from biomass burning in the infrared, simulations have been performed with the MODTRAN band model (v. 3.5) for a nadir viewing observation from space with a spectral resolution of 1 cm^{-1}.

Absorption and emission of the atmospheric constituents, i.e. molecules and aerosols, as well as absorption of the surface radiance, and scattering by aerosols (nb. smoke) are considered by this model. It should be noted that a high resolution radiative transfer model such as FASCODE is used for the actual retrievals exploiting the observations of the image sensor and the Fourier transform spectrometer.

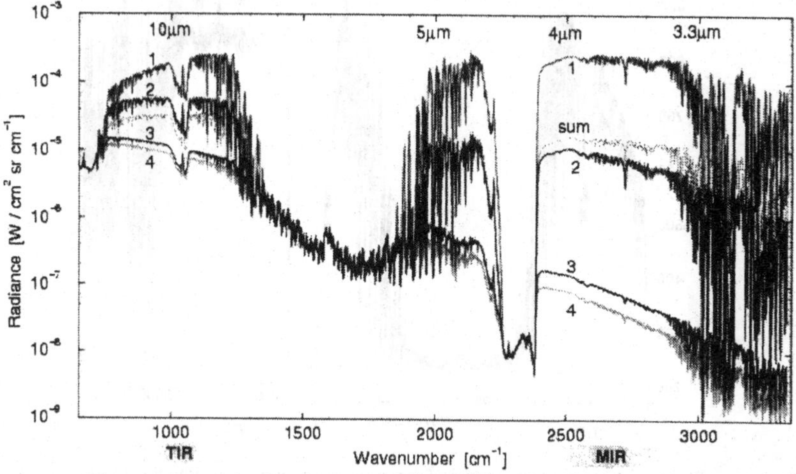

Figure 5. Spectral radiances contributions to forest fires: (1) flaming, (2) smoldering, (3) smoke, (4) standard atmosphere, (sum) summation of 5 % flaming, 30 % smoldering, 30% smoke over unburnt ground and 35 % undisturbed background.

Figure 5 shows the spectral radiance vs. wavenumber for the four types of generic scenes of a forest fire described above. Note that the radiance is dramatically increased for the flaming zone radiance, with up to five orders of magnitudes in the MIR. In the MIR and TIR atmospheric window regions the radiances are dominated by the surface radiance (and hence characterised by the surface temperature) with only minor contributions from atmospheric constituents.

The footprint of the MS-FS will have a diameter of several kilometres. However, even for large wild fires only a small fraction of this large area will be covered by a flaming fire zone, i.e., the MS-FS sees an inhomogeneous ground footprint with a variety of areas with different surface radiance values and correspondingly different atmospheric states. As discussed in [12], the observed spectra can be assumed to have radiance contributions from some combination of the generic fire types. A weighted linear combination of the "generic" spectra, assuming a 5 % contribution of the flaming zone, is shown in *Figure 5* as dashed curve. The increase in radiance for the mixed scene fire spectra is severely reduced in comparison to the generic spectra, but for the forest fire there is still three order of magnitudes increase in the MIR.

The influence of smoke aerosols on the spectra is shown in *Figure 6*. In terms of equivalent blackbody temperatures the difference is as large as about 100 K. The simulations show clear radiation effects from smoke particles in the TIR, hence indicating the feasibility to retrieve aerosol optical depth. With the assumption of a lognormal size distribution for the smoke particles it is possible to derive the median particle radius and the smoke particle mass.

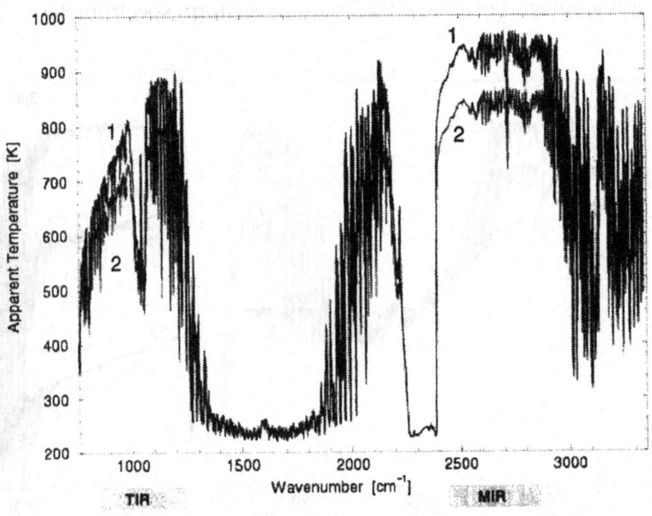

Figure 6. Apparent temperature of forest fire flaming phase, without (1) and with smoke aerosols (2).

5. Retrieval of Biomass Fire Characteristics by Means of Data Fusion

Estimating HTE characteristics is a challenging problem: First, the retrieval of temperature and concentration profiles from spectroscopic measurements obtained by vertical sounding is well known to be an ill-posed problem, and advanced numerical techniques have to be used to find meaningful solutions [17]. Furthermore, due to the limited spatial resolution of Fourier transform spectrometers, HTE scenes observed by MS-FS will be horizontally inhomogeneous. Thus, in order to retrieve more than just average temperature and gas/aerosol concentration profiles for the entire scene additional information such as provided by an image sensor has to be incorporated in the solution process, i.e. data fusion.

Presently our solution to retrieve the state of the atmosphere above fires is based on a sequential approach: Assuming that the distribution of flaming, smoldering, etc. areas is already known from an analysis of the image data, the relative area contribution and corresponding surface temperatures are used as input to the high resolution radiative transfer calculations required to iteratively solve the nonlinear least squares problem to fit the spectral data. We are also discussing a global fit approach where MS-IM images and MS-FS spectra are analysed simultaneously, but because of the significantly higher computational demands and complexity this has not been implemented so far.

In the FOCUS phase A study first retrieval tests have been conducted for a pure gaseous atmosphere: As a first step we have investigated the feasibility to retrieve the temperature and gas concentration profiles above a flaming zone, that, in contrast to standard atmospheric retrieval problems, shows much stronger variability (e.g., temperature gradients of several hundred degrees). For these tests, in addition to temperature the main infrared absorbers water, carbon dioxide, and carbon monoxide have been assumed to be unknown. The tests were performed for a variety of instrument parameter scenarios (i.e., NESR and spectral resolution) representative for spaceborne FTIR spectrometers. Comparisons of the retrieved profiles with the true profiles used to generate the "measurements" revealed that a NESR of some hundred nW/(cm^2 sr cm^{-1}) or better and a spectral resolution of some tens of a wavenumbers or better should be realized in order to allow successful retrievals.

For the inhomogeneous scene retrievals, the observations of the image sensor had to be taken into account in order to retrieve horizontal as well as vertical profiles. Because of computational constraints, a simplified atmospheric scenario had been adopted and only flaming phase and smoldering phase profiles were considered. Assuming NESR = 100 nW/(cm^2 sr cm^{-1}) and a spectral resolution of 0.1 cm^{-1} for the spectrometer, agreement between the true temperature and CO profiles and the retrieved profiles was excellent, for carbon dioxide a slight deviation had to be noticed both for the smoldering and flaming zone lowest column. The water profile retrieved lies between the smoldering and flaming zone profiles, as to be expected.

Simulations also show clear radiation effects from smoke particles in the TIR which makes it feasible to retrieve aerosol optical depth (*Figure 7*). With the assumption of a log-normal size distribution it is possible to derive the median particle radius and smoke particle mass.

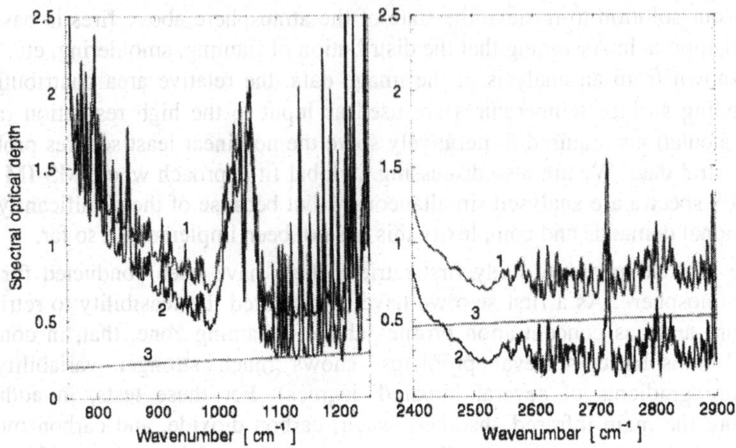

Figure 7. Analysis of optical depth from sum of gas and smoke particles (1), only gas (2) and only smoke (3)

The retrieval model TIMAER (Thermal Infrared Molecule and AErosol Retrieval) currently under development combines both molecules and aerosol properties to retrieve simultaneously temperature, gases, and smoke distribution of biomass burning processes. The data fusion will be based on the sequential approach: Step 1, the surface characterization (temperatures and areas) from MS-IM images, yields the spatial distribution of hot and cold areas within the MS-FS footprint. The classification of areas flaming, smoldering, smoke over unburnt ground and undisturbed background is done in step 2. Step 3 is the retrieval of atmospheric temperatures and gas concentrations for mixture of generic classes, step 4 the estimation of total spectral optical depth of fire plume (gas + smoke) in the atmospheric windows. The last step 5 gives the estimated spectral optical depth of smoke particles as difference to total optical depth using smoke aerosol data base.

Thus, in addition to the gas characterization (temperature and concentrations) we expect to retrieve column amounts of smoke particles (from the optical depth magnitude) and median particle radius for lognormal size distribution.

Acknowledgement

It is gratefully acknowledged that the work was supported by the European Space Agency through ESA Contract 1307/98/NL/JS-.

References

1. Oertel, D., P. Haschberger, V. Tank, F. Schreier, B. Schimpf, B. Zhukov, K. Briess, H.-P. Roeser, E. Lorenz, W. Skrbek, J.G. Goldammer, C. Tobehn, A. Ginati, and U. Christmann (1999) Two dedicated spaceborne fire missions, *Proc. The Joint Fire Science Conference and Workshop, Boise, Idaho/USA* **1**, 254–261

2. Lorenz, E., W. Skrbek and H. Jahn (1997) Design and analysis of a small bispectral infrared push broom scanner for hot spot recognition, *Proc. SPIE* **3122**, 25–34
3. Skrbek, W. and E. Lorenz (1998) HSRS – an infrared sensor for hot-spot-detection, *Proc. SPIE* **3437**, 167–176
4. Lorenz, E., W. Skrbek, and B. Zhukov (1999) Airborne testing of bi-spectral infrared push broom scanner for hot spot detection, *Proc. SPIE* **3759**, 97–105
5. Oertel, D., R. Reulke, R. Sandau, M. Scheele, and T. Terzibaschian (1992) A flexible digital wide-angle optoelectronic stereo scanner, *International Archives of Photogrammetry and Remote Sensing* **29**, B1, 44–49
6. Haschberger P., O. Mayer, V. Tank, and H. Dietl (1991) Ray tracing through an eccentrically rotating retroreflector used fort pathlength alteration in a new Michelson interferometer, *J. Opt. Soc. Am.* **A8**, 1991–2000
7. Haschberger P. and V. Tank (1993) Optimization of a Michelson interferometer with a rotating retroreflector, in optical design, spectral resolution, and optical throughput, *J. Opt. Soc. Am. A* **1.10**, 2338–2345
8. Haschberger P. (1994) Impact of the sinusoidal drive on the instrumental line shape function of a Michelson interferometer with rotating retroreflector, *Appl. Spectrosc.* **48**, 307–315
9. Lindermeir E., P. Haschberger, and V. Tank (1992) Calibration of a Fourier transform spectrometer using three blackbody sources, *Appl. Opt.* **31**, 4527–4533
10. Berk A. and D. Robertson (1989) *MODTRAN: A moderate spectral resolution LOWTRAN7*, Technical Report, Geophysics Laboratory, Mass., USA
11. Clough S.A., F.X. Kneizys, G.P. Anderson, E.P. Shettle, J.H. Chetwynd, L.W. Abreu, L.A Hall, and R.D. Worshamm (1988) FASCOD3: spectral simulation, in J. Lenoble and J.F. Geleyn, editors, IRS'88: *Current Problems in Atmospheric Radiation. A.* Deepak Publ.
12. Worden H., R. Beer, and C.P. Rinsland (1997) Airborne infrared spectroscopy of 1994 western wildfires. *J. Geophys. Res.* **102**, 1287–1300
13. Levoni C., M. Cervino, R. Guzzi, and F. Torricella (1997) Atmospheric Aerosol Optical Properties: A Database of Radiative Characteristics for Different Components and Classes. *Appl. Opt.* **36**, 8031–8041
14. Andreae M.O., J. Fishman, and J. Lindesay (1996) The Southern Tropical Atlantic Region Experiment (STARE): Transport and atmospheric chemistry near the Equator-Atlantic (TRACE A) and Southern African Fire-Atmosphere Research Intitiative (SAFARI), *J. Geophys. Res.***101**, 23519–24164
15. Kaufman Y.J., P.V. Hobbs, V.W.J.H. Kirchhoff, P. Artaxo, L.A. Remer, B.N. Holben, M.D. King, D.E. Ward, E.M. Prins, K.M. Longo, L.F. Mattos, C.A. Nobre, J.D. Spinhirne, Q. Ji, A.M. Thompson, J.F. Gleason, S.A. Christopher, and S.-C. Tsay (1998) Smoke, Clouds, and Radiation – Brazil (SCAR-B) experiment. *J. Geophys. Res.* **103**, 31783–32157
16. Kaufman Y.J., A. Setzer, D. Ward, D. Tanke, B.N. Holben, P. Menzel, M.C. Pereira, and R. Rasmussen (1992) Biomass Burning Airborne and Spaceborne Experiment in the Amazonas (BASE-A). *J. Geophys. Res.* **92**, 14581–14599
17. Schimpf B. and F. Schreier (1997) Robust and Efficient Inversion of Vertical Sounding Atmospheric High-Resolution Spectra by Means of Regularization. *J. Geophys.Res.* **102**, 16037–16055

Simulation of Air Pollution in a Wind Tunnel

S. Civiš, Z. Zelinger, M. Střižík and Z. Jaňour[a]
J. Heyrovský Institute of Physical Chemistry
Academy of Sciences of the Czech Republic
Dolejškova 3, 182 23 Prague 8,CR

[a]*Institute of Thermomechanics*
Academy of Sciences of the Czech Republic
Dolejškova 5, 182 23 Prague 8,CR

Abstract

Laser photoacoustic spectrometry and a line permeation pollution source were used in a study of the dispersion of pollution in an urban agglomerate using simulation in a wind tunnel. Applications of this measuring technique utilize the high sensitivity and broad dynamic range (3 orders of magnitude in this case) of the photoacoustic detection method. The minimum detected absorbance in the photoacoustic detection method employed in this work was at the level of $\cong 4.3 \times 10^{-6}$ ($\cong$ 8 μg/m^3 CH_3OH). The effectiveness and flexibility of the permeation method of generation of various concentrations of gases was verified for simulation of emission pollution sources in a wind tunnel. The line permeation pollution source developed in this work, with a concentration flux of ($8.3 \times 10^{-5} \pm 2 \times 10^{-6}$) g/s at 20 °C, generated a concentration level in the model used from a background value of 80-90 μg/m^3 up to values of $\cong$ 1000 μg/m^3 of methanol. A simple model street canyon together with the pollution source was employed to carry out a number of measurements of spatial profiles. The dispersion of the pollutant was studied at the bottom and on the walls of the street canyon together with the concentration variation with changes in the wind speed. The laboratory model was used to demonstrate the differences in ventilation of the street canyon. Using the laser sheet visualisation method, a video technique was applied to show the real flow in the street canyon for various reference velocities. Spatial measurement of the concentration distribution inside the street canyon was carried out on the model with reference velocity of 1.5 m/s. The dependence of the concentration field on the Reynolds number was estimated from the measurements of concentrations in the vicinity of the line permeation source.

J. Demaison et al. (eds.), Spectroscopy from Space, 275–299.

Introduction

Study of the dispersion of pollutants in the atmosphere forms a basis for implementation of effective methods to prevent environmental damage. The troposphere frequently exhibits dangerous pollutant concentrations, especially in the area of urban agglomerates. Atmospheric pollution has a number of detrimental consequences in the form of acid rain, photochemical smog and global climatic changes. Understanding of the complex processes occurring in the atmosphere permits the creation of models that not only assist in suppressing and eliminating the instantaneous detrimental phenomena, but also provide predictions of how these detrimental consequences can be avoided over longer periods of time.

The dispersion of pollutants in the atmosphere can be studied either directly (*in situ*) by monitoring methods (e.g. [1]), which are often expensive and mostly provide only partial results, or by methods of mathematical and physical modelling. For complicated cases it is useful to use the methods of physical modelling - e.g. simulation of processes in the atmosphere in wind tunnels [2]. This method provides useful information on the global picture of dispersion of pollutants in urban agglomerates, which is frequently very difficult or almost impossible to obtain by direct monitoring methods. Simulation of the dispersion of pollutants in the atmosphere on a model of future construction in a given kind of landscape can prevent urbanistic mistakes and errors that could, for example, cause local accumulation of toxic substances in the air of urban agglomerates and thus directly endanger its future inhabitants.

Qualitative estimates of the processes occurring on a model of the atmosphere in wind tunnels can be made using various methods permitting visualizing of the flow and dispersion of pollutant substances, see e.g. [2]. A quantitative estimate is often made using the Laser-Doppler anemometric method [2]. It is also important to quantitatively estimate imissions in space and time. This kind of concentration measurements is carried out using fast flame-ionization detectors [3], characterized by high sensitivity. High sensitivity of the analytical methods used in simulation in a wind tunnel is very important from the standpoint of the extent of spatial coverage of the monitored area. In addition to sensitivity, it is important in obtaining quantitative concentration profiles to have available an analytical method with a broad dynamic range, where possible on a linear scale. In this work we used the spectroscopic method of laser photoacoustic detection as an analytical tool for monitoring in a wind tunnel [4].

Absorption spectroscopic methods [5] are based on absorption of radiation passing through a sampling cell. In case of the use of infrared radiation sources, it is possible to monitor absorption in the vibration and rotation-vibration modes of selected molecules. Almost all molecular pollutants can be monitored in the infrared region. The application of this method in the atmosphere is complicated by difficulties caused by absorption by atmospheric water and CO_2, which covers an extensive spectral region and coincides with potentially useful radiation sources. However, there are "atmospheric windows" where the absorption of these molecules is negligible. One of these windows is the spectral region around 10 μm, which was used in our measurements [6].

CO_2 Laser Photoacoustic Spectrometry

Absorption spectroscopic methods are based on measuring the intensity difference between the entering and leaving radiation. In monitoring small concentrations of absorbing molecules, small differences are detected between the entering and leaving intensities of the radiation used. When a laser source is used, radiation of high intensity is available; thus small changes caused by the absorption of the molecule are observed on a relatively large detected signal. However, this approach is very disadvantageous from the standpoint of the ratio of the signal to the noise and the dynamic detection range. In these cases, the minimum detectable signal is at the level of 10^{-4} [7]. It then follows for the CO_2 laser that the minimum detectable absorbance is given by this minimum detectable signal and an increase in the sensitivity can be achieved only by lengthening the optical pathway. It has been found that it is more effective to detect small signals in the absence of the large offset background. One of the ways of dealing with this aspect is to abandon the classical scheme of evaluation of the difference between the entering and leaving signals and to measure the absorbed energy by some other method. A molecule that absorbs a photon has several options for ways of losing this energy. At atmospheric pressure the most probable pathway is collision with some other molecule and conversion of the absorbed energy to kinetic energy of molecular motion. If the exciting radiation is in the infrared range, absorption occurs in the vibration modes of the molecule with subsequent vibration-translation relaxation. The increased kinetic energy of the molecule is manifested by a pressure change in the closed volume of the sampling cell ; this change can be very sensitively detected by a microphone The photoacoustic signal obtained S can be simply expressed as:

$$S = CN\sigma P \tag{1}$$

where C is a constant characterizing the properties of the photoacoustic cell (length, cross-section and sensitivity of the microphone), N is a number of absorbing molecules per cm^3, σ an absorption cross-section of an absorbing molecule, and P is the power of the laser radiation. Any offset background is given only by instrumental processing of the signal (e.g. synchronic noise). In contrast to classical absorption spectrometry, the photoacoustic signal is measured practically from zero background. This fact, together with the linear characteristics of the microphone provide photoacoustic methods with a broad dynamic range of 5 orders of magnitude or more [8]. The minimum detectable absorbance $N_{min}\sigma L$ for photoacoustic detection can be estimated:

$$N_{min}\sigma\ L = \frac{S_{min}}{PC} \approx 10^{-8} \tag{2}$$

where N_{min} is the minimum detectable concentration for a given molecule, L is optical path length and S_{min} is the noise-limited minimum microphone signal (e.g. S_{min}/P = 30-50 nV/W and C ≅ 3.5 V.cm/W, see [6]. It follows from relationships (1) and (2) that these small absorbances and thus trace concentrations can be detected by this method even on short optical pathways.

The principle of photoacoustic detection has been known for a number of years [9]; however, this technique is undergoing a renaissance in combination with lasers (e.g. [5]-[8],[10]). The CO_2 laser has a number of advantages that can be utilized for analytical applications: simplicity, high power and minimum interference with the absorption lines of atmospheric H_2O and CO_2. Combination of a CO_2 laser and photoacoustic detection provides high sensitivity and a linear dynamic range in a broad range of up to 6 orders of magnitude.

A broad dynamic range together with the high sensitivity of the photoacoustic method is utilized by this work. In our laboratory, we have carried out a number of spectroscopic and analytical studies using laser photoacoustic detection [4],[11]-[15]. The radiation source was a discrete tunable CO_2 laser. Topics of this work are following:

- a gas analyzer based on photoacoustic detection and its application for concentration monitoring in a wind tunnel,
- development of concentration standards, permitting both the actual concentration calibration and simultaneously simulation of emission sources of pollution in a wind tunnel,
- the actual measurement of the spatial profile of the concentration of pollutant on a model street canyon with line emission sources.

Simulation method

The physical simulation of flow and dispersion inside the urban street canyon must be divided into two subjects:

1) approximate simulation of boundary conditions, i.e. the conditions in the up-wind and down-wind direction in the streets of interest, termed the far fields,
2) approximate simulation of flow in the street canyon.

For simulation of urban buildings up-wind and down-wind of the streets of interest (the far field)- it is assumed that there is a sufficiently large urban built-up area on a flat plain in front of the streets of interest. It is thus possible to introduce the simplifying assumption according to which flow over the urban area is similar to flow over a rough surface, with a given, large value of the roughness parameter z_0 and a given flow of heat on the surface. Simultaneously, we assume that the urban area is sufficiently extensive to show horizontal homogeneity. Thus, processes in the immediate vicinity of the Earth's surface can be described by the Monin-Obukhov theory [16]. According to this theory, the mean velocity is described by the relationship

$$U = \frac{u^*}{\kappa}\left[f(\xi) - f(\xi_0)\right], \tag{3}$$

where:

$$\xi = \frac{z}{L}, \tag{4}$$

u* is the friction velocity, κ (≈ 0.42) is the von Kármán constant, ξ_0 is the dimensionless value of the roughness parameter z_0 and L denotes the Monin-Obuchov scale.
This scale is:

L>0	for a stable layered atmosphere
L<0	for an unstable layered atmosphere
L=0	for an indifferent atmosphere.

For the purpose of simulation it is necessary that the surface layer correspond to these relationships at sites prior to the studied street. The aerodynamic tunnel at the Institute of Thermomechanics does not yet provide modelling of stratification[1], and it is necessary to limit studies to indifferent stratification, for which the velocity profile is described by the known logarithmic profile. As the roughness parameter z_0 is large for urban areas, it is useful to shift the zero level of the velocity profile in the positive direction along the z axis by an amount h_0 and to set:

$$\frac{U}{u^*} = \ln \frac{(z - z_0)}{z_0} \tag{5}$$

It is assumed that the outside pollution level is equal to zero. In the near field case the assumption is made that the flow fields with characteristic scale H and characteristic velocity U are similar for sufficiently large values of the Reynolds number $Re \equiv H\, U/\nu$ [17]. Re generally depends on the shape of the flow boundaries, on the roughness of the surface and especially on other information that is required from simulation. For example Snyder [18] states that:

- for concentrations above the roof of a cubic building, it is necessary for the simulation - see [19] - to set

$$Re \equiv H\, U/\nu > 1.1\ 10^4\ , \tag{6}$$

where H is the height of the building,

- for the wake beyond the building, if the criterion of wake similarity is met, relationship (6) is sufficient - see [20].

For modelling structures with sharp edges, from the standpoint of pressure distribution on the surface, it is necessary that the Reynolds number correspond to the relationship[21]:

$$Re \equiv H\, U/\nu > 4.5\ 10^5\ . \tag{7}$$

Hoydysh et al. [22], fix a boundary value of 3400. It is necessary to verify this assumption because it would seem that a universal criterion does not exist.

A new aerodynamic tunnel was constructed in the Institute of Thermomechanics for research on flow and diffusion in the atmospheric boundary layer

---

[1] A design of an apparatus for simulation of thermal layering has been prepared.

(ABL) using the method of physical modelling in general. This was constructed as a linear tunnel with a constant cross-section of 1.5 x 1.5 m^2 and the working section 25.5 m from the entrance. The entrance is located in a quiescent chamber with dimensions of 6 x 6 x 12 m^3. A dust filter with an area of 64 m^2 is incorporated into its walls. This design limits the effect of external conditions. The entrance is connected to an elbow with blades, turning the flow to the horizontal direction, and a section consisting of a honeycomb flow structure and six removable grids with permeability coefficients β = 0.57 and 0.79. This section ensures equilibration of the velocity profiles and suppression of turbulent fluctuations. It is in turn connected to a 20.5 m long test section, which is connected to the actual working section. The distance of the measuring space from the entrance is thus 25.5 m. The working section with a length of 2 m has removable sidewalls made of 12 mm thick glass, permitting the use of optical and visualisation methods. The models are placed on the bottom or roof. The outlet of the tunnel consists of two elbows, mutually rotated by 90° on which, similar to the elbow at the entrance, are fitted components with blades, turning the flow to the vertical and then to the horizontal plane. The tunnel is driven by a ventilator with a power of 30 kW, with rotation regulation in the interval (0: 1550) r.p.m. and with adjustable blade angle. This configuration permits a velocity between 0 and 13 m/s to be attained in the centre of the working section [23].

Mathematical models

Solution of a system of the equations of motion together with initial and boundary conditions can be referred to as mathematical modelling. According to Atkinson´s [24] estimation the following system can be used for our task

$$\frac{\partial}{\partial x_j}(u_j) = 0 \tag{8}$$

$$\frac{\partial}{\partial t}(u_i) + \frac{\partial}{\partial x_j}(u_i u_j) = -\frac{1}{\rho}\frac{p}{\partial x_i} + \frac{\partial \tau_{ij}}{\partial x_j} \tag{9}$$

where τ_{ij} is the shear stress. Two types of approaches have to be used for the micro-scale flow tasks : Large Eddy Simulation (LES) and Reynolds averaging simulation (RAS). In LES, small-scale motions are filtered out,

$$\left\langle f(\vec{X})\right\rangle = \int f(\vec{X}')G(\vec{X},\vec{X}')d\vec{X}' \tag{10}$$

Here G is a space filter. The large-scale evolution is governed by the following system of equations

$$\langle U_{i,t}\rangle + \left\langle \langle U_i\rangle\langle U_j\rangle\right\rangle_{,j} + \frac{1}{\rho}\langle P\rangle_{,j} = -\left\langle \langle U_i\rangle u'_j + \langle U_j\rangle u'_i + u'_i u'_j\right\rangle_{,j} \tag{11}$$

with a proper modelling of the subgrid scales, e. g. by using the relation

$$\left\langle \langle U_i \rangle u'_j + \langle U_j \rangle u'_i + u'_i u'_j \right\rangle - \frac{1}{3}\delta_{kk} \left\langle \langle U_k \rangle u'_k + \langle U_k \rangle u'_k + u'_k u'_k \right\rangle = \\ = -K\left(\langle U_i \rangle_{,j} + \langle U_j \rangle_{,i} \right) \tag{12}$$

The various method of subgrid modelling and solution are discussed for example in [25].

In the Reynolds averaging approach, independent random functions – velocity vector, pressure, temperature,... are split into their mean value and a turbulent fluctuation f = <f>+ f´. Due to the non-linearity of the Navier-Stokes equations a tensor of the Reynold´s stress $-\rho<u'_i u'_j>$ appears in the Reynolds equations for the mean velocity vector:

$$\frac{\partial}{\partial x_j}(\overline{u}_j) = 0 \tag{13}$$

$$\frac{\partial}{\partial t}(<u_i>) + \frac{\partial}{\partial x_j}(<u_i><u_j>) = -\frac{1}{\overline{\rho}}\frac{\partial <p>}{\partial x_i} + \frac{(<u'_i u'_j>)}{\partial x_j} \tag{14}$$

These terms have to be computed by using a turbulence model. A great number of models have been considered.

The most simple one is based on Boussinesq´s turbulent viscosity assumption

$$\langle u'_i u'_j \rangle = -K\left(\frac{\partial \langle u_i \rangle}{\partial x_j} + \frac{\partial \langle u_j \rangle}{\partial x_i} \right) \tag{15}$$

Here K is the eddy viscosity that should be:

a) parameterized, or determined by Prantl´s mixing length l – Mean Velocity Field Closure;
b) determined from the turbulent field – Mean Turbulent Energy Closure e. g. by the relation

$$K = c_\mu \frac{k^2}{\varepsilon} \tag{16}$$

where k is the turbulent kinetic energy governed by the equation

$$\frac{Dk}{Dt} = \frac{\partial}{\partial y}\left[\left(\nu + \frac{\nu_t}{\sigma_k} \right) \frac{\partial k}{\partial y} \right] + \nu_t \left(\frac{\partial U}{\partial y} \right)^2 - \varepsilon \tag{17}$$

and ε the dissipation rate given by the equation[2]

[2] The equation (15) and (16) forms well known „k-epsilon“ model.

$$\frac{D\varepsilon}{Dt}=\frac{\partial}{\partial y}\left[\left(\nu+\frac{\nu_t}{\sigma_\varepsilon}\right)\frac{\partial\varepsilon}{\partial y}\right]+c_1 f_1\frac{\varepsilon}{k}\nu_t\left(\frac{\partial U}{\partial y}\right)^2-c_2 f_2\frac{\varepsilon^2}{k}+2\nu\nu_t\left(\frac{\partial U}{\partial y}\right)^2 \tag{18}$$

More complicated turbulence models solve the transport equation for Reynold´s stress – Mean Reynolds Stress Closure.

There are other sophisticated turbulence models. We should mention for example the Renormalization-Group techniques, which started from certain similarities between non-linear spin dynamics and Navier-Stokes equations with random forcing. Every commercial code contains a limited number of the models, e.g. models in FLUENT were tested for Hamburg University's experiments [3], and some numerical results [26] are on Fig 1.

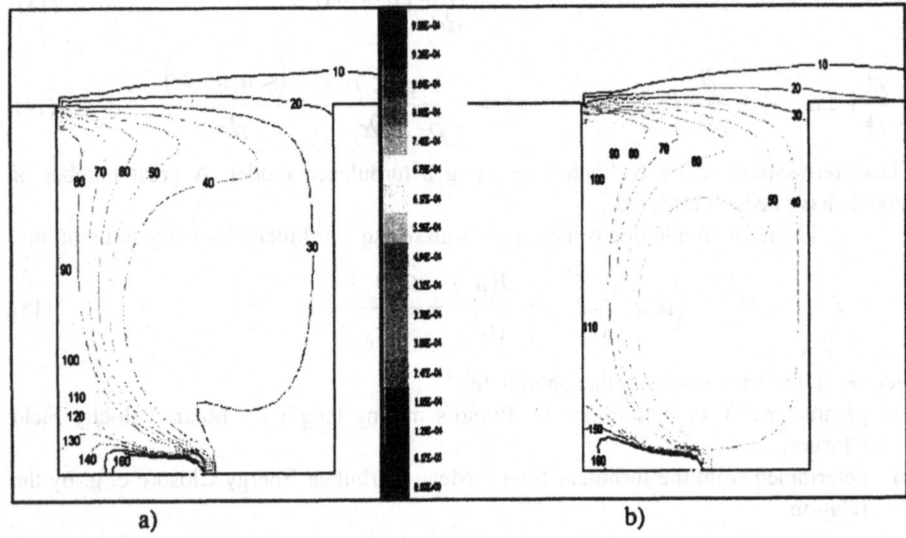

Fig 1.
Numerical simulation of dimensionless concentration K: a) k-ε model. b) RNG k-ε model

Experimental

The experimental system used in this work is depicted in Fig. 2. In a wind tunnel with dimensions of 1.5 x 1.5 m is created a model of a symmetric street canyon with a height and width of 70 cm, oriented perpendicular to the wind direction. A 30 kW ventilator with a speed regulator can be used to simulate the boundary layer of the atmosphere [23] with an external velocity in the interval from 1 m/s to 12 m/s.

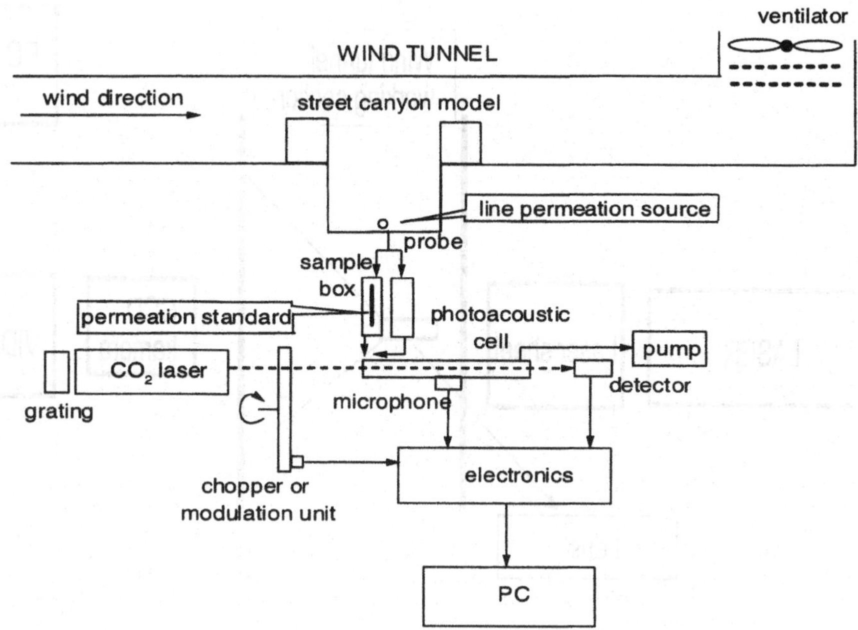

Fig. 2
Experimental set-up for concentration measurements of air pollution in the wind tunnel

The method of visualisation was employed for primary qualitative evaluation of the flow and indication of interesting areas for later local measurements. The principle of the system for visualisation and evaluation of the flow field consists see Fig.3 in capturing light traces of the particles moving in the flow field using a suitable photosensitive element. Illumination of the particles in the system used here is provided by a continuous Argon laser with power of 1 W, used in the "multi-line" regime, whose exit beam is introduced into fibers by the optical system. Cylindrical optics are placed at the end of the fiber to form a thin (about 1 mm thick) light wall. The use of optical fibers as an intermediate between the laser and the cylindrical focusing optics permits simple manipulation of the light wall.

For evolution of particles, a SAFEX smoke generator (DANTEX company) was used. It was found that this apparatus is suitable for visualising a large part of the flow field. An adapter was developed for visualising a small part of the flow field. The smoke is mixed with compressed air and the mixture is forced into selected places. This apparatus was also used for quantitative evaluation of the flow inside the street canyon.

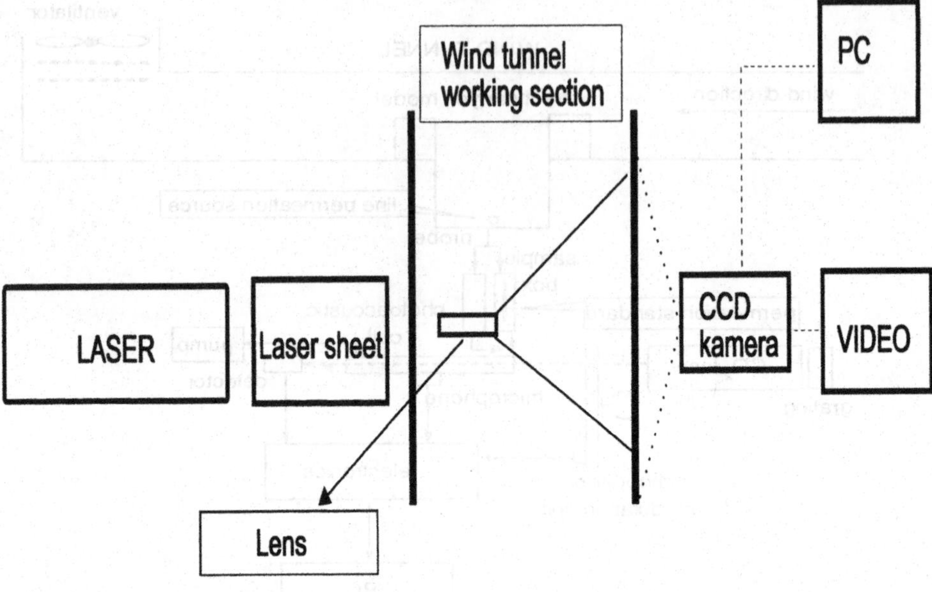

Fig 3
Experimental set-up of laser-sheet

A line pollution source is located in the center of the model street canyon, and is based on the principle of a permeation tube [11],[12], [27],[28]. This permeation source was made from a silicone hose with a length of 150 cm, with an external diameter of ∅ 8 mm and wall thickness of 1.5 mm. Silicone material was selected because it provided an adequate permeation concentration flux of the sample (simulating a series of cars on the street canyon). The hose was filled with 99.5% methanol and closed at both ends by stainless steel balls with a diameter of 8.5 mm. The permeation source could be reproducibly filled under pressure and was weighed over a prolonged period of time, to determine its concentration flux.

The line permeation source developed here has the following advantages:

- high precision of the resultant values for the flux of the pollutant substance (determined by long-term weighing of the source at constant temperature)
- high homogeneity of the flux along the entire source line.

The characteristics of this line source based on the principle of release of CH_3OH at various temperatures are depicted in Fig. 4.

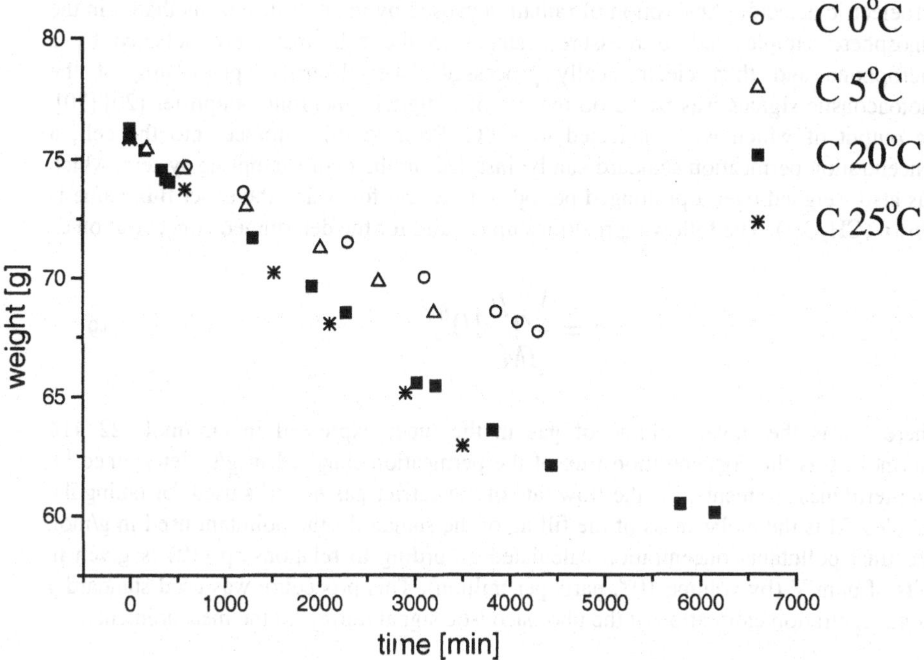

Fig. 4
Time dependence of the mass decrease in the permeation line pollutant source (CH_3OH) at various temperatures

A spatially adjustable sampling probe was introduced into the wind tunnel (a moveable polyethylene tube connected through the photoacoustic cell with a pump), and was used to continuously sample the atmosphere at defined places along the studied street canyon. To a first approximation, we limited the study to monitoring of the transverse central cross-section of the modelled street canyon, to obtain a concept of the transverse concentration profile. Samples of the atmosphere were taken by the probe at a rate in the range 3-4 cm^3/s - these are thus volumes that are negligible from the standpoint of the dynamics of the given wind tunnel. The sampling cell used was a thermally stabilized (30 ± 1 °C) brass tube with a length of 38 cm and diameter of 8 mm, fitted with an electret microphone (TPR 175 E, Japan), and windows for infrared optics (KBr) with a diameter of 5 cm and thickness of 7 mm, next to which were located the buffer spaces for suppressing parasitic acoustic signals caused by absorption on the windows of the cell. The infrared radiation of the discretely tunable CO_2 laser (Edinburgh Instruments WL-8-GT) passed through the cell and impinged on a pyroelectric detector. Absorption of radiation caused by the presence of methanol in the atmosphere samples led to pressure changes in the cell, that were detected by a microphone and then electronically processed. The electronic processing of the photoacoustic signals was based on the use of a digital synchronic amplifier [29],[30], the output of which was connected to a PC. Prior to the entrance into the cell, a concentration permeation standard can be inserted in the entire sampling system, which was also weighed over a prolonged period of time and for which the exact flux value is known (CH_3OH). The following relationship is valid for the determined concentration c:

$$c = \frac{V_m p}{fM} 10^6 \tag{19}$$

where V_m is the molar volume of gas in the tube, expressed in cm^3/mol (22 414 cm^3/mol), p is the concentration flux of the permeation standard in g/s, determined by long-term measurements, f is the flow rate of the carrier gas in cm^3/s used for taking the samples, M is the molar mass of the filling of the standard - the pollutant used in g/mol. The final pollutant concentration calculated according to relationship (19) is given in units of ppm V (by volume 10^{-6}, parts per million). This procedure was used standardly for concentration calibration of the photoacoustic signal during all the measurements.

Results and Discussion

The assumptions of the Monin-Obuchov theory [16] for the exit flow were verified for various values of the mean velocity U_0 obtained at the centre of the tunnel in front of the measuring space. Thus, good agreement was demonstrated between the measured values of the mean velocity vector at the wall in front of the measuring space and a logarithmic profile (5). Fig. 5, demonstrates such a good agreement for a reference velocity of $U_0 = 5$ m/s.

For quantitative estimation of the flow inside the street canyon, the visualization method was first employed for a model with H = 0.7 m. A mixture of smoke was emitted into the air stream in the centre of the edge of a lee building so that a major part of the mixture was drawn into the street canyon. The flow field was illuminated by the light wall, perpendicular to the axis of the street from a lens located

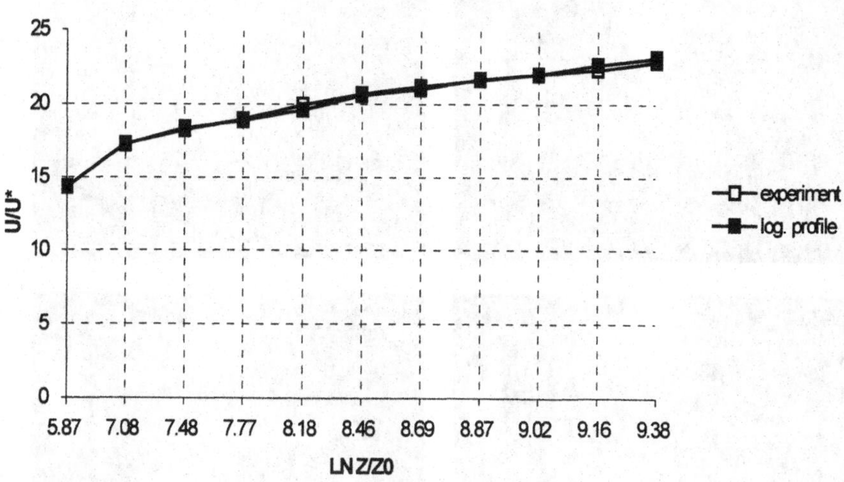

Fig. 5
Comparison of mean velocity profile over urban area with logarithmic one. (z_0 = 0.04 m, u^* = 0.2134 m/s)

in the roof of the tunnel above the model. The camera was located on the side glass wall to capture the developments in the illuminated 2-D flow field in the centre of the street canyon. In this way, a sequence of "instantaneous" pictures of the flow field inside the street canyon were obtained for values of the external velocity in the interval $U_0 \in (0.5; 5.0)$. The recordings from the video camera are available at the Institute of Thermomechanics of the Academy of Sciences of the Czech Republic and were presented at EUROMECH 391 Col. [31]. For illustration, instantaneous flow fields are taken from the recorded sequence for cases with external velocity U_0 in the interval (0.5; 5.0) m/s, i.e. for values of the Reynolds number of $Re \equiv U_0H/\nu \in (2.3 \times 10^4; 2.3 \times 10^5)$, see Fig. 6, where views are given from the side into the street canyon.

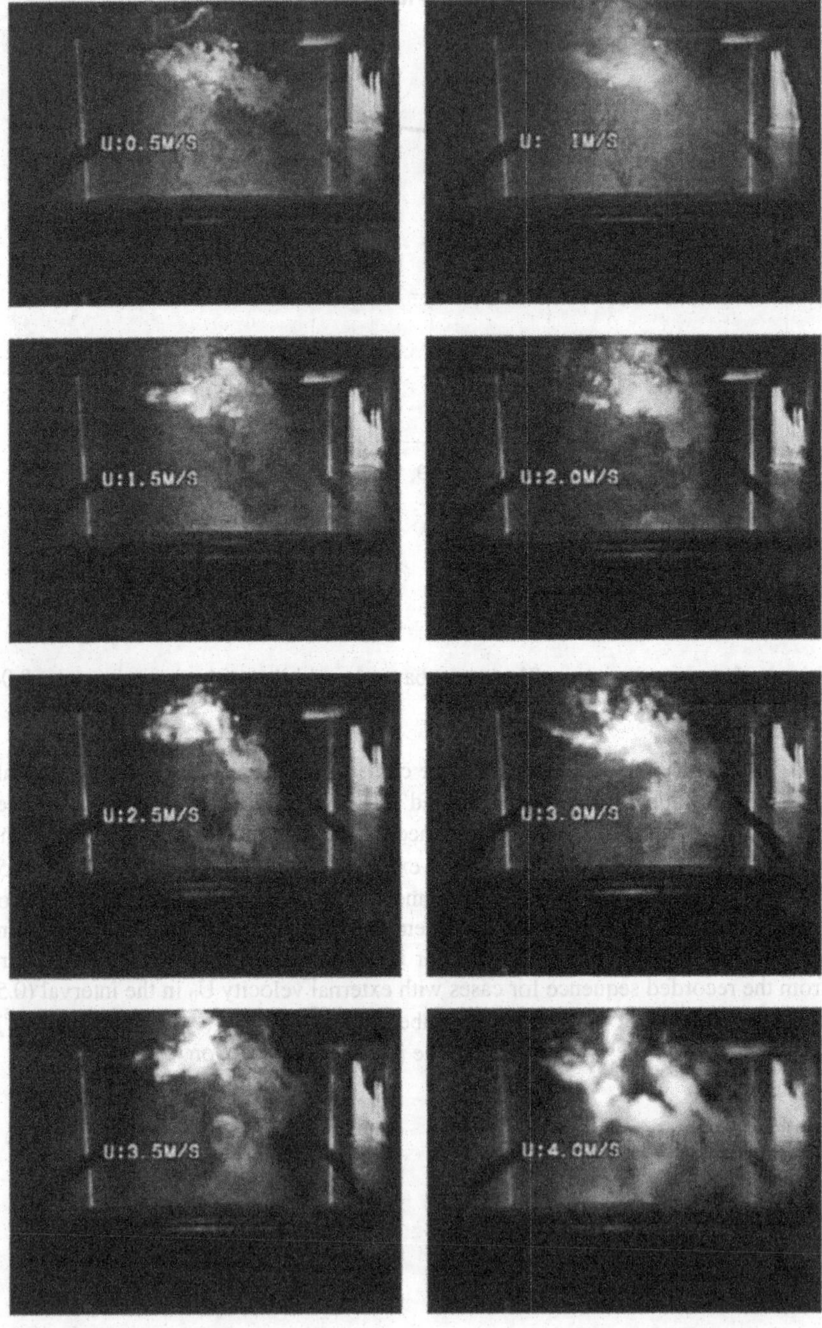
U:0.5M/S
U: 1M/S
U:1.5M/S
U:2.0M/S
U:2.5M/S
U:3.0M/S
U:3.5M/S
U:4.0M/S

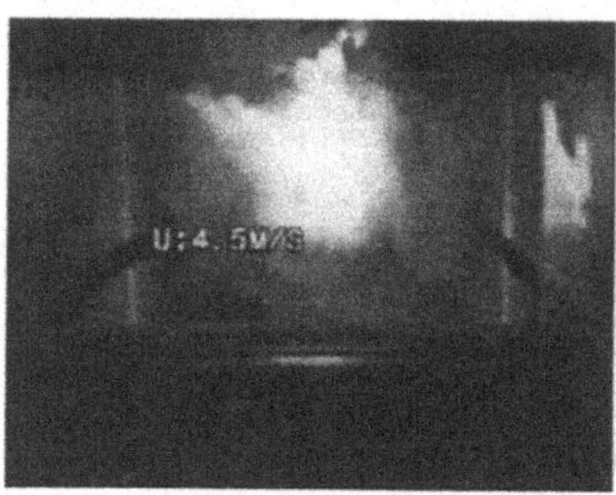

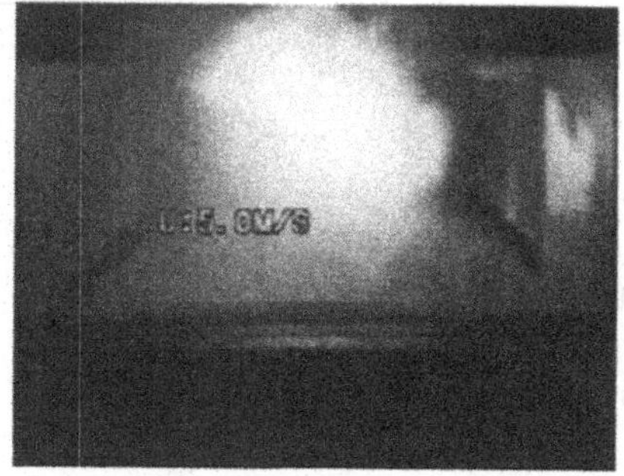

Fig. 6
Visualisation of the flow inside a street canyon with the mean external velocity U_0 = 0.5, 1.0, 1.5, 2.0, 2.5, 3.0, 3.5, 4.0, 4.5 and 5.0 m/s, (Re = 2.3 10^4, 4.6 10^4, 7 10^4, 9.3 10^4, 1.1 10^5, 1.4. 10^5, 1.6 10^5, 1.9 10^5, 2.1 10^5 a 2.3 10^5)

Amongst other things, it was estimated from the recording that:

- the flow inside the street, which forms a characteristic eddy, is highly nonstationary and, especially for lower velocities, liquid is drawn from the cavern into the external stream,
- the flow fields with Reynolds number Re>5 . 10^4 seem to be qualitatively similar
- however, with increasing external velocity U_0, i.e. with increasing Re, the centre of the eddy area inside the canyon is shifted in the up-wind direction.

Work has started on verification of the validity of the assumption on the independence of the flow field on the Reynolds number - see the previous discussion on this subject. At selected sites in the vicinity of the linear source at the bottom of the street, concentration measurements were carried out as a function of the reference velocity and the measured concentration values were converted into dimensionless form on the basis of equation (21), where the yield of the source Q/L = 5.5 . 10^{-05} g/s. The results are given in Fig. 7. On the basis of these results, it can be suggested that, for the configuration used in the vicinity of a linear source, the mean concentration values are independent of the Reynolds number for approx.

$$Re \equiv L_b\, U/\nu \geq 1.5 \times 10^4 \qquad (20)$$

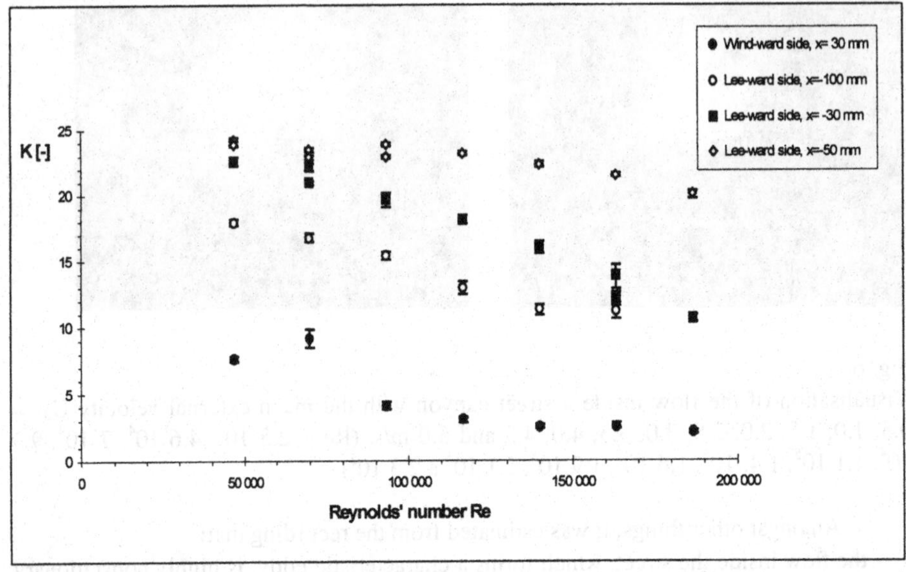

Fig.7
Dependence of the nondimensional concentrations $K = CHU_0\,L/Q$ around the line source on Reynold´s number.

The actual sensitivity of detection by the laser photoacoustic spectrometric method was tested using a generated concentration of 1.76 ppm V (2513 µg/m^3). This concentration level was detected with a signal to noise ratio of ≅ 300. The absorption cross-section σ of CH_3OH is 72 x 10^{-20} cm^2 [32] for a line of the CO_2 laser (9 P(34)) used. From Eq. (2) we can determine that the minimum detectable absorbance $N_{min}\sigma L$ of the apparatus used was equal to approx. 4.3 x 10^{-6}, corresponding to a concentration of ≅ 8 µg/m^3. The sensitivity of the adjusted apparatus was not the maximum possible that can be attained using the analyzer, but was quite sufficient for monitoring the concentration level in the wind tunnel. We used a moveable sampling probe to carry out spatial profile measurements of the dispersion of a pollutant from a line source located in the centre of the model street canyon for one value of the wind speed of 1.5 m/s. The results are depicted in Fig. 8.

The values of the concentration measured C [g/cm^3] were converted to dimensionless concentration K using the height of the buildings H [cm], the transverse length of the street L [cm], wind speed outside of the street canyon U [cm/s] and the concentration flux of the line pollutant source Q [g/s] according to the formula:

$$K = \frac{CHUL}{Q} \qquad (21)$$

The distance from the linear source is standardized in relation to the overall distance from the source to the building, where a negative sign is used to distinguish

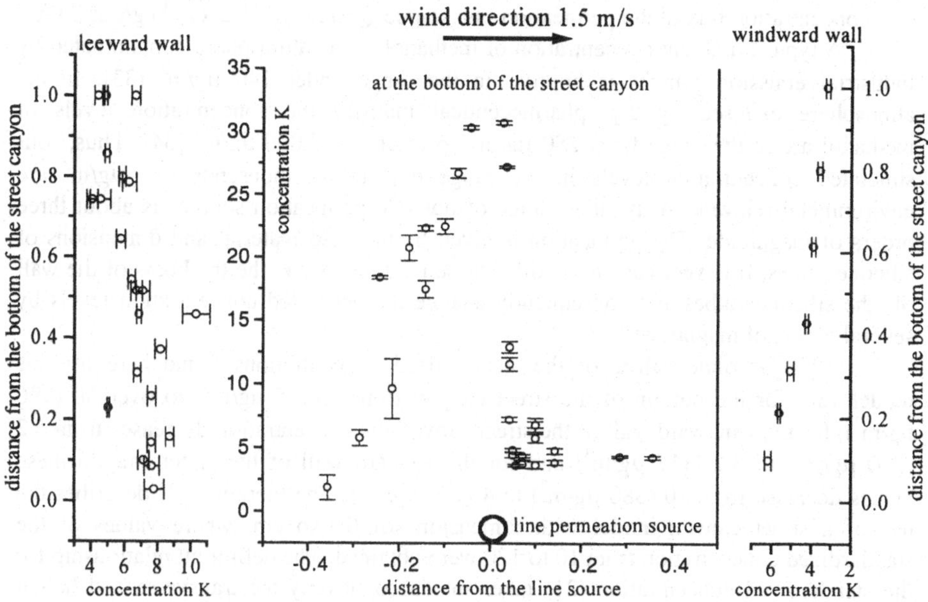

Fig. 8
Concentration profile of dispersion of pollutants on the bottom of a street canyon and on the wall of the street canyon

the leeward side of the street from the windward side of the street. The sampling position on the walls of the building is again standardized in relation to the height of the model building. The measurement is carried out on the bottom and on the walls of the street canyon. The resultant profiles clearly show a substantially higher concentration level on the leeward side of the street, while the opposite windward side of the street is ventilated with surprising efficiency. This phenomenon is closely connected with the formation of circulation currents which shift the pollutants from the windward to the leeward side of the street, where they are accumulated. The concentration profiles on the walls of the buildings exhibit a clear decreasing tendency and the windward side of the street is once again ventilated better than the leeward side. The error values of the individual measurements contain clearly visible areas where there is a greater concentration variation, probably caused by rapid circulation of the air at these places (area -0.2 for the bottom of the street and area 0.4 for the leeward wall of the street). These phenomena could be elucidated in detail only using measurements with a much smaller time constant (in this experiment we used a time constant of 1s). The model used had the following parameters:

- transverse length L = 150 cm,
- distance from the line source to the walls of the street canyon on both sides, 70 cm,
- height of the walls of the street canyon H = 70 cm,
- wind speed U = 150 cm/s,
- concentration flux of the permeation line source $Q=(8.3x10^{-5} \pm 2 x10^{-6})$ g/s at 20 °C.

A typical ambient concentration of methanol in the atmosphere contaminated by industrial emission sources is approx. in the range under 140 $\mu g/m^3$ [33]. In the atmosphere exposed by e.g. pharmeceutical industry the concentration levels of methanol are in the range from 280 $\mu g/m^3$ up to above 22000 $\mu g/m^3$ [34]. Thus, our simulated concentration levels in the range tens up to thousends of $\mu g/m^3$ are environmetal relevant. A dynamic range of our line permeation source is about three orders of magnitude. The permeation is given by the used material and dimensions of silicone tubes. It is very easy to modify these parameters e.g. the thickness of the wall of the silicone tubes and consequently change the generated concentration levels by several orders of magnitude.

The absolute values of the standardized concentrations found here for this model vary for the bottom of the street canyon from 4 (135 $\mu g/m^3$) to over 30 (990 $\mu g/m^3$); for the windward wall of the street canyon the concentration decreases from 6.2 (230 $\mu g/m^3$) to 3.2 (117 $\mu g/m^3$) and for the leeward wall of the street canyon these values decrease from 10 (380 $\mu g/m^3$) to 4 (150 $\mu g/m^3$). The literature [3] describes the use of a street canyon model with dimensions of 6x6x6 cm, where values of the standardized concentration from 10 to 120 were obtained. The definition relationship for the standardized concentration (21) takes into account only the transverse and height dimensions of the street canyon and the longitudinal dimension is an independent variable. The order of magnitude difference in the geometry of the two models could be the reason for the differences in the measured concentrations.

In addition to the spatial concentration profiles, we also carried out concentration measurements in dependence on the wind speed, see Fig. 9. These measurements were carried out at three places on the leeward side and at one place on the windward part of the street canyon. The dependences show how the pollutant concentration decreases with increasing wind speed. We employed the range from 1 m/s to 4 m/s. For speeds of greater than 2 m/s the values on the windward side of the street in the ventilated space are decreased down to the background value. A similar dependence was found for the leeward side of the street canyon for measurements at distances greater than 10 cm (< -0.143). However, the values do not decrease down to the background value. Areas in the leeward side of the street canyon, in the vicinity of the line source - distance of less than 10 cm (> -0.143) exhibit a much slower decrease the pollutant concentration with increasing wind speed; however, this effectiveness of ventilation increases with each, even minimal, increase in distance from the line source (compare "left wall 3 cm, -0.043" and "left wall 5 cm, -0.071" in Fig. 9). It follows from the studies carried out for the leeward side of the street canyon that the steepness of ventilation in the tested distance range rapidly increases with increasing distance from the linepollution source (from steepness $\cong$ -0.8 $m^{-1}s$ for 3 cm to $\cong$ -4.1 $m^{-1}s$ for 5 cm). At greater distances from the line sources, the steepness begins to decrease (see steepness ≅ -3.1 $m^{-1}s$ for 10 cm). The large difference between the concentration levels on the windward and leeward sides of the street canyon is illustrated very well on the concentration levels at the sampling point 3 cm from the source in the

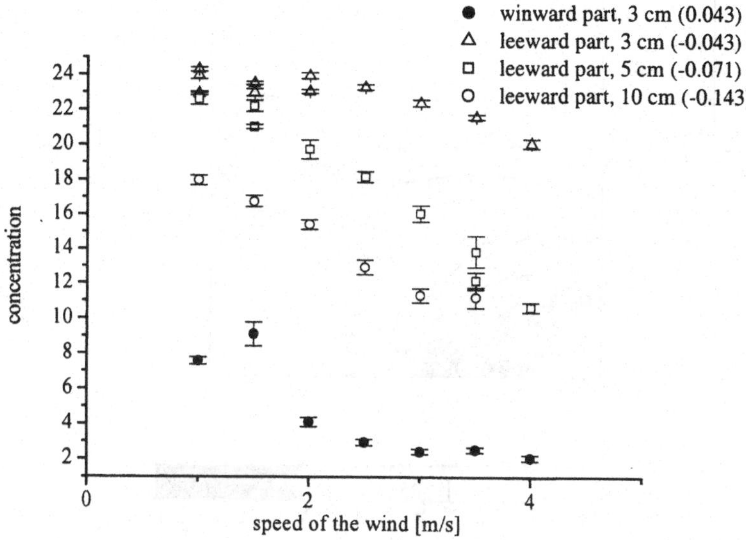

Fig. 9
Dependence of the concentration level of pollution at various sites on the bottom of the street canyon on the wind speed.

windward part and 3 cm from the source in the leeward part (See Fig. 6). At a wind speed of 4 m/s the value on the windward side is almost equal to the background concentration of 87 $\mu g/m^3$, while at a symmetrical distance on the opposite side of the source on the leeward side of the street canyon the concentration level is an order of magnitude higher (850 $\mu g/m^3$).

Model of a simple road with one lane

The spatial measurements of methanol concentration levels were performed in the whole volume of the street canyon for a reference velocity 1.5 m/s (Fig. 10).

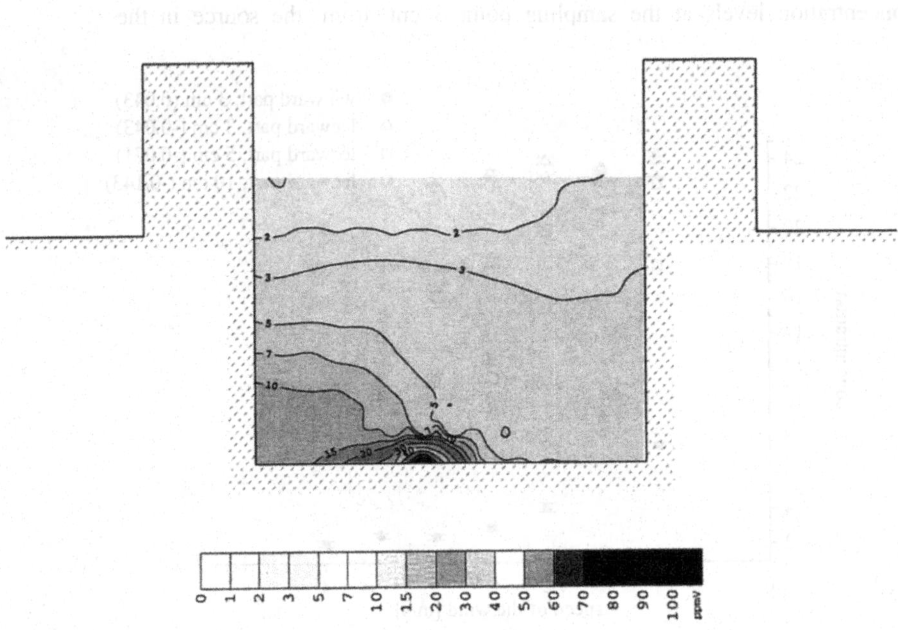

Fig. 10
Mean concentration across the street-canyon distribution in units ppmV. Model of a simple road with one lane.

Horizontal distances between individual sampling points were 5 cm (at a level of 0, 5, 7, 10 and 20 cm from the bottom) and near to the top of canyon varied from 5 to 20 cm (at heights of 30, 40 and 50 cm). The highest concentrations found at the bottom of the street canyon in the near of the permeation source reached values of about 90 -100 ppmV. A minimum was found in the concentration profile at a height of approximately 10% - 20% of the height of the street above the linear source. Increasing concentration on the leeward side of the street is also obvious from Fig. 8, 10 and 11. By contrast, the opposite windward side is ventilated with greater efficiency. This fact relates to the

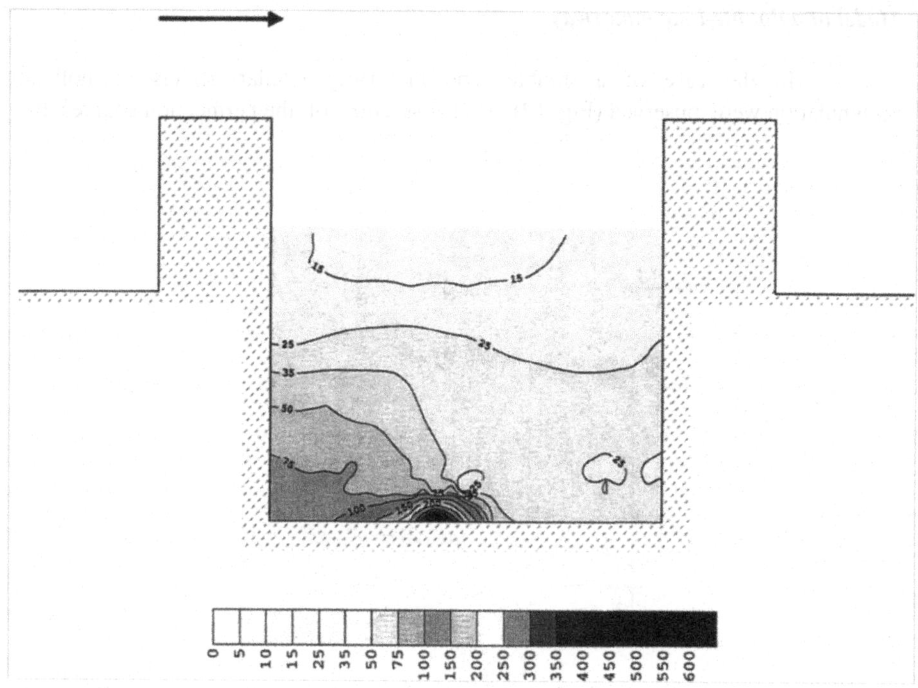

Fig. 11
Dimensionless mean concentration K across the street-canyon distribution. Model of a simple road with one lane.

formation of circulation currents, which move the contamination into the leeward side of the street where it accumulates. This is in good agreement with the results of mathematical simulations [35]. The simulations (see Fig. 1) predict that the concentration decreases with increasing wind speed and, furthermore, concentrations at the leeward side of the street are higher than those at the windward side. These are the most essential features of pollutant dispersion in the street canyons, and therefore the street model, with some minor modifications, is still widely used (especially for engineering applications). A more detailed description cannot be made on the basis of simple mathematical modelling. An essential drawback of the model is its very rough parameterisation of the wind direction dependence. Moreover, with decreasing wind speed, uniform concentration distribution across the street canyon may be assumed. Mathematical models generally do not describe these properties. Therefore it is not recommended to use them for wind speeds lower than 1 ms^{-1}. Under these conditions, experimental modelling plays the dominant role.

Model of a double-lane motorway

In the case of a double lane motorway similar effects of pollutant accumulation were observed (Fig. 12). In the location of the permeation sources, the

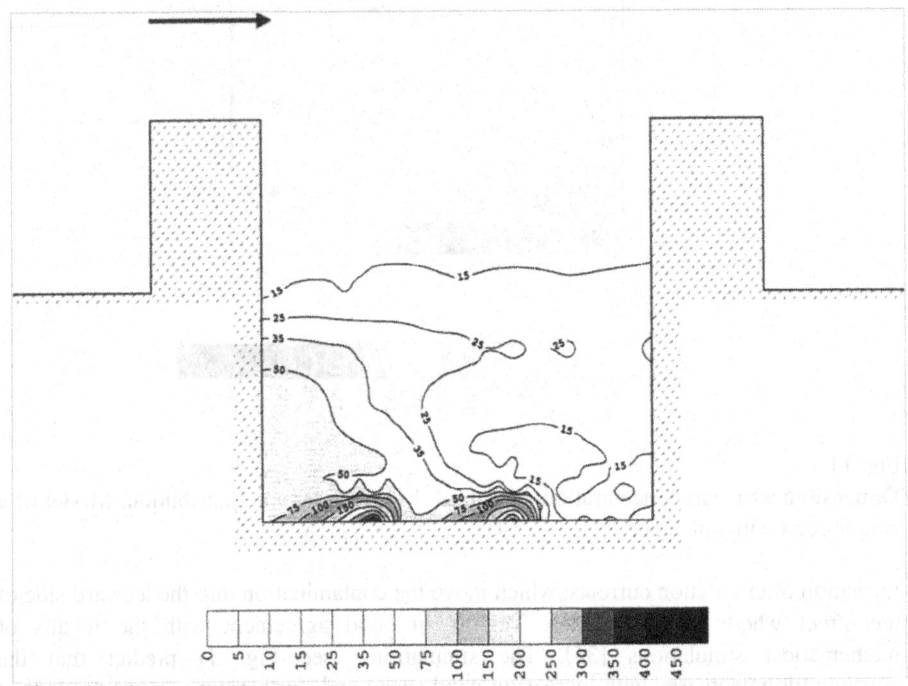

Fig. 12
Dimensionless mean concentration K across the street-canyon distribution. Model of a double-lane motorway.

methanol concentration is the highest and reaches values exceeding 100 µgm^{-3} with permeation rate of the used source of 6.07x10^{-4} g s^{-1} (±10%). The figure 12 clearly suggests the great magnitude of contamination that the drivers of slowly moving cars in narrow motorway lanes are exposed to as well as the people walking on the leeward side of the street canyon. Only 10 % of the ground-level concentration remains at half the vertical distance on the leeward side of the street canyon.

Conclusion

This article describes the results of physical simulation of diffusion in a street canyon. A model street canyon was proposed and produced, with variable characteristic scale, on which it is possible to study the effect of the Reynolds number, in addition to measuring the flow and concentration fields. For qualitative evaluation of the flow fields, the apparatus was modified for visualization of the flow. The technique of laser photoacoustic detection was used for measuring mean concentration values. It has the advantages of high sensitivity and a broad dynamic range, permitting monitoring of concentrations from trace amounts to saturation values. The use of this method permitted proposal of a simple model of a linear permeation pollutant source, functioning on the principle of concentration standards and ensuring high precision and homogeneity of the concentration flow on a permeation basis.

The visualization method yielded a video recording depicting the flow in the street canyon for various reference velocities. On the basis of this recording, it is possible to make a preliminary estimate that the flow inside the canyon is independent of the Reynolds number for $Re \equiv L_b\ U/\nu \geq 5 \times 10^4$. Spatial measurements were carried out on the street model of the concentration field inside the street canyon for a reference velocity of 1.5 m/s. The measurement demonstrated that the maximum concentration is located at the bottom of the street in the vicinity of the source in the direction towards the lee wall. Finally, experiments were carried out on verification of the dependence of the concentration field on the Reynolds number. This measurement was carried out in the vicinity of the linear source. It was estimated from the results that the mean value of the concentrations in the vicinity of the source are independent of Re for $Re \equiv L_b\ U/\nu \geq 1.5 \times 10^4$.

It is apparent that further detailed measurements are necessary. It will also be necessary to estimate the dependence on the Reynolds number for areas more distant from the source and to carry out detailed measurements of the concentration fields, where possible for cases where the criterion of independence of Re is fulfilled. It further seems necessary to carry out detailed measurements of the concentration in the vicinity of its minima above the source and to attempt to explain this phenomenon

This work describes the use of the laser photoacoustic detection technique, proposed for simulation of the dispersion of atmospheric pollution in urban agglomerates in a wind tunnel. The advantages of this method consist in its high sensitivity and broad dynamic range, which permit monitoring of the concentration from trace amounts to saturation values. A model street canyon with a line permeation pollution source was employed to carry out a number of measurements of spatial profiles; the dispersion of the pollutant was studied at the bottom and on the walls of the street canyon together with the concentration variation with changes in the wind speed. The model was used to demonstrate the differences in ventilation of the bottom of the street canyon. It was clearly shown that, in contrast to the windward side of the street canyon, the leeward side is never ventilated down to the background value of the detected pollution source, even at a wind speed of 4 m/s.

Spatial measurements of the concentrations were carried out on the models inside the street canyon for a reference velocity of 1.5 m/s. The measurement demonstrates that the maximum concentration is located at the bottom of the street in the vicinity of the source in the direction towards the leeward wall.

For simulation in a wind tunnel, a new type of line permeation pollution source has been proposed, that has been adopted from the field of permeation concentration standards, ensuring high precision and homogeneity of the concentration flux. Both of the applied methods - laser photoacoustic detection and the use of a permeation standard - contribute substantially to the development of new analytical methods of study and simulation of the dispersion of pollution in urban agglomerates.

Acknowledgment

This work was carried out in the framework of projects supported by the Grant Agency of the Academy of Sciences of the Czech Republic (A3076801/1998) and COST 715.

References

1. Meyers R. A.(ed.)(1998): *Encyclopedia of Environmental Analysis and Remediation*, ed. John Wiley & Sons, Inc., New York.
2. Adrian, R. J.(1991): Particle-Imaging techniques for fluid mechanics. *Annu. Rev. Fluid Mech.*, **23**, 261.
3. Pavageau M.(1996): *Concentration Fluctuations in Urban Street Canyons*, Meteorologisches Institut der Hamburg Universität, Hamburg.
4. Zelinger, Z., Civiš, S., Jaňour, Z. (1999): *Analyst,* **124**, 1205.
5. Sigrist M.W.(1998): Air Monitoring, Optical Spectroscopic Methods, in *Encyclopedia of Environmental Analysis and Remediation*, ed. R. A. Meyers, John Wiley & Sons, Inc., p. 84.
6. Sigrist, M.W.(1994): Air Monitoring by Laser Photoacoustic Spectroscopy, in *Air Monitoring by Spestroscopic Techniques*, ed. M. W. Sigrist, Chemical Analysis Series, Vol.**127**, p. 163 John Wiley & Sons, Inc.
7. Patel, C.K.N(1978):. *Science*, **202**, 157.
8. Sigrist, M.W., Bernegger, S., and Meyer, P.L.(1989): Atmospheric and exhaust air monitoring by laser photoacoustic spectroscopy. In *Photoacoustic, Photothermal and Photochemical Processes in Gases* (P. Hess, Ed.), Top. Curr. Phys., vol. **46**, chapter 7. Springer-Verlag, Berlin.
9. Bell A.G.(1880): *Am. J. Sci.*, **20**, 305.
10. Sigrist, M.W., Bernegger, S., Meyer, P.L.(1989): *Infrared Phys.*, **29**, 805.
11. Zelinger, Z., Papoušková, Z., Jakoubková, M., Engst, P.(1988): *Coll. Czech. Chem. Commun.*, **53**, 749.
12. Zelinger, Z., Engst, P., Papoušková, Z., Jakoubková M. (1988): *Springer Ser. Opt. Sci.*, **58**, 131, Heidelberg.
13. Steiner, V., Engst, P., Zelinger, Z., Horák, M.(1989): *Coll. Czech. Chem. Commun.*, **54**, 2667.
14. Zelinger, Z., Jančik, I., Engst, P. (1992): *Applied Optics*, **31**, 6974.
15. Zelinger, Z., Střižík, M., Kubát, P., Civiš, S. (2000): *Anal. Chim. Acta*, **422**, 179.
16. Monin A. S., Obukhov A. M. (1953): Dimensionless characteristics of turbulence in the atmospheric surface layer, *Doklady AN SSSR* **93**, 223-226.

17. Townsend A. A. (1976): *The Structure of Turbulent Shear Flow,* Second Edition, Cambridge University Press, Cambridge, 429.
18. Snyder W. H.(1972): *Boundary Layer Meteoroplogy* **3**, 113-134.
19. Golden J.(1961): *Scale Model Techniques*, M. S. Thesis, College of Eng., New York Univ.
20. Smith E. G. (1951): The Feasibility of Using Models for Predetermining Natural Ventilation, *Res. Rep. Tex. Eng. Exp.*, 26.
21. Fisher O., Koloušek V., Pirner M.(1977): *Aeroelasticita stavebních konstrukcí,* ACADEMIA, Praha,495.
22. Hoydysh W. G., Griffiths R. A., Ogawa Y.(1974): A Scale Model Study of the Dispersion of Pollution in Street Canyon, *67 th Annual Meeting of the Air Poll. Control Assoc.*, Denver, CO, 24.
23. Jaňour Zb.(1995): A new atmospheric boundary layer wind tunnel at the Institute of Thermomechanics- *EUROMECH Col. 338*, Bologna.
24. Atkinson B. W.(1995): *Introduction to the fluid mechanics of meso-scale flow field,* in Gyr A., Franz-S. Rys (ed.): *Diffusion and Transport of Pollutants in Atmospheric Mesoscale Flow Fields*, Kluwer Academic Publishers, Dotrecht Boston, London, 1-22.
25. Lesieur M.(1997): *Turbulence in Fluids*, Kluwer Academic Publishers, Dortrecht Boston London, 515.
26. Drábková S., Jaňour Z., Kozubková M., Šťáva P.(1997): *Kolokvium Dynamika tekutin '97*, 11.
27. O'Keefe A.E., Ortman G. C.(1966): *Anal. Chem.*, **38**, 760.
28. Stellmack M.L., Street Jr. K.W(1983): *Anal. Lett.*, **16** (A2), 77.
29. Tanarro, I., Campos, J.(1986): *J. Phys. E: Sci. Instrum.*, **19**, 125.
30. Probst, P.A., Collet, B.(1985): *Rev. Sci. Instrum.*, **56**, 466.
31. Jaňour Z., Holpuch J. Střižík M.(1999): in *EUROMECH Col. 391*, Book of Abstracts, Prague 1999, 17.
32. Loper, G.L., Calloway, A.R., Stamps, M.A., and Gelbwachs(1980): J.A., *Appl. Opt.*, **19**, 2726.
33. Meyer P.L. and Sigrist M.W.(1990): Rev. Sci. Instrum., **61**, 1779.
34. Sigrist, M.W.(1994b): *Analyst*, **119**, 525.
35. Berkowitz R. (1998): in *Urban Air Pollution – European Aspects*, J. Fenger, O. Hertel, F. Palmgren (Eds.), Kluwer Academic Publishers, London, p. 223-251.

ANHARMONICITY IN THE VIBRATIONAL SPECTRA OF C_{60} AND ITS IMPLICATIONS IN LABORATORY SPECTROSCOPY AND ASTROPHYSICS

LÁSZLÓ NEMES
Research Laboratory for Materials and Environmental Chemistry, Chemical Research Center, Hungarian Academy of Sciences, Pusztaszeri út 59-67, H-1025 Budapest, Hungary

DANIEL A. JELSKI
Department of Chemistry, Rose-Hulman Institute of Technology, 5500 Wabash Avenue, Terre Haute, Indiana 47803 U.S.A.

1. Introduction

Following the discovery and laboratory preparation of fullerenes [1] there has been explosive scientific activity concerning the chemical and physical properties of C_{60}, C_{70} and other carbon clusters. Fullerenes continue to be at the center of fundamental inquiry. Perhaps the first of originally posed questions was whether fullerenes occur in interstellar space and in circumstellar shells (e.g. [2,3]) . This is a logical expectation as long, linear carbon chains, the cyanopolyynes had previously been found in interstellar space by means of microwave rotational transitions in radio astronomy, first HC_3N [4] and later much longer chains, up to $HC_{11}N$ [5]. In the laboratory, sensitive spectroscopic methods, such as cavity ringdown spectroscopy, are used for the detection and characterization of these long carbon molecules [5]. In fact, it was the detection of these chain molecules that started the query that ended in the discovery of fullerenes (see, e.g. [6]).

There have been proposals for the detection of C_{60} in space. As C_{60} has icosahedral symmetry, it has no permanent dipole moment, so radio astronomy cannot be used for its detection. Alternative possibilities are electronic and vibrational transitions. Fullerene properties important for searches in circumstellar and interstellar sources were summarized already in 1992 [7]. Looking for electronic spectra of C_{60} and C_{60}^+ was suggested in connection with diffuse interstellar bands and simulations of rotational band contours have been published [8-10]. The finding of two interstellar absorption bands coincident with near infrared spectral features of C_{60}^+ was reported [11]. In addition a test of finding C_{60}^+ in the interstellar medium by means of infrared vibrational emission from C_{60}^+ in the mid infrared range was proposed [12]. A summary of the relevance of C_{60} and its derivatives for diffuse interstellar bands was provided in 1995 [13].

Potential sources for fullerenes and their chemical derivatives are carbon-rich astronomical objects, such as R. Coronae Borealis (RCB) stars and circumstellar shells of

J. Demaison et al. (eds.), Spectroscopy from Space, 301–316.

some cool giant stars [14-16]. A mid-infrared search for C_{60} in RCB stars and in IRC+10216 using the NASA Infrared Telescope Facility in the 8.6 micrometer spectral region was published in 1995 [17].

All the spectroscopic searches for C_{60}, C_{70} and other fullerenes in astronomical objects are based either on calculated spectral features, as referred above, or on a very limited basis of laboratory spectroscopy on fullerenes in the gaseous phase. In the ultraviolet range resonant two-photon ionization (R2PI or REMPI) spectra were measured for C_{60} and C_{70} using supersonic beams [18-21]. The ultraviolet absorption spectrum of C_{60} has been remeasured recently using the helium nanodroplet technique providing a gas-phase analogue for solid phase matrix isolation spectroscopy [22]. Infrared resonance enhanced multiphoton ionization spectra for C_{60} and C_{70} were also obtained [23,24]. In the infrared range there is a scarcity of data for gas-phase measurements. As far as we know there have only been two papers published on the gas-phase infrared emission spectra of C_{60} and C_{70} [25,26].

2. Vibrational characteristics of C_{60}

The C_{60} fullerene belongs to the icosahedral point group (I_h), it has 180 cartesian degrees of freedom, of which 174 motions correspond to genuine vibrations. Most of these modes are grouped into degenerate symmetry species. The distribution of the 46 normal modes among the I_h species is:

$$\Gamma_{vib} = 2A_g + 3F_{1g} + 4F_{2g} + 6G_g + 8H_g + 1A_u + 4F_{1u} + 5F_{2u} + 6G_u + 7H_u \qquad (1)$$

There are four infrared active modes (F_{1u}), ten Raman active modes ($2A_g$+$8H_g$) and 32 optically inactive, silent modes. Optical spectroscopy probes only 14 modes, vibrational information for the rest comes mainly from various diffraction and scattering methods, such as neutron diffraction, and electron energy loss spectroscopy. A detailed overview of the vibrational spectroscopy of fullerenes is given in a recent book chapter [27].

Approaches aiming at the determination of force-constants using experimental data are rendered difficult by the fact that there are many more harmonic force constants for C_{60} than there are data to use, even if full icosahedral symmetry for C_{60} is utilized. This is easy to show by calculating the number of harmonic force constants. For an n by n symmetry block, the number of independent force constants (f.c.) is n(n+1)/2. For C_{60} there are 3, 36 and 10 harmonic f.c.'s for the A_g, H_g and F_{1u} species, resp., and there are 14 fundamental frequencies available for these modes from Raman and infrared spectra. The total number of harmonic f.c.'s for the optically silent 32 modes is 6+10+21+1+15+21+28 =102 from Eq.(1). So the f.c.'s are seriously underdetermined.

In principle it would be possible to use other force-field dependent experimental data, such as isotopic shifts in fundamental frequencies and vibration-rotation interaction data to improve upon this situation. Such experimental data are, however, not available for C_{60} or for any other fullerene. Thus the only solution to this indeterminacy problem is to limit the number of symmetry f.c.'s by using some harmonic force-field models. From the above numerical comparison it is clear that the model to be used should contain as few f.c. parameters as possible.

As a starting point one has to resort to an harmonic vibrational analysis. The results can then be extended to involve vibrational anharmonicity. Following the earliest approaches [28,29] using the Wilson GF matrix technique [30], many empirical and theoretical approaches to the vibrational problem have been applied. The book chapter by Dresselhaus, Dresselhaus and Eklund [27] provides a comparison for works up to 1996. Empirical approaches are typically using valence force-field models (e.g. [31,32], and potentials that are transferable among diamond, graphite and C_{60} [33]. Theoretical approaches include the QCFF/PI method [34,35], tight binding and quantum molecular dynamics studies [36,37], the use of the bond-charge model [38], molecular mechanics methods [39], various semi-empirical quantum-chemical calculations (e.g. [40]) using the MNDO method), and 'ab initio' quantum-chemical calculations (e.g., [40,41]). One of the latest 'ab initio' papers on the harmonic force field of C_{60} addresses another important problem arising from the cyclical connectedness of atoms in C_{60}, the redundancy of force-constants [42]. A review of the existing literature on fullerene vibrational analyses reveals a staggering number of works.

3. The approach taken in this work

Our method for the study of the vibrational problem follows closely an earlier computational work [43]. In this work the vibrational self-consistent field (vscf) method was used to construct a computer program that can be applied to large molecules for assessing anharmonic couplings among normal coordinates. The vscf method has been developed by the Bowman group [44]. A semi-classical version was described by Ratner and Gerber [45]. A short description for the Bowman vscf method is given in [43], while more details are available in [44].

The first step of the calculations is to obtain the harmonic normal mode frequencies for C_{60} using a force-field model with the fewest possible force-field parameters and preferably such parameters that possess maximum physical meaning. The requirement of physical meaning is dictated by chemical intuition that leads one to expect the valence force-field parameters for C_{60} to be similar to those of graphite.

In the previous vscf calculations for C_{60} [43] the force-field model was taken from [31], having eight valence parameters describing bond-stretching and angle-bending motions for a carbon atom surrounded by first and second neighbors. This potential was modified for the first-neighbor stretch terms by using the Simons-Parr-Finlan (SPF) method [46,47]. The essence of this is to modify the stretching potential term from the form $(1/2)*\{r-r_e\}^2$ to $(1/2)*r_e^2*\{(r-r_e)/r\}^2$. Thus anharmonicity for the stretching motions was introduced into the calculation of the potential energy hypersurface. Although in [43] the vscf method was successfully used to show that anharmonic couplings among the vibrational normal coordinates amount to a maximum of 15 cm^{-1}, so are significant, no spectroscopic predictions for specific vibrational levels of C_{60} were given. We have therefore undertaken a new vscf calculation to extend the results from [43] with the final purpose to obtain such predictions. In the present calculations we used a different valence force-field model, taken from Feldman et al. [32]. The force parameters in this model are chemically more significant, using this model a direct comparison among C_{60}, graphite and diamond force-constants is possible (see Table 1. in [32]).

We have not introduced SPF-type displacement coordinates for the stretching motions, but used the original potential energy construction in [32] as follows:

$$U = (1/2)\Sigma\, k_r(\Delta r)^2 + \Sigma k_{r,r'}\,\Delta r \Delta r' + (1/2)\Sigma\, k_\Theta\, rr'(\Delta\Theta)^2 + + (3/2)\,\Sigma\beta\{\mathbf{r}.\mathbf{r}' - \mathbf{r}_0.\mathbf{r}_0')^2/rr'\} + (1/2)\, k_p\{h(a/a_0) - h_0\}^2 \qquad (2)$$

where the summation includes all distinct bonds and angles, Δ indicates displacement, e.g. ($\Delta r = r - r_e$), and the bold products indicate vector products. The first two terms contain force constants for bond stretching motions, the third term describes bond bending ($\Delta\Theta$), the fourth term is basically a bond-stretch/bond-bending interaction term (coming from the so-called Keating term (β) proposed for diamond [48]), and the final term corresponds to a puckering motion. Each carbon atom and its three nearest neighbors constitute a pyramid. Moving the carbon atom radially against the basis of this pyramid (the equilibrium height is denoted by h_0, so the displacement is $h-h_0$, and the basal area of the pyramid is denoted by 'a') corresponds to this puckering displacement. This type of a force-constant was found necessary by Feldman et al. [43] to obtain a good simultaneous fit to infrared and (especially low) Raman frequencies. When the differences among force-constants due to the two different bonds in the C_{60} hexagons are taken into account, there result eleven distinct force-constants. Two of those were taken to vanish in [32], so nine force-constants were finally derived from a fit to the four infrared and ten Raman experimental frequencies (see Table 1. in [32]). Upon inspecting Eq.(2) we see that the Keating term is formally similar to the structure of the SPF stretching coordinates. Thus it is expected that this term will introduce anharmonicity into the potential energy calculated, except that in this case the anharmonicity is not of a bond stretching type.

We have used the Feldman f.c. parameters in Eq.(2) to carry out the usual normal mode analysis for C_{60}. I_h symmetry was not used, the calculations were done on a cartesian coordinate basis, resulting in 180 frequencies, the six non-genuine motion frequencies were found to be very small, arising from rounding errors. The remaining 174 frequencies come out in groups of degenerate components, corresponding to the dimensions of symmetry species in the I_h point group. The frequency groups so obtained were assigned to the experimental frequencies on the basis of the inversion symmetry behavior of the corresponding eigenvectors. Our results were very close to those in Table 2. of [32], providing assurance that the corresponding eigenvectors from our calculations would yield normal coordinate forms corresponding to the true vibrational forms of C_{60}.

As the next step potential energy surfaces were created using pairs of normal coordinate displacements (q_i, q_j) over a 7 by 7 grid (49 energy points), and these surfaces were fit by a non-linear least-squares procedure to a polynomial up to fourth order:

$$U= c_1*q_i^2+c_2*q_j^2+c_3*q_i^3+c_4*q_i^2*q_j+c_5*q_i*q_j^2+c_6*q_j^3+c_7*q_i^4+$$

$$+c_8*q_i^3*q_j+c_9*q_i^2*q_j^2+c_{10}*q_i*q_j^3+c_{11}*q_j^4 \qquad (3)$$

The above expansion corresponds to the usual Taylor potential energy expansions, omitting factorial coefficients. There are terms in Eq.(3) containing only one type of normal coordinate (uncoupled anharmonic oscillator terms: uao), and coupling terms. The fit contained the first two harmonic terms, thus provided a check on the fitting procedure, as one should recover the harmonic frequencies of the normal coordinates used in the fit. All binary combinations (two-dimensional cuts of the hypersurface) out of the 174 normal coordinates should be covered. As the Cartesian force-field calculations lead to degenerate sets of normal coordinates, in principle only one component in each degenerate set may be used for forming binary combinations. The vscf program does not contain symmetry restrictions, so all higher order c_i coefficients for all the combinations may be used for the description of the anharmonic force-field. This constitutes one part of the input to the vscf program.

The 'a posteriori' use of icosahedral symmetry for the c_i 'force constants' is, however, straightforward. Only those coefficient should be non-vanishing for which the symmetrized direct product of the species of the normal coordinates involved in that term contain the totally symmetric A_g representation. A necessary but not sufficient requirement is that the direct product should be 'gerade' relative to the inversion operation. This symmetry requirement may be used to constrain some c_i coefficients to zero, for the uao and the coupling terms as well. To predict the symmetrically non-vanishing number of various orders of vibrational operators (thus force constants) the theory of the symmetric n-th powers of representations may be used. A recent paper contains these results [49]. For C_{60} there are 151 harmonic, 7519 cubic and 330468 quartic force constants. Many of the symmetrically allowed terms may vanish because of the vibrational mechanics of C_{60}, however a great number of them will contribute cubic and quartic c_i coefficients. The implementation of such a symmetry screening would involve generating the symmetric product for each combination in Eq.(3) and its reduction to find whether it contains the A_g species.

In our present calculations we have not used symmetry restrictions, but we present here two numerical examples, one for the cubic and quartic coupling of two infrared active modes, and another for the coupling of two Raman-active modes. The harmonic coefficients in Tables 1. and 2. are given in mdyn.A^{-1}.amu^{-1}, the cubic ones in mdyn.A^{-2} $amu^{-3/2}$ and the quartic ones in mdyn.A^{-3} amu^{-2} units. These units are called here 'Feldman' units.

TABLE 1. The coupling coefficients for two F_{1u} (IR) modes

Expansion term	Coupling coefficients in 'Feldman' units	Harmonic fitted terms in cm^{-1} units
q_i^2	0.99754	531
q_j^2	1.15428	571
q_i^3	0.00000	
$q_i^2q_j$	0.00000	
$q_iq_j^2$	0.00000	
q_j^3	0.00000	
q_i^4	0.06841	
$q_i^3q_j$	0.00000	
$q_i^2q_j^2$	-0.01295	
$q_iq_j^3$	0.00000	
q_j^4	0.00498	

Our results show that the use of the Feldman potential energy construction does indeed introduce vibrational anharmonicity. As the direct product of three F_{1u} modes should result in an 'ungerade' mode, while the product of four F_{1u} modes is of 'gerade' symmetry, one expects the cubic coefficients to vanish, while the quartic coefficients may be different from zero. Table 1. entries bear this out nicely, although some quartic coefficients vanish. This could be due to the special nature of the coupled normal modes. In future work error margins will be provided in order to estimate the statistical significance of the coupling coeffcients derived.

In the next table analogous coupling coefficients are shown for two Raman-active modes. For 'gerade' modes all higher order direct products should be symmetric to the inversion center, so symmetry does not require them to vanish. In fact, again, the numerical values from the fit conform to this symmetry rule, all coupling coefficients are different from zero.

TABLE 2. The coupling coefficients for two H_g (Raman) modes

Expansion term	Coupling coefficients in 'Feldman' units	Harmonic fitted terms in cm^{-1} units
q_i^2	0.65708	431
q_j^2	0.25422	268
q_i^3	0.04777	
$q_i^2 q_j$	-0.00131	
$q_i q_j^2$	-0.00906	
q_j^3	-0.00151	
q_i^4	0.02857	
$q_i^3 q_j$	-0.00008	
$q_i^2 q_j^2$	0.00961	
$q_i q_j^3$	0.00002	
q_j^4	0.00075	

4. Manifestation of anharmonicity in laboratory spectra

In solid state (film) infrared and Raman spectra there are far more spectroscopic features than expected solely on the basis of the harmonic vibrational picture. One explanation for these features is that they originate from symmetry breakdown in the solid state, in other words, the individual C_{60} molecules are not totally isolated and there are coupled vibrations involving lattice modes. Another source for symmetry breaking is that the vibrational motions in fullerenes cannot be taken as independent normal vibrations, vibrational anharmonicity couples these and leads to vibrational states containing multiple vibrational excitation. Then there is the possibility of ^{13}C –> ^{12}C isotopic substitution in the fullerene sample, as the natural abundance of the ^{13}C isotope (about 1%) results in roughly half of the C_{60} molecules being singly or multiply isotopically substituted, and only half of them containing pure ^{12}C fullerenes. Finally other fullerenes might contaminate the sample, or chemically altered forms might be present. In case of very pure and fresh C_{60} samples this latter possibility can be excluded so we do not consider it in the following.

The infrared spectrum of a thick C_{60} film shows many more features than just the four fundamental bands, as Figure 1. illustrates.

Figure 1. The infrared spectrum of a 1mm thick C_{60} film at 300 K *

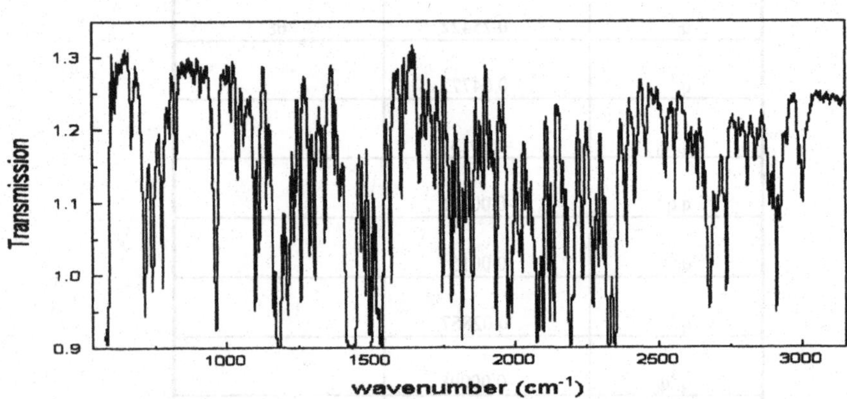

The most important source of spectral complexity is vibrational anharmonicity. Combination and difference tones, as well as overtones are signatures of the mechanical anharmonicity of vibrational motions. A number of papers are devoted to this subject [42, 50-57]. In these works most of the weaker infrared and Raman bands were assigned to second-order combination bands, that is to vibrational transitions from the ground state to vibrational states containing quanta from two vibrational modes. Difference modes, i.e. vibrational transitions that arise from a vibrationally excited level, have generally not been considered, as their intensity is expected to be much smaller than those of combination bands. The intensity of the infrared combination tones $(\nu \approx \nu_1+\nu_2)$ and difference tones $(\nu \approx \nu_1-\nu_2)$ is proportional to $(n_i + ½) \pm (n_j + ½)$, where n_i (the so-called Bose factor) is $½\{\coth(h\nu_i/2k_BT) - 1\}$.

Not all transitions to vibrationally multiply excited states are allowed, either in the infrared or Raman spectra. The group theoretical requirement for combination bands to appear in the infrared spectrum is that the direct product of symmetry species of the two combining upper vibrational states (state i and state j) should contain the F_{1u} species (there are 380 such allowed combinations in C_{60} (see: e.g. [26]):

$$\Gamma_i \otimes \Gamma_j \supset F_{1u} \tag{4}$$

* The authors are grateful to Prof. Laszlo Mihaly, SUNY, U.S.A., for sending the spectral data

Similarly second-order Raman bands are only allowed when the analogous direct product contains the A_g and/or H_g species (the number of extra Raman modes due to this source is 151 A_g and 661 H_g modes [51]:

$$\Gamma_k \otimes \Gamma_l \supset A_g, H_g \tag{5}$$

First overtones are not infrared active, since the direct product of any species with itself is always symmetric to the inversion center, so cannot contain 'ungerade' representations, such as F_{1u}. This simplifies the infrared spectrum.

As was pointed out in [53] the appearance of Raman active modes in the infrared spectrum shown in Fig.1. suggests that icosahedral symmetry is reduced in the solid state. A further possible explanation of the weak infrared and Raman bands is the presence of the naturally occurring ^{13}C isotope. Although the exchange of a single ^{12}C atom by ^{13}C reduces icosahedral symmetry to C_s symmetry, thus rendering all optically silent modes infrared and Raman active, the change of the kinetic energy is so small that the newly arising vibrational transitions will cluster close to the infrared or Raman transitions of the unsubstituted C_{60} sample. In other words, if we were able to record the spectra of an isotopically pure $^{12}C_{60}$ sample, it would essentially look identical to that of the commonly available C_{60}. Thus this source for increased spectral complexity can, in the first approximation, be neglected. This conclusion is born out by spectra of C_{60} samples isotopically enriched in ^{13}C [56,57]. The presence of one or more ^{13}C atoms in C_{60} should thus lead only to a broadening of the spectral features, and not to extra spectral lines. Notwithstanding, several of published assignments (e.g. [52,53]) list features activated by ^{13}C isotopic substitution.

The intensity of combination tones and difference tones is not predicted by a mechanical anharmonicity model, one needs to consider electrical anharmonicity as well. Electrical anharmonicity appears in higher (than linear) order terms in the Taylor expansion of the electric dipole moment in terms of the harmonic normal coordinates: $q_k, q_l, \ldots$

$$\mu = \mu_e + \Sigma_k \mu_k q_k + \tfrac{1}{2} \Sigma_{kl} \mu_{kl} q_k q_l + \ldots \tag{6}$$

where μ_e is the equilibrium permanent dipole moment (zero in the case of icosahedral symmetry), μ_k is first derivative of the dipole moment μ_e w.r.t. the normal coordinate q_k, μ_{kl} is the second derivative w.r.t. q_k and q_l, etc. Band intensities are proportional to the square of μ in Eq.(6). In the harmonic approximation the third and higher terms in Eq.(6) vanish. Mechanical anharmonicity determines the shift of overtones, combination and difference tones from their simple frequency sum or difference values, resp., while electrical anharmonicity determines their intensity (zero in the harmonic approximation). Fabian [55] carried out such calculations for C_{60} and found that for the description of intensities of combination and difference tones electrical anharmonicity is decisive. In the present work we do not consider electrical anharmonicity.

There are a number of works (e.g. [42,52-54]), in which weak second-order features in the infrared and Raman spectra of C_{60} were assigned to combination tones. There is also a very recent work giving the latest assignments for the fundamentals [42]. Thus now it is possible to compare second-order spectral assignments in [52-54] to fundamental mode assignments in [42]. A selective comparison is given in Table 3.

TABLE 3. Comparisons among different spectral assignments for second-order spectral lines in infrared and Raman spectra of C_{60} (in cm^{-1})

Assignments in [52]	Assignments in [53]	Assignments in [54]	Assignments from [42]
$G_u(1)+H_g(1)=672.5$			621
$H_g(2)+H_u(1)=775$			834
$F_{1u}(1)+H_g(1)=799$			793
$H_u(1)+G_g(1)=828.5$			888
$G_g(2)+G_u(1)=1020.5$			946
		$H_g(2)+H_u(4)=1260$	1168
$G_u(2)+H_g(4)=1535$			1482
		$H_g(3)+F_{2u}(4)=1747$	1853
$F_{2g}(3)+G_u(3)=1838$			1628
	$G_g(1)+G_g(5)=1842$		1833
	$F_{1g}(3)+H_g(2)=1790$		1721
$F_{2g}(2)+G_u(3)=1789$			1465
		$H_g(1)+G_u(6)=1792$	1696
	$H_g(4)+H_g(5)=1876$		1876
		$F_{2g}(3)+H_u(3)=1876$	1525
$G_g(4)+H_u(4)=1876.5$			1777

As the few examples in Table 3. show (in particular the last six entries) there is still ample room for the revision of second-order spectral band assignments. This is one of our main task that we plan to achieve using the newly calculated anharmonic coefficients. It should, however, be noted that one important reason for the different assignments in Table 3. is the difference in fundamental mode assignments used in the above listed works.

Gas-phase spectra are free from all possible solid state effects, so the low intensity features found in high temperature infrared emission spectra of C_{60} [26] should correspond mainly to binary combination tones. Figure 2. shows a part of the high temperature gas-phase emission spectrum of C_{60} at a temperature 960 K from [26]. In the Figure the strong infrared fundamental at 1410 cm^{-1}, and two weak features at 1497 and 1538 cm^{-1} are visible.

Figure 2. The gas-phase infrared emission spectrum of C_{60} at 960 K

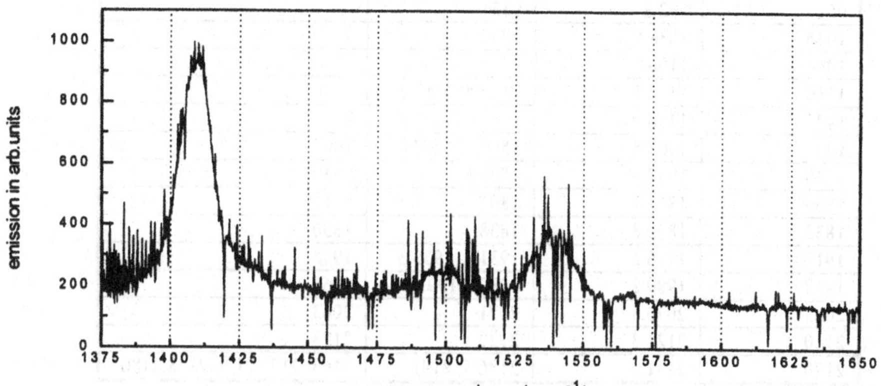

5. Temperature effects in infrared spectra

In order to obtain spectroscopic information for the detection of C_{60} in astrophysical sources, and to better understand the vibrational and rotational characteristics of C_{60} at very low laboratory temperatures, the changes in the infrared emission or absorption spectrum of C_{60} upon cooling should be studied. In [26] temperature effects were studied, in the temperature range 880 – 1212 K for C_{60}, and 897 – 1234 K for C_{70} in infrared emission. The changes in band position and band width were examined. These quantities were extrapolated to 0 K, even though the extrapolation was very long.

In Table 4. a comparison is shown for weak features in C_{60} spectra found in [26] and in [50,52,54]. Only those features are listed from [52-54] that were assigned to 'ungerade' combination tones, or explicitly to infrared active modes, as there is no reason to expect Raman-active vibrations to appear in a gas-phase infrared spectrum. For most of the features in the gas-phase spectrum, corresponding features can be found in the solid state spectra, suggesting that the weak gas-phase bands arise from binary combination tones (in [26] three assignments were given based on [50]). Weak bands for C_{70} can be assigned on the basis of solid state spectra [58,59].

TABLE 4. Comparison among gas-phase and solid state weak infrared features for C_{60} (in cm^{-1})

Gas-phase [26]	Solid phase [50]	Solid phase [52]	Solid phase [54]
951	962.8	953	-
1035	1038.5	1033	-
1094	1100.2	-	-
1110	1115.1	1112.5, 1117.5	-
1425	1430.0	-	-
1497	1502.8	1500.5, 1502	1497
1538	1538.9	1538.5	1539
1800	1817.1	1801	1811
1833	1852.8	1838	1830
1919	1936.2	1931.2, 1933.3	1938
1967	1989.2	1989.5,1991.5	1968
2065	2076.5	2071	2063
2110	2121.3	2120	2123
2170	2191.2	2186.5, 2191	2168, 2172,2176,2178,2180

The results showed that upon cooling the hot gas, bands got narrower and shifted as a linear function of T towards higher energies. The 0 K extrapolated band centers moved very close to previous matrix values [60] - but remained significantly different from those -, and the temperature dependent band widths dropped basically as a function of $T^{1/2}$. This behavior is explained partly by the narrowing of the rotational structure, as described by the Darling-Dennison formula [61], also by a special study of the rotational band shapes of C_{60} [62]. An additional important source for temperature dependence is that hot transitions given rise by anharmonicity cool out effectively, but not in a symmetrical fashion, hot bands red-shift the band maxima, thus cooling results in a blue-shift. Actually the finding in [26] that the band width temperature dependence had also a linear component in T shows that at high temperatures both vibrational and rotational cooling influence band-widths.

To study the vibrational effects in temperature dependence, one has to calculate the vibrational partition function (v.p.f.) for C_{60}. In former work on band envelope calculations for C_{70} [63] the Haarhoff algorithm for a collection of harmonic oscillators [64] was used for the v.p.f. Convoluting the vibrational level density obtained from the Haarhoff algorithm by the vibrational Boltzmann factor, and integrating it over an appropriately wide wavenumber range one obtains the v.p.f. Using the assignments for C_{60} fundamentals in [42] the v.p.f. value for T=950 K is obtained as $7.45 \times 10^{+25}$. A similar, very high value was obtained in [26], using a simpler formula for calculating the v.p.f. Such high values for the v.p.f. mean that the ground vibrational state population is vanishingly small at elevated temperatures. The vibrational energy content of hot C_{60} molecules at 950 K peaks at about 60,000 cm^{-1}, as shown in Figure 3.

The rotational partition function for C_{60} changes much less with the temperature, it is proportional to $T^{3/2}$, so the cooling out of the rotational level population is very slow, and the rotational level populations extend up to very large J values, since the rotational constant of C_{60} is exceedingly small: 0.00278 cm^{-1}. Thus to remove the majority of rotational energy content of C_{60} cooling to near absolute zero temperature is needed. An estimate was given in [62]; for T= 30 K the Boltzmann maximum in the P or R branch occurs at J=85.

Figure 3. Internal energy distribution for C_{60} at 950 K

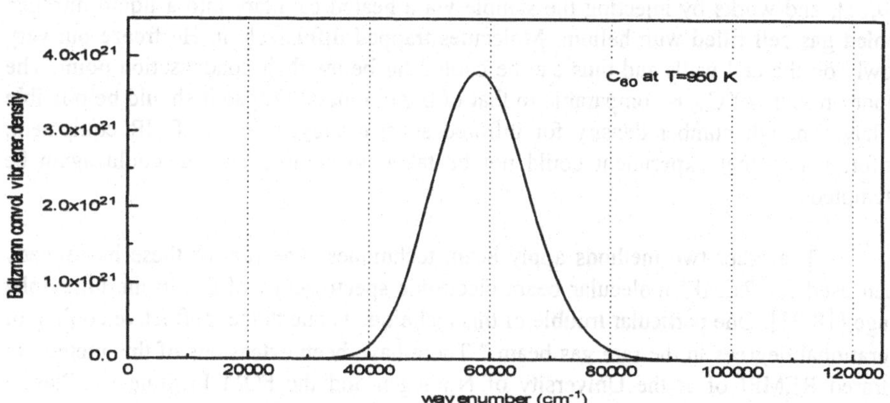

The calculated rotational contours in [62] took Coriolis vibration-rotation coupling into account and resulted in P-R separation values between 1 and 1.5 cm^{-1}, as a measure of infrared band width, at T=30 K (this value was chosen to correspond to conditions in the Egg Nebula (CRL 2688) [65]). Band widths for the infrared fundamentals of C_{60} extrapolated to 2.7 K in [26] are between 0.4 – 0.8 cm^{-1}, and 0.6 cm^{-1} was obtained using band contour simulation in [63] for C_{70}. These values are sufficiently small to detect ^{13}C-^{12}C isotopic substitutions, as, e.g. for a single isotopic substitution the shift should be around 0.6 cm^{-1}, while for double and triple substitution, 1.2 and 1.8 cm^{-1}, resp. [66].

Cold infrared spectral data would presumably help in disentangling various components in spectral broadening (anharmonic red-shifts, rotational structure and isotopic shifts), and in astrophysical searches for fullerenes. Furthermore it would open the way for such interesting laboratory spectroscopy projects as the experimental analysis of Coriolis interactions in C_{60}.

6. Proposal for cold infrared gas-phase spectroscopy

There are at least three different experimental approches to record infrared gas-phase spectra of stable fullerenes (C_{60}, C_{70}, perhaps also C_{36} [67]). Of course, there might be other techniques to use.

Experiments were started on C_{60} at the Department of Chemistry, University of Waterloo, Canada in 1991, under the initiative of Prof. Peter Bernath, using the technique of diffusive cooling, as applied for infrared spectroscopy [68,69]. The method was originally developed for microwave spectroscopy of low vapor pressure substances [70,71], and works by injecting the sample via a heated capillary into a liquid nitrogen cooled gas cell filled with helium. Molecules trapped diffusively in He freeze out very slowly on the cell walls and thus can be cooled far below their condensation point. The vapor pressure of C_{60} is comparable to that of barium metal [72] so it should be possible to have enough number density for infrared spectroscopy, using a FTIR equipment. Unfortunately that experiment could not be taken to completion, but could again be attempted.

The other two methods apply beam techniques. The first of these has already been used for REMPI molecular beam electronic spectroscopy of C_{60} in the ultraviolet range [18-21]. One particular trouble of this technique is due to the ineffective cooling of vibrational degrees in the rare gas beam.* There have been extensions of this method to infrared REMPI of at the University of Nijmegen and the FOM Institute for Plasma Physics, Rijnhuizen, using an infrared free electron laser [23,24]. These emission spectra show very large band widths, corresponding to high vibrational excitation, in the range of 10-15 cm^{-1} and an asymmetric line shape due to vibrational anharmonic couplings. Otherwise they are similar to the infrared emission spectra in [26].

The other beam method is absorption spectroscopy of molecules isolated in helium droplet beam. This technique is utilizing nanometer size He droplets enveloping individual molecules or clusters, and cooling them near absolute zero K by subsequent steps of evaporation upon photon absorption. This method has been used to study C_{60} electronic spectra [22] in the MPI für Strömungsforschung, Göttingen, but high-resolution spectroscopy is also carried out by this means at the University of Bochum [73], and at Priceton University [74] for the study of single small molecules at high resolution. As vibrational cooling in this method is very effective due to reaching thermal equilibrium with the He droplets at 0.37 K, there is close resemblance to gas-phase data. In case the appropriate tunable infrared laser may be found for application to C_{60} this technique could be perhaps the most suitable for achieving the goal set in this report.

---

* L.N. is grateful to Prof. Gerard Meijer, University of Nijmegen, Holland, for pointing this out

Acknowledgements

L.N. acknowledges the scientific support given to him by Professor Peter F. Bernath, University of Waterloo, Canada in fullerene spectroscopic work (in 1992-1993), as well as financial support from the Hungarian Research Fund (OTKA) for the period 1998 – 2001 under contract numbers # T026068 and #T032549.

References

1. Kroto, H.W., Heath, J.R., O'Brien, S.C., Curl, R.F. and Smalley, R.E. (1988), *Nature* **318**, 162. Krätschmer, W., Lamb, L.D., Fostiropoulos, K. and Huffman, D.R. (1990) *Nature* **347**, 354.
2. Kroto,H.W. (1988), *Science* **242**, 1139.
3. Kroto, H.W. (1994), *Nature* **369**, 274.
4. Turner, B.E. (1971) *Astrophys. J.* **163**, L35.
5. McCarthy, M.C. (1999) Laboratory and Astronomical Spectroscopy of New Carbon Chains and Rings, in: *Conf. Proc. Spectroscopy for the Year 2000*, Oct. 30-Nov. 3, Cornwall, Canada, p. 5.
6. Hargittai, I. (1995), *The Chemical Intelligencer*, **1**, 6.
7. Kroto, H.W. and Jura, M. (1992), *Astron.Astrophys*., **263**, 257.
8. Scarrott, S.M., Watkin, S., Miles, J.R. and Sarre, P.J. (1992), *Mon.Not.R. astr. Soc.* **255**, 11.
9. Edwards, S.A. and Leach, S. (1994) in Nenner, I. (ed.), *Molecules and Grains in Space, AIP Conf. Proc.* No. 132., Am.Inst.Phys., New York, p. 589.
10. Edwards, S.A. and Leach, S. (1993), *Astron. Astrophys.* **272**, 533.
11. Foing, B.H. and Ehrenfreund, P. (1994), *Nature* **369**, 296. Foing, B.H. and Ehrenfreund, P. (1997) *Astron. Astrophys*., **319**, L59.
12. Moutou, C., Leger, A., d'Hendecourt L. and Maier, J.P. (1996), *Astron.Astrophys*., **311**, 968.
13. Herbig, G.H. (1995), *Annu.Rev.Astrophys.* **33**, 19.
14. Hecht, J.H. (1992), *Astrophys.J*., **367**, 635.
15. Webster, A.S. (1992), *Astron. Astrophys*., **257**, 750.
16. Whitney, B.A., Balm, S.P., Clayton, G.C. (1993), , Luminous high latitude stars, in Sasselov, D. (ed.), *ASP Conf. Ser.* No. 45, p. 115.
17. Clayton, G.C., Kelly, D.M., Lacy, J.H., Little-Marenin, I.R., Feldman P.A. and Bernath, P.F. (1995), *Astron. J*., **109**, 2096.
18. Heath, J.R., Curl, R.F. and Smalley, R.E. (1987), *J. Chem.Phys*., **87**, 4236.
19. Haufler, R.E. Chai Y., Chibante, L.P.F., Fraelich, M.R. Weisman, R.B., Curl R. F. and Smalley, R.E. (1991), *J.Chem.Phys*., **95**, 2197.
20. Haufler, R.E., Wang, L.S., Chibante, L.P.F., Jin, C., Conceicao, J., Chai, Y. and Smalley, R.E. (1991), *Chem.Phys.Letters* **179**, 449.
21. Hansen, K. Muller, R., Brockhaus, P., Campbell, E.E.B. and Hertel, I.V. (1997), *Z. Phys.* **D 42**, 153.
22. Close, J.D., Federmann, F., Hoffmann, K. and Quaas, N. (1997), *Chem. Phys. Letters* **276**, 393.
23. von Helden, G., Holleman, I., Knippels, G.M.H., van der Meer, A.F.G. and Meijer, G. (1997), *Phys.Rev. Letters* **79**, 5234.
24. von Helden, G., Holleman, I., van Roij, A.J.A., Knippels, G.M.H.,van der Meer, A.F.G. and Meijer, G. (1998), *Phys.Rev.Letters* **81**, 1825.
25. Frum C.I., Engleman, Jr., R., Hedderich, H.G., Bernath, P.F., Lamb L.D. and Huffman, D.R. (1991),*Chem. Phys. Letters* **176**, 504.
26. Nemes, L., Ram, R.S., Bernath, P.F., Tinker, F.A., Zumwalt, M.C., Lamb L.D. and Huffman, D.R. (1994), *Chem. Phys. Letters* **218**, 295
27. Dresselhaus, M.S., Dresselhaus, G. and Eklund, P.C. (1996) *Science of Fullerenes and Carbon Nanotubes*, Chapter 11, Academic Press, New York.
28. Cyvin, S.J., Brendsdal, E., Cyvin, B.N. and Brunvoll, J. (1988), *Chem.Phys.Letters* **143**, 377.
29. Brendsdal, E., Cyvin, B.N., Brunvoll, J. and Cyvin, S.J. (1988), *Spectr.Letters* **21**, 313.
30. Wilson, E.B. Jr., Decius, J.C. and Cross, P.C. (1955) *Molecular Vibrations*, McGraw-Hill, New York.
31. Jishi, R.A., Mirie, R.M. and Dresselhaus, M.S. (1992), *Phys.Rev.* **B45**, 13685.

32. Feldman, J.L., Broughton, J.Q., Boyer, L.L., Reich, D.E. and Kluge, M.D. (1992), *Phys.Rev.* **B46**, 12731.
33. Burgos, E., Halac, E. and Bonadeo, H. (1998), *Chem.Phys.Letters* **298**, 273.
34. Negri, F., Orlandi, G. and Zerbetto, F. (1991), *J.Am.Chem.Soc.*,**113**, 6037.
35. Negri, F and Orlandi, G (1996), *J.Phys. B: At.Mol.Opt.Phys.* **29**, 5049.
36. Adams, G.B., Page, J.B., Sankey, O.F., Sinha, K., Menendez, J. and Huffman, D.R. (1991), *Phys.Rev.* **B44**, 4052.
37. Wang, C.Z., Chan, C.T. and Ho, K.M. (1992), *Phys.Rev.* **B46**, 9761.
38. Onida, G. and Benedek, G. (1992), *Europhys.Letters* **18**, 403.
39. Murry, R.L.,Colt, J.R. and Scuseria, G.E. (1993), *J.Phys.Chem.*, **97**, 4954.
40. Raghavachari, K., Rohlfing, C.M. (1991), *J.Phys.Chem.*, **95**, 5768.
41. Häser, M., Almlöf, J. and Scuseria, G.E. (1991), *Chem.Phys.Letters* **181**, 497.
42. Choi, C.H., Kertesz M. and Mihaly, L. (2000), *J.Phys.Chem.*, **A104**, 102.
43. Jelski, D.A., Haley, R.H. and Bowman, J.M. (1996), *J.Comp.Chem.*, **17**, 1645.
44. Bowman, J.M. (1986), *Acc.Chem.Res.*, **19**, 202.
45. Ratner, M.A. and Gerber, R.B. (1986), *J.Phys.Chem.*, **90**, 20.
46. Simons, G., Parr, R.G. and Finlan, J.M. (1973), *J.Chem.Phys.*, **59**, 3229.
47. Simons, G. (1974), *J.Chem.Phys.*, **61**, 369.
48. Keating, P.N. (1966), *Phys. Rev.*, **145**, 637.
49. Varga, F., Nemes, L. and Watson, J.K.G. (1996), *J.Phys.B: At.Mol.Opt.Phys.*, **29**,5043.
50. Chase, B., Herron, N. and Holler, E. (1992), *J.Phys. Chem.*, **96**, 4262.
51. Dresselhaus, G., Dresselhaus, M.S. and Eklund, P.C. (1992), *Phys.Rev.* **B45**, 6923.
52. Wang, K.A., Rao, A.M., Eklund, P.C., Dresselhaus, M.S., Dresselhaus, G. (1993), *Phys.Rev.*, B **48**, 11375.
53. Dong, Z.H., Zhou, P., Holden, J.M., Eklund, P.C., Dresselhaus, M.S. and Dresselhaus, G. (1993), *Phys. Rev.* **B48**, 2862.
54. Martin, M.C., Du, X., Kwon, J. and Mihaly, L. (1994), *Phys.Rev.* **B50**, 173.
55. Fabian, J. (1996), *Phys.Rev.* **B53**, 13864.
56. Martin, M.C., Fabian, J., Godard, J., Bernier, P., Lambert, J.M. and Mihaly, L., *Phys.Rev.* **B51**, 2844 (1995)
57. Holleman, I., Boogaarts, M.G.H., van Bentum, P.J.M. and Meijer, G. (1995) in Kuzmany, H., Fink, J., Mehring, M., Roth, S. (eds.) *Physics and Chemistry of Fullerenes and Derivatives*, , World Scientific, Singapore, , p. 55
58. Bethune, D.S., Meijer, G., Tang, W.C., Rosen, H.J., Golden, W.G., Seki, H., Brown, C.A. and de Vries, M.S. (1991), *Chem.Phys.Letters* **179**, 181.
59. Jishi, R.A., Dresselhaus, M.S., Dresselhaus, G., Wang, K.A., Zhou, P., Rao, A.M. and Eklund, P.C. (1993) *Chem.Phys.Letters* **206**, 187.
60. Haufler, R.E., Conceicao, J., Chibante, L.P.F., Chai, Y., Byrne, N.E., Flanagan, S., Haley, M.M., O'Brien, S.C., Pan, C., Xiao, Z., Billups, W.E., Ciufolini, M.A., Hauge, R.H., Margrave, J.L., Wilson, L.J., Curl, R.F. and Smalley, R.E. (1990), *J. Phys. Chem.*, **94**, 8634.
61. Gerhard, S.L. and Dennison, D.M. (1933), *Phys.Rev.*, **43**, 197.
62. Weeks, D.E. and Harter, W.G. (1991), *Chem. Phys. Letters* **176**, 209.
63. Nemes, L. (1997), *J. Mol. Structure*, **436-437**, 25-34.
64. Haarhoff, P.C. (1963), *Mol. Phys.*, **7**, 101.
65. Rieu, N.Q., Winnberg, A. and Bujarrabal, V. (1986), *Astron. Astrophys.*, **165**, 204.
66. Guha, S., Menendez, J., Page, J.B., Adams, G.B., Spencer, G.S., Lehman, J.P., Giannozzi, P. and Baroni, S. (1994), *Phys. Rev. Letters* **72**, 3359.
67. Piskoti, C., Yarger, J. and Zettl, A. (1998) *Nature* **393**, 771.
68. Ewing, G.E. and Sheng, De T. (1988), J.*Phys. Chem.*, **92**, 4063.
69. Dunder, T. and Miller, R.E. (1990), *J. Chem. Phys.*, **93**, 3693.
70. Messer, J.K. and De Lucia, F.C. (1984), *Phys. Rev. Letters* **53**, 2555.
71. Wiley, D.R., Crownover, R.L., Bittner, D.N. and De Lucia, F.C. (1988), *J. Chem. Phys.*, **89**, 1923.
72. Abrefah, J. and Olander, D.R., Balooch, M. and Siekhaus, W.J., *Appl. Phys. Letters* **60**, 1313.
73. Hartmann, M., Miller, R.E., Toennies, J.P. and Vilesov, A. (1995), *Phys. Rev. Letters* **75**, 1566.
74. Lehmann, K.K. (1999) Helium Nanodroplet Isolation Spectroscopy, ,in *Proceedings for the Conference: Inspired by Herzberg, Spectroscopy for the Year 2000*, Cornwall, Canada, , p. 11.

DREAM OR REALITY: COMPLETE BASIS SET FULL CONFIGURATION INTERACTION POTENTIAL ENERGY HYPERSURFACES

Attila G. Császár, György Tarczay
Department of Theoretical Chemistry, Eötvös University
P. O. Box 32, H-1518 Budapest 112, Hungary

Matthew L. Leininger
Sandia National Laboratory, MS 9217, Livermore, CA 94551-0969

Oleg L. Polyansky, Jonathan Tennyson
Department of Physics and Astronomy, University College London
Gower St., London WC1E 6BT, U.K.

Wesley D. Allen
Center for Computational Quantum Chemistry, Department of Chemistry,
University of Georgia, Athens, GA 30602, U.S.A.

1. OVERVIEW

All knowledge about molecules outside of our solar system comes from study of their spectra. Although laboratory measurements are usually the prime source for the data needed to interpret astronomical observations, there are a number of reasons why theory has played, and will continue to play, a central role in the study of molecules in space.

Astronomical environments, such as those found in the interstellar medium, are very different from those on earth. This dissimilarity leads to fundamentally different chemistry and to the production of species

J. Demaison et al. (eds.), Spectroscopy from Space, 317–339.

that can be hard to create in the laboratory. Theory can play an important role in both predicting main features of spectra of such species or in looking for possible spectral matches.

Even where laboratory spectra have been recorded for a particular species, this data may only be partial. One such situation, which is particularly common for unstable or reactive species, is that wavelengths can be measured to high accuracy but there is no information on transition probabilities or line strengths. Line strength data are essential for obtaining useful information from astronomical spectra. Without them observation of a spectrum simply indicates that a particular species is present. Use of the line strengths gives column densities as well as the local temperature if thermodynamic equilibrium prevails, or other environmental information if a non-equilibrium environment is being probed. The H_3^+ molecule provides a good example of this situation. The laboratory infrared spectrum of H_3^+ is well known. Its original measurement [1], and indeed subsequent experiments, relied heavily on *ab initio* theory. H_3^+ has now been observed in giant planets, diffuse and dense clouds in the interstellar medium [2], and the remnants of a supernova [3]. The infrared spectrum of H_3^+ has provided a particularly powerful handle on ionospheric activity in giant planets [4], giving detailed information on a whole variety of effects including, for example, the electrojets which power the auroral activity on Jupiter. Yet while the characteristic wavelengths used for observations of H_3^+ all come from experiment, all line strengths used to model the spectra are theoretical and are the result of *ab initio* calculations [5].

Some astronomical applications, stellar models being a classic example, are particularly demanding on spectroscopic data. The stellar opacity problem for stars hotter than our own Sun was only solved by the systematic calculation of large quantities of spectral data for atoms and atomic ions by the international Opacity Project Team [6]. The spectroscopic properties of cooler stars are dominated by molecular absorptions [7]. However, to model the role of a triatomic molecule such as H_2O or HCN, which are important components of oxygen- and carbon-rich cool stars respectively, it may be necessary to consider up to a billion vibration-rotation transitions. The laboratory measurement and analysis of a dataset of transitions of this size is completely impractical. Recently this task has been addressed by theory, where large linelists of rotation-vibration transitions have been produced for HCN [8] and H_2O [9, 10]. It has been found that including these transitions can fundamentally alter conclusions about the structure of the stellar atmosphere [8]. Despite the use of very high level theory and large datasets, recent stellar models have shown, for example, that the best water linelists are still not adequate for reproducing the observed spectra [11]. Large datasets of energy levels produced by theory can also be used to model other astronomically important properties such as temperature-dependent partition functions [12] or equilibrium constants [13].

The vibration-rotation spectrum of water is perhaps the most important and intensely studied of all molecular spectra. For instance, water vapour is thought to be responsible for absorbing 70% of the sunlight lost in a cloudless atmosphere [14]. Many decades of work have focused on measuring, analysing, and modeling the spectrum of water. In something of a theoretical triumph, Polyansky *et al.* [15, 16] have used the combination of high level *ab initio* electronic structure and variational nuclear motion calculations to assign 1687 transitions to water in the sunspot spectra recorded in the 10 – 20 μm region. Yet these transitions represent only about 15% of the clearly resolved features observed in sunspots in this spectral region. It is likely that nearly all the unassigned features are also due to water. Further significant progress in assigning these features will require corresponding theoretical developments. It is therefore important to consider all possible factors which influence the *ab initio* calculation of transition frequencies. The major factor determining the accuracy of a computed rotation-vibration spectrum is the potential energy hypersurface (PES) employed. State-of-the-art *ab initio* electronic structure calculations have become able to predict vibrational band origins (VBOs) of triatomics, like water, to within a few wavenumbers, and other spectroscopic properties with similar ($\sim 0.1\%$) accuracy [10, 15–21]. Standard treatments of molecular electronic structure theory tacitly neglect several physically significant factors which become important in high-accuracy theoretical work: core-valence electron correlation, coupling between electronic and nuclear motion, part of which is considered in the so-called diagonal Born–Oppenheimer correction (DBOC), and relativistic corrections. A number of groups have recently started exploring the validity of the Born–Oppenheimer approximation when calculating vibration-rotation spectra, using water as the test molecule [10, 22]. Electronic relativistic effects [23] are also receiving considerable interest, although attention is only starting to be paid to the possible spectroscopic consequences of the relativistic correction for light molecules [24–30].

2. INTRODUCTION

Ever since the Egyptians invented them, maps (*pinax* for Greeks, *orbis pictus* or *tabula* for Romans) have helped mankind to find directions in the known world and, at the same time, to make explorations of the unknown world. In fact, the history of maps is intertwined with the history of mankind.

The concept of potential energy hypersurfaces [31, 32] is as fundamental to the understanding of most modern branches of chemistry and physics including almost the whole of spectroscopy and kinetics, as maps are for our everyday life. Nevertheless, PESs exist only within the so-called Born–Oppenheimer (BO) separation of electronic and nuclear motion [33, 34]. Following this separation, potential energy hypersurfaces for nuclear motion arise describing the variation in the total electronic energy of a chemical system as a function of its geometry (internal coordinates of the N constituent nuclei). Usually, attention is focused on cases where a single BO–PES is sufficiently uncoupled from other surfaces (electronic states) that their interaction may be ignored.

Despite their great importance, production of detailed and accurate maps took quite a long time to achieve. In a similar manner, due to methodological problems and insufficient computational power, *ab initio* quantum mechanical study of PESs remained unrealistic until about 1970.

Recent developments in electronic structure theory and computer technology have facilitated the computations necessary for the theoretical determination of high accuracy potential energy hypersurfaces. Theory is now capable, over a rather large range of geometries, of obtaining near chemical accuracy (±1 kcal/mol) for the relative energies of small and medium-sized molecular systems. In favorable cases this translates into an accuracy of a few cm^{-1} for rovibrational eigenstates computed from these PESs, as shown below.

The computational errors in today's approximate solutions to the time-independent non-relativistic electronic Schrödinger equation result from the truncation of the atomic orbital one-electron basis and the truncation of the n-electron basis of all Slater determinants that constitute the full configuration interaction (FCI) wave function. Techniques which aim to achieve chemical accuracy for energetic quantities include the Gaussian-n [35–38], the complete basis set (CBS) [39, 40], and the W-n [41] model chemistries. These approaches may rely on empirical parameters which are optimized to minimize the errors for a given test set of molecules. An alternative parameter-free method is the focal-point [42, 43] approach (see Section 4 below), which seeks to achieve the complete one-electron basis set (CBS) and n-electron (FCI) limit by performing a series of electronic structure computations employing convergent basis sets and correlation methods. The previously mentioned model chemistries may be considered as approximations to the focal-point approach, and are therefore more computationally efficient but may be less accurate for certain troublesome cases. Furthermore, the focal-point approach allows efficient estimation of the remaining computational error [43–45] for the given problem, while in the case of model chemistries one needs to rely on average error estimates.

The determination of the complete one-electron basis set limit has received considerable attention in recent years [41,43–51]. Formulas have been advanced providing an estimate of the CBS energy limit from calculations employing systematically constructed families of basis sets (*e.g.*, the correlation-consistent (cc) basis sets of Dunning and co-workers [52–54]. These studies show that different levels of electronic structure theory follow distinct basis set extrapolation patterns; most notably, Hartree–Fock (HF) energies converge almost exponentially toward the CBS limit [49], while correlation energies seem to follow an X^{-3} dependence [43, 46], where X is the cardinal number of the cc basis sets [52]. Therefore, the expressions $E^X = E_{\rm CBS} + a\ \exp(-bX)$ and $E^X = E_{\rm CBS} + cX^{-3}$, where $E_{\rm CBS}$ is the extrapolated energy, E^X denotes energies obtained from correlation-consistent basis sets with cardinal number X, while a, b, and c are fitting constants, are now commonly employed for estimating the complete basis set ($E_{\rm CBS}$) Hartree–Fock and correlation energies, respectively. Accordingly, the above formulas are employed throughout this study for CBS extrapolations. Note that certain correlation contributions may behave differently, such as the relativistic 2-electron Darwin correction, which scales as slowly as X^{-1} [50].

Of equal importance is the determination of the FCI limit in a given one-particle basis from approximate n-electron methods. Numerous studies have shown the diminishing energy contribution of higher excitation levels. Coupled-cluster (CC) methods [55] including triple excitations [56–58] or configuration interaction with up to quadruple substitutions [59] typically provide accurate and reliable approximations to the FCI energy, in contrast to Møller–Plesset perturbation methods [60]. In particular, the CCSD(T) [56] approach represents a good compromise between cost and accuracy for predicting high-quality energies and properties [61]. Nevertheless, to compute energies to better than chemical accuracy it is necessary to consider the energy contributions from higher-order excitations. For the estimation of higher-order correlation effects neglected in the CCSD(T) and CCSDT approaches, we present here a technique called scaled higher-order correlation energy (SHOC) ([62]; see Section 5). It is important to note that if SHOC is carried out at the extrapolated CBS limit, the penultimate (CBS FCI) solution of the non-relativistic electronic Schrödinger equation may be closely approximated.

Reproduction of spectroscopic observations provides the most stringent test for checking the intrinsic accuracy of any *ab initio* PES. Therefore, in this paper we primarily focus on the *ab initio* generation of PESs and on the accuracy of the subsequent prediction of the rovibrational spectra of two molecules of recent interest for us and to astronomers: H_2O and H_2S.

3. COMPUTATIONAL DETAILS

Detailed documentation of the computational methods described in the text can be found in the original papers [27–29, 43–45]. When excerpts from previous work are presented, the notation utilized in the source is not modified. A brief summary of computational details and programs is provided to aid readers already familiar with most of the notation.

3.1. ELECTRONIC STRUCTURE CALCULATIONS

The configuration interaction (CI) calculations, including full CI (FCI), utilized the DETCI code [59] interfaced with the PSI3 program system [63], while the ACESII code [64] has been employed for the coupled-cluster (CC) calculations. Kinetic relativistic effects have been gauged by first-order perturbation theory applied to the one-electron mass-velocity and the one- and two-electron Darwin terms [23], as implemented via the recipe of Klopper [65] within the DALTON program system [66]. Calculation of the Gaunt correction surface utilized the program package MOLFDIR [67, 68].

The correlation-consistent (cc-pVXZ) basis sets of Dunning and coworkers [52–54], as well as their augmented (aug-cc-pVXZ), core-valence (cc-pCVXZ), and uncontracted variants, have been employed almost exclusively in our studies.

3.2. NUCLEAR MOTION CALCULATIONS

The rovibrational energy states have been determined using an exact kinetic energy (EKE) operator expressed in Radau coordinates augmented with the fitted *ab initio* PESs. The nuclear motion calculations utilized the DVR3D program suite [69] and optimized basis sets [70, 71]. Rovibrational calculations are only reported here for the $H_2^{16}O$ and $H_2^{32}S$ isotopomers. All calculations presented used a hydrogen mass midway between the atomic and nuclear value, as recommended by Zobov *et al.* [22].

4. THE FOCAL-POINT APPROACH (FPA)

Characteristics of the focal-point approach [42, 43], comprising the dual extrapolation to the one- and n-particle *ab initio* limits, are as follows: (a) use of a family of basis sets which approaches completeness in a systematic and well controlled way (*e.g.*, the cc-pVXZ, aug-cc-pVXZ, and cc-pCVXZ sets of Dunning and co-workers [52–54] based on principal expansion); (b) application of low levels of electronic structure theory with basis sets as large as possible (typically direct Hartree–Fock (HF) and MP2 computations with hundreds of basis functions); (c) higher-order valence correlation treatments [CCSD(T), CCSDT, BD(TQ), MP4, MP5, and preferably FCI] with the largest possible basis sets; (d) construction of a two-dimensional extrapolation grid based on the assumed additivity of correlation *increments* to energy differences; and (e) avoidance of empirical corrections. Focal-point investigations [42–45, 72–74] and numerous other theoretical studies have shown that even in systems without particularly heavy elements, account may also be needed for core correlation and relativistic phenomena, as well as the diagonal Born–Oppenheimer correction (DBOC). Auxiliary shifts for these effects are appended to valence-only focal-point analyses not only to ensure the highest possible accuracy but also to enhance our understanding of the manifestation of these effects for different chemical applications (*e.g.*, conformational energy and dissociation energy calculations).

The focal-point scheme assumes that the correlation energy increments have rather different convergence characteristics, with the higher-order correlation increments showing diminishing basis set dependence. This assumption has allowed efficacious estimation of molecular barriers at the CBS FCI limit for H_2O, H_2S, NH_3, SiH_3^-, C_2H_6, HCOOH, HNCO, and SiC_2 [43–45, 72, 73]. These calculations, employing two reference structures on the same PES, represent the simplest way to judge the performance of *ab initio* techniques for the calculation of semiglobal PESs. Indeed, they provide plentiful instruction in the pursuit of the *ab initio* limit. Following Table VII of Ref. [43], the diverse (basis set, correlation) convergence may be characterized as follows: NH_3 and SiH_3^- inversion barriers and H_2O barrier to linearity (poor, good); HNCO inversion barrier (good, poor); C_2H_6 and HCOOH torsional energetics (good, good); and SiC_2 barrier to linearity (poor, poor).

In Table 1 we extend our previous valence-only focal-point studies [43–45, 72–74] with a problem in the (good, good) category, and report results for the rotational barrier of CH_3SiH_3, a prototypical symmetric top molecule with one torsional degree of freedom. CH_3SiH_3 and its isotopomers are ideal candidates for the study [75–77] of internal rotation

Table 1. Valence focal-point analysis of the torsional barrier of methyl silane[a]

Series CC	ΔE_e (RHF)	δ [MP2]	δ [CC2]	δ [CCSD(T)]	δ [CC3]	δ [BD(TQ)]	ΔE_e (CC)
VDZ(62)	588	+10	−33	+3	−1	−1	566
a-VDZ(104)	554	−44	−22	−6	−1	[−1]	[480]
VTZ(148)	568	−2	−23	0	[−1]	[−1]	[541]
VTZ+d(153)	577	−3	−22	−1	[−1]	[−1]	[549]
CVTZ(210)	577	−10	−22	−1	[−1]	[−1]	[542]
a-VTZ(234)	576	−28	−23	−4	[−1]	[−1]	[519]
a-VTZ+d(239)	583	−27	−23	−4	[−1]	[−1]	[527]
VQZ(294)	574	−19	−21	−3	[−1]	[−1]	[529]
a-VQZ(440)	575	−25	[−21]	[−3]	[−1]	[−1]	[524]
V5Z(516)	579	−22	[−21]	[−3]	[−1]	[−1]	[531]
a-V5Z(738)	579	−26	[−21]	[−3]	[−1]	[−1]	[527]
CBS	583	−27	−21	−3	−1	−1	530
Series MP	ΔE_e (RHF)	δ [MP2]	δ [MP3]	δ [MP4]	δ [MP5]	δ [MP∞]	ΔE_e (MP)
VDZ(62)	588	+10	−23	0	−5	−1	568
a-VDZ(104)	554	−44	−18	−4	[−5]	[−1]	[482]
VTZ(148)	568	−2	−18	−2	[−5]	[−1]	[540]
VTZ+d(153)	577	−3	−17	−2	[−5]	[−1]	[549]
a-VTZ(234)	576	−28	−19	−4	[−5]	[−1]	[519]
a-VQZ(440)	575	−25	−17	−4	[−5]	[−1]	[523]

[a] For each basis set (VXZ ≡ cc-pVXZ and a-VXZ ≡ aug-cc-pVXZ) the total number of contracted Gaussian functions is given in parentheses. For correlated-level calculations (note that CC2 ≡ CCSD and CC3 ≡ CCSDT) the symbol δ denotes the increment in the relative energy (ΔE_e) with respect to the preceding level of theory as given by the hierarchy RHF → MP2 → CCSD → CCSD(T) → CCSDT → BD(TQ) and RHF → MP2 → MP3 → MP4 → MP5 → MP∞ for Series CC and Series MP, respectively. Brackets signify assumed increments from smaller basis set results. All values are given in cm^{-1}. Extrapolations to the CBS limit were performed according to the formulas given in Section 2.

as well as couplings between vibrational modes during large-amplitude internal motion. The rotational barrier represents the difference between the energies of the eclipsed and staggered conformations. RHF theory recovers most of the rotational barrier; specifically, the RHF barrier is less than 55 cm^{-1} from the true valence-only electronic barrier. Since there are no stereochemically active lone pairs in CH_3SiH_3, diffuse functions are only moderately important. The valence-only CBS FCI electronic barrier is 530 cm^{-1} (note the excellent agreement between extrapolated Series CC and Series MP results). We estimate, at the CVTZ CCSD(T) level, the core-valence and the relativistic corrections

Table 2. A comparison of predicted barriers to linearity for water ($H_2^{16}O$ when adiabatic effects are considered)

Reference	Barrier Height	Comments
Carter and Handy [78]	11493 cm^{-1}	Spectroscopic Empirical
Jensen [79]	11246 cm^{-1}	Spectroscopic Empirical
Polyansky *et al.* [80]	10966 cm^{-1}	Spectroscopic Empirical
Lanquetin *et al.* [81]	11154 cm^{-1}	Effective Hamiltonian
Partridge and Schwenke (PS) [10]	11155 cm^{-1}	*Ab initio*
Partridge and Schwenke [10]	11128 cm^{-1}	Spectroscopic Empirical
PS + adiabatic + relativistic corr.	11192 cm^{-1}	*Ab initio*
Császár *et al.* [43]	11046 ± 70 cm^{-1}	Extrapolated *ab initio*
Tarczay *et al.* [44]	11127 ± 35 cm^{-1}	High accuracy *ab initio*
Kain *et al.* [82]	11105 ± 5 cm^{-1}	Semitheoretical
Valeev *et al.* [83]	11119 ± 12 cm^{-1}	*Ab initio* (MP2-R12)

as +6 and +1 cm^{-1}, respectively. These small values, similar to those found for ethane [43], are indicative of the lack of rehybridization during the rotational motion. Contribution of the DBOC effect to the barrier, found to be -0.5 cm^{-1} for ethane [43], should be negligible. Zero-point vibrational effects, estimated at the cc-pVTZ MP2 level, increase the barrier by 54 cm^{-1}. These corrections bring the calculated effective barrier to $530 + 6 + 1 + 54 = 591$ cm^{-1}, in perfect agreement with the best one-dimensional value of 591.4 cm^{-1} [75].

In the various focal-point studies we looked critically at the auxiliary corrections to standard valence-only *ab initio* electronic structure calculations. For water, for example, inclusion of the kinetic relativistic correction raises the all-electron extrapolated barrier height by 50 cm^{-1} or about 0.5%. After this correction the extrapolated *ab initio* result for the barrier to linearity of water [43] was in good but not perfect agreement with the best empirical value available in 1998 [80]. In Table 2 we give a compilation of older and recent (semi)theoretical and (semi)empirical estimates of the barrier to linearity on the ground-state PES of water. Using the focal-point approach Császár *et al.* [43] deduced a converged estimate of 11046 ± 70 cm^{-1} for the barrier, while later research [44], extending the previous work, determined 11127 ± 35 cm^{-1}. In an attempt to reproduce available empirical rovibrational levels supported by the ground electronic state of water, Kain *et al.* [82] modified existing high-quality PESs [10, 22, 27] and arrived at a value of 11105 ± 5 cm^{-1} for the barrier, which is bracketed by the previous, nearly-converged *ab initio* estimates. Important conclusions of the first *ab initio* investigations were that, as usual, it was easiest to estimate the

Hartree–Fock (HF) limit. the well-established HF value being 11247 ± 2 cm^{-1}; it proved hardest to arrive at the MP2 limit; and corrections beyond MP2 were much less dependent on the quality of the basis set. Using the MP2-R12/A level and very large basis sets, Valeev *et al.* [83] have most recently obtained a converged estimate of –357 ± 5 cm^{-1} for the MP2 increment, significantly different from the best previous *ab initio* estimate of –348 cm^{-1} [44]. The improved estimate of the MP2 limit yielded an improved (lower) estimate for the uncertainty of the *ab initio* barrier. Relativistic corrections beyond the kinetic term also influence the barrier. For example. the Gaunt correction increases the barrier by 6 cm^{-1}, while consideration of the Lamb-shift effect [29] lowers the barrier by almost 4 cm^{-1}. These and some other high-quality *ab initio* results led Valeev *et al.* [83] to surmise that the best present *ab initio* estimate of the barrier to linearity of H_2O is 11119 ± 12 cm^{-1}. In summary, consideration of all possible computational effects on the barrier has finally resulted in excellent accord between theory and experiment.

5. SCALED HIGHER-ORDER CORRELATION ENERGIES (SHOC)

Since FCI computations are very expensive, even with small basis sets, there is little hope that larger basis set FCI calculations will become routine in the near future. Even with the explosive growth in computing power, FCI benchmarks are typically limited to small basis sets and molecules with up to two heavy atoms [60, 84–96]. However, large basis set coupled-cluster (CC) computations, which include through triple excitations (CCSD(T) [56] and CCSDT [57, 58]), are viable for a large number of molecular systems where small basis set FCI computations are feasible.

Therefore, in a recent study Császár and Leininger [62] explored a simple multiplicative approach, termed scaled higher-order correlation (SHOC), for approximating the full configuration interaction (FCI) limit at the complete basis set (CBS) limit from small basis set FCI and coupled cluster (most notably CCSD(T) and CCSDT) calculations. In the SHOC approach an attempt is made to scale correlation energies obtained at lower levels of theory to correct for neglected excitations and arrive at the FCI limit within a particular one-particle basis set. Note that linear scalings, though with different goals in mind, have been employed before, see, *e.g.*, the PCI-X method of Siegbahn and co-workers [97], the G3S method of Pople and co-workers [38], as well as the scaled external correlation (SEC) scheme [98], and the scaling all correlation (SAC) method [99] of Truhlar and co-workers. The utility of the SHOC

Table 3. Hartree–Fock and correlation energies (in E_h) of the N(^4S) atom and the N_2 molecule (at $r = 1.0996$ Å) calculated at different levels of theory, and the resulting valence-only total atomization energy (TAE, in kcal/mol) of N_2^a

	(U)HF	ΔE_{corr}[(U)CCSD]	ΔE_{corr}[(U)CCSDT]	ΔE_{corr}[FCI]
N atom				
cc-pVDZ	-54.391115	-0.086760	-0.087517	-0.087436
cc-pVTZ	-54.400686	-0.111746	-0.114341	-0.114183
cc-pVQZ	-54.403718	-0.118297	-0.121440	-0.121294
CBS	-54.405124	-0.123077	-0.126620	-0.126483
N_2				
cc-pVDZ	-108.953856	-0.309691	-0.321684	-0.323350
cc-pVTZ	-108.983089	-0.372249	-0.390516	-0.392539
cc-pVQZ	-108.990687	-0.393463	-0.413286	-0.415426
CBS	-108.993355	-0.408943	-0.429902	-0.432127
TAE	HF	CCSD	CCSDT	FCI
cc-pVDZ	107.7	193.1	200.3	200.9
cc-pVTZ	114.0	207.4	215.6	217.0
cc-pVQZ	115.0	213.4	221.9	223.4
CBS	114.9	217.1	225.8	227.3

[a] Finite-basis HF, CCSD, CCSDT, and FCI results are taken from Refs. [101, 102]. The FCI results reported for N_2 utilize CCSDT results and the SHOC scheme (see text). Basis set extrapolations were performed with formulas given in the text.

scaling approach is best understood by examples. In this review two examples are given. The first concerns the total atomization energy of N_2. The second example concerns the PES of H_2O.

5.1. HOC EFFECTS ON THE TOTAL ATOMIZATION ENERGY OF N_2

The scaled higher-order correlation (SHOC) correction scheme has been demonstrated [62] to extrapolate CCSD(T) correlation energies for the global minima of the ground electronic states of BH and AlH, as well as $\tilde{X}^3B_1$ and $\tilde{a}^1A_1$ CH_2 [100], to the FCI limit with almost μE_h accuracy. The SHOC approach is used here to investigate the HOC correction on the total atomization energy of N_2, employing data computed at the CCSD, CCSDT, and FCI levels mostly by Feller and Sordo [101–103]. As it is clear from Table 3, containing Hartree–Fock and correlation energies for the N(^4S) atom and the N_2 molecule, the SHOC correction for the total atomization energy of N_2, with a SHOC factor of 1.005182 obtained from our own cc-pVDZ CCSDT and FCI computations, is +1.5

kcal/mol. The resulting extrapolated CBS FCI value is 227.3 kcal/mol. Core-valence correlation and relativistic effects, not considered during calculation of the valence-only CBS FCI value, have been estimated by Feller and Sordo [102] to be +0.8 kcal/mol. This last correction brings our computed value very close to the best experimental value, 228.5 ± 0.4 kcal/mol [104].

5.2. HOC EFFECTS ON POTENTIAL ENERGY SURFACES

The proposed SHOC scheme is multiplicative, and it can readily be applied to the study of PESs by performing small basis FCI calculations at each of the grid points. The effectiveness of a simple alternative procedure, which employs a single, average SHOC factor, has been tested on the ground-state PES of H_2S in Refs. [62, 71]. The single-factor SHOC energy correction resulted in an order of magnitude *uniform* reduction of the valence-only correlation energy error of the CCSD(T) calculation. Here the analysis is extended to H_2O. Four geometries [105] have been selected, covering an energy range of 0–23000 cm^{-1}. Valence-only cc-pVDZ FCI and CCSD(T) computations have been performed at these geometries, resulting in estimates of valence-only correlation energies. The FCI − CCSD(T) energy differences before and after the SHOC scaling, using the arithmetical average of the scale factors determined, 1.002811, are as follows: [-636, -625, -706, -419] and [-34, -26, -97, +147] μE_h, respectively. It is clear from these numbers that the SHOC scale factors change relatively little over the PES and that the single-factor SHOC correction results again in an order of magnitude reduction in the valence-only correlation energy error of the CCSD(T) calculation.

6. THE VIBRATIONAL SPECTRUM OF H_2S

Several valence-only potential energy hypersurfaces have been determined at the CCSD(T) level for H_2S in Ref. [71], with aug-cc-pVTZ+d, aug-cc-pVQZ+d, and aug-cc-pV5Z basis sets. As discussed in some detail in Refs. [45, 106], augmentation of the original aug-cc-pVXZ basis sets of Dunning [52, 53] for S with tight polarization functions (+d) is necessary in order to obtain reliable RHF energies. This so-called core polarization effect has been dealt with through extension of the d-space (+d) of the aug-cc-pVXZ (X = T and Q) basis sets with an additional function whose exponent is the same as the largest d exponent in the aug-cc-pV5Z basis. Another surface, denoted CBS CCSD(T), results when the electronic energies obtained with the finite basis sets are ex-

Table 4. Differences between observed (obs.) vibrational band origins of $H_2^{32}S$ and those computed at the CCSD(T) level[a]

Label	Obs.	TZ+d	QZ+d	5Z	CBS	CBS FCI
010/00 1	1182.57	-0.3	-1.7	-2.1	-2.5	-3.0
020/00 2	2353.96	+5.4	+2.2	+0.9	+0.3	-0.7
$100/10^+0$	2614.14	-5.0	+0.9	+1.8	+3.1	+2.3
$001/10^-0$	2628.46	-6.1	+0.2	+1.3	+2.5	+1.8
030/00 3	3513.79	+11.3	+6.3	+4.1	+3.5	+1.9
040/00 4	4661.68	+14.3	+7.6	+5.1	+4.2	+2.2
$120/10^+2$	4932.70	+2.1	+5.0	+4.2	+5.0	+3.2
$021/10^-2$	4939.10	-0.2	+3.4	+3.6	+4.4	+2.6
$200/20^+0$	5144.99	-10.2	+2.9	+4.6	+6.9	+5.4
$101/20^-0$	5147.22	-10.8	+2.5	+4.0	+6.5	+5.0
002/11 0	5243.10	-10.7	+1.6	+3.6	+5.8	+4.4
050/00 5	5797.24	+12.8	+4.7	+2.8	+1.5	-1.1
$210/20^+1$	6288.15	-9.4	+2.7	+3.9	+5.8	+3.7
$111/20^-1$	6289.17	-10.0	+2.2	+3.4	+5.5	+3.5
$300/30^+0$	7576.38	-16.4	+5.3	+7.8	+11.4	+9.0
$102/21^+0$	7752.26	-16.2	+5.4	+8.0	+11.5	+9.0
$003/21^-0$	7779.32	-16.2	+3.4	+6.3	+9.5	+7.2

[a] All values are given in cm^{-1}. All differences are reported as calculated − observed. TZ+d = aug-cc-pVTZ+d CCSD(T) PES; QZ+d = aug-cc-pVQZ+d CCSD(T) PES; 5Z = aug-cc-pV5Z CCSD(T) PES; CBS = complete basis set CCSD(T) PES; CBS FCI = CBS full configuration interaction PES. Normal mode $(v_1 v_2 v_3)$/local mode$(v_{r1} v_{r2} v_\theta)$ labeling. Observed VBOs from Ref. [107]. Average absolute errors, in cm^{-1}, for the entries of this table are as follows: TZ+d = 9.3, QZ+d = 3.4, 5Z = 4.0, CBS = 5.3, and CBS FCI = 3.9.

trapolated, using the equations given in Section 2, to the complete basis set (CBS) limit. The most accurate valence-only PES, at least within the present approach, is denoted as CBS FCI and is obtained when the CBS CCSD(T) correlation energies are scaled, as described in Section 5, to the FCI limit. All these surfaces have been employed to calculate VBOs for $H_2^{32}S$. The results obtained are presented in Table 4.

It is clear from Table 4 that extension of the basis set ($X = 3$, 4, 5, and ∞) shifts the calculated VBOs more or less systematically. The least accurate results are obtained with $X = 3$, followed by the CBS CCSD(T) results. The $X = 4$ results [aug-cc-pVQZ+d CCSD(T)] have the smallest mean deviation. This observation is not surprising, in fact it is a result of fortuitous error cancellation which is often utilized in quantum chemical calculations. Considerably better agreement between theory and experiment is achieved when core-valence correlation, relativistic, and DBOC effects are properly included in the *ab initio* treatment [71].

7. THE ROVIBRATIONAL SPECTRUM OF H_2O

There is a very high quality *ab initio* ground-state PES available for H_2O, determined by Partridge and Schwenke [10]. Nevertheless, neither basis set extrapolations nor SHOC energy corrections have been performed to reach the CBS FCI limit. For this reason and due to remaining computational errors, small empirical adjustment [10] of the *ab initio* PES resulted in a much better quality semitheoretical PES, which reproduces a large number of rovibrational levels up to about 10000 cm^{-1} with an average accuracy of about 0.2 cm^{-1}, and levels up to 18000 cm^{-1} with an average accuracy of about 1.2 cm^{-1}. From the detailed *ab initio* studies [43, 44] of the barrier to linearity of water, it became clear that corrections due to the special theory of relativity should influence substantially the calculated VBOs. Over the last two years we calculated several relativistic correction surfaces for the ground-state PES of H_2O, including a one-electron mass-velocity and Darwin (MVD1) surface [27], a two-electron Darwin (D2) surface, a Gaunt correction surface, and correction surfaces due to the Lamb-shift effect [29].

Our results, presented in Tables 5 and 6, clearly show that the relativistic corrections have a significant influence on the calculated behavior of both the vibrational and rotational states of water. This should be compared with inclusion of the DBOC, which only has a minor influence [22]. As expected from the changes in the barrier to linearity found upon inclusion of relativistic effects [43, 44, 29], the relavistic corrections to the PES can either raise or lower the rovibrational bands. For example, the kinetic relativistic effect (MVD1) raises both the barrier and the band origins of the bending states. Empirical test calculations [82] which augmented the BO potential with a simple term proportional to the bending coordinate alone resulted in a decrease in the barrier height and a simultaneous decrease in bending band origins. These observations are particularly interesting because of difficulties encountered in representing bending excitations both in water [80] and in H_2S [107] by fitting to spectroscopic data. While the average discrepancy between *ab initio* theory and observation for the VBOs of water is not significantly changed by inclusion of the relativistic corrections in the calculation, there is a marked shift in the error. The error in all band origins using the nonrelativistic BO potential surface of Partridge and Schwenke [10] is approximately constant at $0.1-0.2\%$. Inclusion of the relativistic correction in the PES greatly improves predictions for the stretching states at the expense of worsening (doubling) the error for the pure bending modes. It is also clear from the calculated VBOs that, in accord with

Table 5. Independent contributions of selected relativistic correction surfaces to the vibrational ($J = 0$) band origins of $H_2^{16}O^a$

(v_1, v_2, v_3)	Observed	+MVD1	+D2	+Gaunt	+L1	+L2
(010)	1594.75	+1.29	+0.09	−0.09	−0.09	+0.00
(020)	3151.63	+2.73	+0.18	−0.15	−0.18	+0.01
(030)	4666.79	+4.38	+0.28	−0.17	−0.29	+0.02
(040)	6134.01	+6.40	+0.40	−0.13	−0.43	+0.02
(050)	7542.44	+8.96	+0.54	−0.01	−0.60	+0.03
(060)	8869.95	+12.72	+0.73	+0.27	−0.86	+0.04
(100)	3657.05	−2.80	−0.05	−0.79	+0.18	+0.00
(200)	7201.54	−5.60	−0.09	−1.57	+0.36	−0.01
(300)	10599.69	−8.38	−0.14	−2.34	+0.54	−0.01
(400)	13828.28	−11.06	−0.18	−3.09	+0.71	−0.01
(500)	16898.40	−13.0	−0.20	−3.73	+0.83	−0.01
(600)	19782.00		−0.26	−4.47	+1.01	−0.01
(700)	22529.30		−0.34		+1.19	−0.02
(101)	7249.82	−5.68	−0.09	−1.69	+0.37	−0.01
(201)	10613.35	−8.43	−0.14	−2.31	+0.54	−0.01
(301)	13830.94	−11.05	−0.18	−1.83	+0.71	−0.01
(401)	16898.84	−12.98	−0.20	−4.66	+0.83	−0.01
(501)	19781.10	−15.80	−0.26		+1.01	−0.01
(601)	22529.44	−19.31	−0.34		+1.19	−0.02
(701)	25120.28		−0.39		+1.29	−0.02

[a] All VBOs and corrections are given in cm^{-1}. Observed VBOs, provided only for guidance, are taken from Refs. [108, 109, 110]. The one-electron mass-velocity and Darwin (MVD1) results are taken from Ref. [27]. L1 = one-electron Lamb-shift correction surface. L2 = two-electron Lamb-shift correction surface. The Lamb-shift correction surfaces were obtained using the procedure of Ref. [29].

expectation, the kinetic relativistic correction has the largest effect followed by the Gaunt correction and the Lamb shift. Note that the full Breit correction [28] is somewhat smaller than the Gaunt correction. It is also notable how small the influence of the 2-electron correction terms (2-electron Darwin, D2, and 2-electron Lamb, L2) is on the VBOs, though for some VBOs, especially for the bends, the D2 correction, increasing with increasing excitation, is not negligible.

Table 6 shows the $J = 20$ rotational term values calculated using the same models analysed above for the VBOs. Results are only presented for the vibrational ground state [109]. The effect of the inclusion of relativistic corrections on the rotational term values is strongly dependent on K_a. For low values of K_a, the relativistic correction has almost no effect. For mid K_a values, about $K_a = 5$ to 8 for $J = 20$, it raises rota-

Table 6. Independent contributions of selected relativistic correction surfaces to the $J = 20$ rotational $[(v_1 v_2 v_3)=(000)]$ energy levels of $H_2^{16}O$[a]

J_{K_a,K_c}	Observed	+MVD1	+D2	+Gaunt	+L1	+L2
$20_{0,20}$	4048.250	-0.455	+0.067	-1.051	+0.079	+0.003
$20_{1,19}$	4412.316	-0.481	+0.073	-1.144	+0.085	+0.004
$20_{2,18}$	4738.622	-0.426	+0.083	-1.214	+0.085	+0.004
$20_{3,17}$	5031.794	-0.297	+0.094	-1.268	+0.078	+0.006
$20_{4,16}$	5292.103	-0.045	+0.111	-1.297	+0.064	+0.006
$20_{5,15}$	5513.236	+0.526	+0.141	-1.282	+0.027	+0.007
$20_{6,14}$	5680.788	+1.400	+0.183	-1.218	-0.032	+0.009
$20_{7,13}$	5812.074	+1.397	+0.186	-1.247	-0.032	+0.010
$20_{8,12}$	5966.823	+0.206	+0.137	-1.421	+0.052	+0.007
$20_{9,11}$	6170.832	-1.023	+0.087	-1.609	+0.140	+0.004
$20_{10,10}$	6407.443	-1.968	+0.051	-1.771	+0.207	+0.002
$20_{11,9}$	6664.173	-2.820	+0.018	-1.924	+0.268	+0.001
$20_{12,8}$	6935.428	-3.646	-0.013	-2.075	+0.328	-0.001
$20_{13,7}$	7217.562	-4.463	-0.043	-2.226	+0.386	-0.002
$20_{14,6}$	7507.545	-5.274	-0.072	-2.375	+0.445	-0.005
$20_{15,5}$	7802.709	-6.084	-0.103	-2.523	+0.502	-0.005
$20_{16,4}$	8100.291	-6.895	-0.132	-2.669	+0.560	-0.007
$20_{17,3}$	8397.648	-7.712	-0.162	-2.814	+0.617	-0.009
$20_{18,2}$	8691.927	-8.538	-0.193	-2.956	+0.676	-0.011
$20_{19,1}$	8979.881	-9.382	-0.225	-3.095	+0.735	-0.012
$20_{20,0}$	9257.459	-10.260	-0.258	-3.230	+0.797	-0.015

[a] All rotational data are given in cm^{-1}. The observed rotational term values, provided only for guidance, are taken from Refs. [108, 109, 111, 112]. Many rotational transitions come in quasi-degenerate pairs (cf. Ref. [27]). To save space, only one member in each pair is reported in this table. The one-electron mass-velocity and Darwin (MVD1) results are taken from Ref. [27]. L1 = one-electron Lamb-shift correction surface. L2 = two-electron Lamb-shift correction surface. The Lamb-shift correction surfaces were obtained using the procedure of Ref. [29].

tional energies by 0.2 – 1.4 cm^{-1} bringing the calculated and observed values into reasonably good agreement [27]. For high values of K_a, the relativistic correction lowers the rotational term values leading to significant disagreement with the observed levels [27]. This is consistent with the large increase observed in the VBOs of the bending overtones. Similarly to the VBOs, the order of the magnitude of the different corrections to the rotational term values is kinetic > Gaunt > Lamb. The two-electron relativistic corrections (D2 and L2) are again rather small, the two-electron Lamb-shift correction can safely be ignored. This is not true, however, for the one-electron Lamb shift, L1. For example, the $20_{20,0}$ rotational level is shifted, due to L1, by more than 0.7 cm^{-1}, some 500 times more than the present experimental accuracy [108], which in this case could be improved by up to three orders of magnitude using current technology. The Lamb-shift effect increases with increasing excitation both for the vibrations and for the rotations.

8. SUMMARY

High-resolution rovibrational spectra provide a wealth of information on molecular properties and vibration-rotation dynamics provided that the "inverse eigenvalue dilemma" of molecular spectroscopy can be unravelled to some degree. First-principles techniques do not suffer from the inverse eigenvalue dilemma and thus provide both complementary and competitive approaches to the understanding of chemical phenomena. A big step in this direction is the *ab initio* determination of potential energy hypersurfaces of "spectroscopic" accuracy. In this review a snapshot of some aspects of the state of evolution of *ab initio* methodologies in the determination of accurate PESs has been presented. The focus of the attention has been on the focal-point approach involving a dual extrapolation to the complete basis set (CBS) and full configuration interaction (FCI) asymptotes. The numerical results presented show again the intrinsic accuracy and utility of *ab initio* techniques in general and the focal-point approach in particular. It is shown that scaling the higher-order correlation energy (SHOC) increases the dissociation energy, D_e, of N_2, and reduces the error with respect to experiment, and it also results in an order of magnitude error reduction in the FCI – CCSD(T) correlation energy of H_2O. A pragmatic and sophisticated *ab initio* approach, based on the focal-point and SHOC schemes, has been utilized for the construction of valence-only PESs for the ground electronic state of H_2S. Convergence analysis of the VBOs computed using these PESs reveals the usefulness of the approach. Small corrections, not considered in the valence-only treatment, including core-valence correlation, relativistic effects, and the diagonal Born–Oppenheimer correction (DBOC), are known to be important for the accurate prediction of rovibrational levels. In this review the relativistic effects are quantified for the rovibrational spectrum of H_2O. It is shown that among the relativistic effects the dominant one is the MVD effect, while, in general, the Gaunt correction and quantum electrodynamics (QED) corrections make smaller and smaller contributions to the rovibrational levels. Nevertheless, at about 10000 cm^{-1} and above the QED effect may result in changes as large as 1 cm^{-1} for some rovibrational states.

Acknowledgments

The work of Attila G. Császár and György Tarczay has been supported by the Hungarian Ministry of Culture and Education (FKFP 0117/1997) and by the Scientific Research Fund of Hungary (OTKA T024044 and T033074). The work of Matthew L. Leininger was supported by Sandia National Laboratories. Sandia is a multiprogram laboratory operated by Sandia Corporation, a Lockheed Martin Company, for the United States Department of Energy under Contract DE-AC04-94AL85000. The work of Jonathan Tennyson and Oleg L. Polyansky was supported by the UK Engineering and Physical Science Research Council (grant GR/K47702). The work of Oleg L. Polyansky was also supported by the Russian Fund for Fundamental Studies. Scientific exchanges between Budapest and London received support from the Hungarian-British Joint Academic and Research Programme (project no. 076). Scientific exchanges between Budapest and Athens, GA received support from a NATO Linkage Grant (CRG.LG 973892).

References

[1] Oka, T. (1980) *Phys. Rev. Lett.*, **45**, 531–534.

[2] Geballe, T.R. (2000) *Phil. Trans. Royal Soc. London A*, **358**, 2503–2512.

[3] Miller, S., Tennyson, J., Lepp, S. and Dalgarno, A. (1992) *Nature*, **355**, 420–422.

[4] Miller, S., Achilleos, N., Ballester, G.E., Geballe, T.R., Joseph, R.D., Prange, R., Rego, D., Stallard, T.S., Tennyson, J., Trafton, L.M. and Waite Jr., J.H. (2000) *Phil. Trans. Royal Soc. London A*, **358**, 2485–2501.

[5] Kao, L., Oka, T., Miller, S. and Tennyson, J. (1991) *Astrophys. J. Suppl.*, **77**, 317–329.

[6] The Opacity Project Team (1995) *The Opacity Project*, vol. 1 and (1996) *The Opacity Project*, vol. 2, IOP Publishing: Bristol, UK.

[7] Allard, F., Hauschildt, P. H., Alexander, D. R. and Starrfield, S. (1997) *Ann. Rev. Astron. Astrophys.*, **35**, 137–177.

[8] Jorgensen, U.G., Almlöf, J., Gustafsson, B., Larsson, M. and Siegbahn, P. (1986) *J. Chem. Phys.*, **83**, 3034–3042.

[9] Allard, F., Hauschildt, P.H., Miller, S. and Tennyson, J. (1994) *Astrophys. J.*, **426**, L39–L41.

[10] Partridge, H. and Schwenke, D.W. (1997) *J. Chem. Phys*, **106**, 4618–4639.

[11] Allard, F., Hauschildt, P.H. and Schwenke, D.W. (2000) *Astrophys. J.*, **540**, 1005–1015.

[12] Neale, L. and Tennyson, J. (1995) *Astrophys. J.*, **454**, L169–L173.

[13] Sidhu, K.S., Miller, S. and Tennyson, J. (1992) *Astron. Astrophys.*, **255**, 453–456.

[14] Ramanathan, V. and Vogelmann, A.M. (1997) *Ambio*, **26**, 38–46.

[15] Polyansky, O.L., Zobov. N.F., Viti. S., Tennyson. J. Bernath, P.F. and Wallace, L. (1997) *Science*, **277**. 346–348.

[16] Polyansky, O.L., Zobov. N.F., Viti, S., Tennyson. J., Bernath, P.F. and Wallace, L. (1997) *J. Molec. Spectrosc.*, **186**. 422–447.

[17] Kedziora, G.S. and Shavitt, I. (1997) *J. Chem. Phys.*, **106**, 8733–8745.

[18] Császár, A.G. and Mills, I.M. (1997) *Spectrochimica Acta*, **53A**, 1101–1122.

[19] Császár, A.G., Allen, W.D., Yamaguchi, Y. and Schaefer III, H.F. (2000) in *Computational Molecular Spectroscopy*, Eds. P. Jensen and P. R. Bunker, Wiley: New York.

[20] Searles, D. and Nagy-Felsobuki, E. (1993) *Ab Initio Variational Calculations of Molecular Vibrational-Rotational Spectra*, Springer-Verlag: Berlin.

[21] Tennyson, J. (2000) in *Computational Molecular Spectroscopy*, Eds. P. Jensen and P. R. Bunker, Wiley: New York.

[22] Zobov, N.F., Polyansky, O.L., Le Sueur. C.R. and Tennyson, J. (1996) *Chem. Phys. Lett.*, **260**, 381–387.

[23] Balasubramanian, K. (1997) *Relativistic Effects in Chemistry, Part A: Theory and Techniques and Part B: Applications*, Wiley: New York.

[24] Cencek, W., Rychlewski, J., Jaquet, R. and Kutzelnigg, W. (1998) *J. Chem. Phys.*, **108**, 2831–2836.

[25] Jaquet, R., Cencek, W., Kutzelnigg, W. and Rychlewski, J. (1998) *J. Chem. Phys.*, **108**, 2837–2846.

[26] Polyansky, O.L. and Tennyson, J. (1999) *J. Chem. Phys.*, **110**, 5056–5064.

[27] Császár, A.G., Kain, J.S., Polyansky, O.L., Zobov, N.F. and Tennyson, J. (1998) *Chem. Phys. Lett.*, **293**, 317-323; (1999) *ibid*, **312**, 613–616 (E).

[28] Quiney, H.M., Barletta, P., Tarczay, G., Császár, A.G., Polyansky, O.L. and Tennyson, J. (2001) *Chem. Phys. Lett.* to be submitted.

[29] Pyykkö, P., Dyall, K.G., Császár, A.G., Tarczay, G., Polyansky, O.L. and Tennyson, J. (2000) *Phys. Rev. A* in print.

[30] Tarczay, G., Császár, A.G. and Klopper, W. (2001) *J. Chem. Phys.* to be submitted.

[31] Murrell, J.N., Carter, S., Farantos, S.C., Huxley, P. and Varandas. A.J.C. (1984) *Molecular Potential Energy Surfaces*, Wiley: New York.

[32] Hirst, D.M. (1985) *Potential Energy Surfaces*, Taylor and Francis: London.

[33] Born, M. and Oppenheimer, J.R. (1927) *Ann. Physik*, **84**, 457.

[34] Born, M. and Huang, K. (1954) *Dynamical Theory of Crystal Lattices*. Oxford University Press: London.

[35] Curtiss, L.A. and Raghavachari, K. (1998) *ACS Symp. Ser.*, **677**, 176.

[36] Curtiss, L.A., Raghavachari, K., Trucks, G.W. and Pople, J.A. (1991) *J. Chem. Phys.*, **94**, 7221–7230.

[37] Curtiss, L.A., Raghavachari, K., Redfern, P.C., Rassolov, V. and Pople, J.A. (1998) *J. Chem. Phys.*, **109**, 7764–7776.

[38] Curtiss, L.A., Raghavachari, K., Redfern, P.C. and Pople, J.A. (2000) *J. Chem. Phys.*, **112**, 1125–1132.

[39] Montgomery Jr., J.A., Ochterski, J.W. and Petersson, G.A. (1994) *J. Chem. Phys.*, **101**, 5900–5909.

[40] Montgomery Jr., J.A., Frisch, M.J., Ochterski. J.W. and Petersson, G.A. (1999) *J. Chem. Phys.*, **110**, 2822–2827.

[41] Martin, J.M.L. and Oliveira, G. (1999) *J. Chem. Phys.*, **111**, 1843–1856.

[42] Allen, W.D., East, A.L.L. and Császár, A.G. (1993) in *Structures and Conformations of Non-Rigid Molecules*, NATO ASI Series C, Eds. J. Laane, M. Dakkouri, B. van der Veken, and H. Oberhammer, Kluwer: Dordrecht, p.343.

[43] Császár, A.G., Allen, W.D. and Schaefer III, H.F. (1998) *J. Chem. Phys.*, **108**, 9751–9764.

[44] Tarczay, G., Császár, A.G., Klopper, W., Szalay, V., Allen, W.D. and Schaefer III, H.F. (1999) *J. Chem. Phys.*, **110**, 11971–11981.

[45] Tarczay, G., Császár, A.G., Leininger, M.L. and Klopper, W. (2000) *Chem. Phys. Lett.*, **322**, 119–128.

[46] Klopper, W., Bak, K.L., Jørgensen, P., Olsen, J. and Helgaker, T. (1999) *J. Phys. B*, **32**, 103–130.

[47] Feller, D. (1992) *J. Chem. Phys.*, **96**, 6104–6114.

[48] Martin, J.M.L. (1996) *Chem. Phys. Lett.*, **259**, 679–682.

[49] Halkier, A., Helgaker, T., Jørgensen, P., Klopper, W. and Olsen, J. (1999) *Chem. Phys. Lett.*, **302**, 437–446.

[50] Halkier, A., Helgaker, T., Klopper, W. and Olsen, J. (2000) *Chem. Phys. Lett.*, **319**, 287–295.

[51] Halkier, A., Helgaker, T., Klopper, W., Jørgensen, P. and Császár, A.G. (1999) *Chem. Phys. Lett.*, **310**, 385–389.

[52] Dunning Jr., T.H. (1989) *J. Chem. Phys.*, **90**, 1007–1023.

[53] Kendall, R.A., Dunning Jr., T.H. and Harrison, R.J. (1992) *J. Chem. Phys.*, **96**, 6796–6806.

[54] Wilson, K.A., v. Mourik, T. and Dunning Jr., T.H., (1997) *J. Mol. Struct. (THEOCHEM)*, **338**, 339–349.

[55] Bartlett, R.J. (1997) *Recent Advances in Coupled Cluster Methods*, World Scientific: Singapore.

[56] Raghavachari, K., Trucks, G.W., Pople, J.A. and Head-Gordon, M. (1989) *Chem. Phys. Lett.*, **157**, 479–483.

[57] Noga, J. and Bartlett, R.J. (1987) *J. Chem. Phys.*, **86**, 7041–7050; (1988) *ibid.*, **89**, 3401(E).

[58] Scuseria, G. and Schaefer III, H.F. (1988) *Chem. Phys. Lett.*, **152**, 382–386.

[59] Sherrill, C.D. and Schaefer III, H.F. (1999) *Adv. Quant. Chem.*, **34**, 143–269.

[60] Leininger, M.L., Allen, W.D., Schaefer III, H.F. and Sherrill, C.D. (2000) *J. Chem. Phys.*, **112**, 9213–9222.

[61] Lee, T.J. and Scuseria, G.E. (1995) in *Quantum Mechanical Electronic Structure Calculations with Chemical Accuracy*, Ed. Langhoff, S.R., Kluwer: Dordrecht, pp. 47–108.

[62] Császár, A.G. and Leininger, M.L. (2001) *J. Chem. Phys.* submitted for publication.

[63] PSI 3.0, T. D. Crawford, C. D. Sherrill, E. F. Valeev, J. T. Fermann, M. L. Leininger, R. A. King, S. T. Brown, C. L. Janssen, E. T. Seidl, Y. Yamaguchi, W. D. Allen, Y. Xie, G. Vacek, T. P. Hamilton, C. B. Kellogg, R. B. Remington and H. F. Schaefer III (PSITECH Inc., Watkinsville, GA, 1999).

[64] Stanton, J.F., Gauss, J., Lauderdale, W.J., Watts, J.D. and Bartlett, R.J. ACES II. The package also contains modified versions of the MOLECULE Gaussian integral program of J. Almlöf and P. R. Taylor, the ABACUS integral derivative program written by T. U. Helgaker, H. J. As. Jensen, P. Jorgensen, and P. R. Taylor, and the PROPS property evaluation integral code of P. R. Taylor.

[65] Klopper, W. (1997) *J. Comp. Chem.*, **18**, 20–27.

[66] DALTON, an *ab initio* electronic structure program, Release 1.0 (1997), written by Helgaker, T., Jensen, H.J.Aa., Jørgensen, P., Olsen, J., Ruud, K., Agren, H., Andersen, T., Bak, K.L., Bakken, V., Christiansen, O., Dahle, P., Dalskov, E.K., Enevoldsen, T., Fernandez. B., Heiberg, H., Hettema, H., Jonsson, D., Kirpekar, S., Kobayashi, R., Koch, H., Mikkelsen, K.V., Norman, P., Packer, M.J., Saue, T., Taylor, P.R. and Vantras, O.

[67] MOLFDIR, Aerts, P.J.C., Visser. O., Visscher, L., Merenga, H., de Jong, W.A. and Nieuwpoort, W.C., University of Groningen, The Netherlands.

[68] Visscher, L., Visser, O., Aerts, P.J.C., Merenga, H. and Nieuwpoort, W.C. (1994) *Comp. Phys. Commun.*, **81**, 120–144.

[69] Tennyson, J., Henderson, J.R. and Fulton, N.G. (1995) *Comp. Phys. Comms.*, **86**, 175–198.

[70] Polyansky, O.L., Jensen, P. and Tennyson, J. (1994) *J. Chem. Phys.*, **101**, 7651–7657.

[71] Tarczay, G., Császár, A.G., Polyansky, O.L. and Tennyson, J. (2001) *J. Chem. Phys.* to be submitted.

[72] Nielsen, I.M.B., Allen, W.D., Császár, A.G. and Schaefer III, H.F. (1997) *J. Chem. Phys.*, **107**, 1195–1211.

[73] Aarset, K., Császár, A.G., Sibert III, E.L., Allen, W.D., Schaefer III, H.F., Klopper, W., Noga, J. (2000) *J. Chem. Phys.*, **112**, 4053–4063.

[74] Valeev, E.F., Allen, W.D., Schaefer III, H.F., Császár, A.G. and East, A.L.L. (2001) *J. Phys. Chem.*, submitted.

[75] Moazzen-Ahmadi, N. and Ozier, I. (1987) *J. Mol. Spectrosc.*, **123**, 26–36.

[76] Moazzen-Ahmadi, N. and Ozier, I., McRae, G. A. and Cohen, E. A. (1996) *J. Mol. Spectrosc.*, **175**, 54–61.

[77] Ozier, I. (1999) personal communication.

[78] Carter, S. and Handy, N.C. (1987) *J. Chem. Phys.*, **87**, 4294–4301.

[79] Jensen, P. (1989) *J. Mol. Spectrosc.*, **133**, 438–460.

[80] Polyansky, O.L., Jensen, P. and Tennyson, J. (1996) *J. Chem. Phys.*, **105**, 6490–6497.

[81] Lanquetin, R., Coudert, L.H. and Camy-Peyret, C. (1999) *J. Mol. Spectrosc.*, **195**, 54–67.

[82] Kain, J.S., Polyansky, O.L. and Tennyson, J. (2000) *Chem. Phys. Lett.*, **317**, 365–371.

[83] Valeev, E.F., Császár, A.G., Allen, W.D. and Schaefer III, H.F. (2001) *J. Chem. Phys.*, accepted for publication.

[84] Evangelisti, S., Bendazzoli, G.L. and Gagliardi, L. (1994) *Chem. Phys.*, **185**, 47–56.

[85] Evangelisti, S., Bendazzoli, G.L., Ansaloni, R. and Rossi, E. (1995) *Chem. Phys. Lett.*, **233**, 353–358.

[86] Evangelisti, S., Bendazzoli, G.L., Ansaloni, R., Durí, F. and Rossi, E. (1996) *Chem. Phys. Lett.*, **252**, 437–446.

[87] Christiansen, O., Koch, H., Jørgensen, P. and Olsen, J. (1996) *Chem. Phys. Lett.*, **256**, 185–194.

[88] Olsen, J., Christiansen, O., Koch, H. and Jørgensen, P. (1996) *J. Chem. Phys.*, **105**, 5082–5090.

[89] Olsen, J., Jørgensen, P., Koch, H., Balková, A., and Bartlett, R.J. (1996) *J. Chem. Phys.*, **104**, 8007–8015.

[90] Sherrill, C.D., Van Huis, T.J., Yamaguchi, Y. and Schaefer III, H.F. (1997) *J. Mol. Struct. (THEOCHEM)*, **400**, 139–156.

[91] Ben-Amor, N., Evangelisti, S., Maynau, D. and Rossi, E. (1998) *Chem. Phys. Lett.*, **288**, 348–355.

[92] Rossi, E., Bendazzoli, G.L. and Evangelisti, S. (1998) *J. Comp. Chem.*, **19**, 658–672.

[93] Sherrill, C.D., Leininger, M.L., Van Huis, T.J. and Schaefer III, H.F. (1998) *J. Chem. Phys.*, **108**, 1040–1049.

[94] Van Huis, T.J., Leininger, M.L., Sherrill, C.D. and Schaefer III, H.F. (1998) *Coll. Czech. Chem. Comm.*, **63**, 1107–1142.

[95] Leininger, M.L., Sherrill, C.D., Allen, W.D. and Schaefer III, H.F. (1998) *J. Chem. Phys.*, **108**, 6717–6721.

[96] Rossi, E., Bendazzoli, G.L., Evangelisti, S. and Maynau, D. (1999) *Chem. Phys. Lett.*, **310**, 530–536.

[97] Siegbahn, P.E.M., Blomberg, M.R.A. and Svensson, M. (1994) *Chem. Phys. Lett.*, **223**, 35–45.

[98] Brown, F.B. and Truhlar, D.G. (1985) *Chem. Phys. Lett.*, **117**, 307–313.

[99] Gordon, M.S. and Truhlar, D.G. (1986) *J. Am. Chem. Soc.*, **108**, 5412–5419.

[100] Leininger, M.L. and Császár, A.G. (2001) *Mol. Phys.* to be submitted.

[101] Feller, D. (1999) *J. Chem. Phys.*, **111**, 4373–4382.

[102] Feller, D. and Sordo, J.A. (2000) *J. Chem. Phys.*, **113**, 485–493.

[103] Feller, D. and Sordo, J.A. (2000) *J. Chem. Phys.*, **112**, 5604–5610.

[104] Chase Jr, M.W. (1998) *J. Phys. Chem. Ref. Data*, **9**, 1.

[105] The following geometries ([r_{OH}/Å, r_{OH}/Å, θ_{HOH}/degree]), in order of increasing energy, have been selected: [0.9583, 0.9583, 104.418], [0.9583, 0.9583, 148.870], [0.7784, 1.1959, 104.418], and [0.7784, 0.7784, 104.418].

[106] Martin, J.M.L. (1998) *J. Chem. Phys.*, **108**, 2791–2800.

[107] Polyansky, O.L., Jensen, P. and Tennyson, J. (1996) *J. Mol. Spectrosc.*, **178**, 184–188.

[108] Tennyson, J., Zobov, N.F., Williamson, R., Polyansky, O.L. and Bernath, P.F. (2001) *J. Chem. Phys. Ref. Data*, to be submitted.

[109] Polyansky, O.L., Zobov, N.F., Tennyson, J., Lotoski, J.A. and Bernath, P. (1997) *J. Molec. Spectrosc.*, **184**, 35–50.

[110] Rothman, L.S., Gamache, R.R., Tipping, R.H., Rinsland, C.P., Smith, M.A.H., Benner, D.C., Malathy Devi, V., Flaud, J.-M., Camy-Peyret, C., Perrin, A., Goldman, A., Massie, S.T., Brown, L.R. and Toth, R.A. (1992) *J. Quant. Spectrosc. Radiative Transf.*, **48**, 469–507.

[111] Flaud, J.-M., Camy-Peyret, C. and Maillard, J.-P. (1976) *Mol. Phys.*, **32**, 499–521.

[112] Polyansky, O.L., Busler, J.R., Guo, B., Zhang, K. and Bernath, P. (1996) *J. Mol. Spectrosc.*, **176** 305–315.

[113] Polyansky, O.L., Zobov, N.F., Viti, S. and Tennyson, J. (2000) *J. Mol. Spectrosc.*, in press.

OPTICALLY ACTIVE HYDROGEN BONDED COMPLEXES IN THE ATMOSPHERE

N.A. ZVEREVA
Optics and Atmosphere Inst., Akademicheskii av.1, 634055, Tomsk, Russia
e-mail: zvereva@phys.tsu.ru

1. Introduction

The study of intermolecular interactions and their spectral manifestation has attracted much attention recently. Significant progress has been achieved in the experimental investigations of systems with hydrogen bonds in the gas phase [1-8]. However, low concentration of complexes and interference from highly intense rotational spectra occurring in the far-IR (infrared) region makes interpretation of molecular complexes spectra difficult. A complete set of intermolecular vibrations has been determined for only a very limited number of hydrogen complexes. The main attention of researchers has concentrated on hydrogen bond systems of high and medium strength, for instance, H_2O...HF [9], where IR spectra in the gas phase were completely recorded. Weak hydrogen bond systems, such as H_2O...HCl, are much less completely understood in spite of their chemical importance. Geometry of this complex was determined by rotational spectroscopy in the gas phase [10], however data on vibration spectra are limited to intramolecular modes in solid matrices [11-13]. In Ref. [14] the calculations of H_2O...HCl complex were carried out by ab initio methods and vibration frequencies and geometry of the complex were determined. The aim of the present work is a quantum mechanical study of stability and vibration spectra of 1:2 and 2:1 $(H_2O)_n(HCl)_m$ and $(H_2O)_n(HF)_m$ (n,m $\geq$ 2) complexes. The correct quantum mechanical studies place rather demanding requirements on the choice of an optimum method of calculation. The most consistent and reliable method is ab initio method of restricted Hartree-Fock-Roothaan (RHF) [15]. For structurally nonrigid molecules, such as intermolecular complexes, the displacements in motions having anomalously large amplitudes have been shown [16-17] not to violate the Born-Oppenheimer approach for the case of far-off located electronic terms. This follows from nonadiabatic matrix element evaluations.

2. Methods of calculation

Ab initio calculations have been carried out using the Monstergauss code package [18]. For intermolecular 1:2 and 2:1 $(H_2O)_n(HCl)_m$ and $(H_2O)_n(HF)_m$ (n, m $\geq$ 2) complexes,

J. Demaison et al. (eds.), Spectroscopy from Space, 341–350.

split-valence basis set 6-31G** was chosen, containing polarization functions on all atoms. Addition of diffuse sp-shells to heavy atoms and additional set of d-functions to atoms O and Cl has slightly affected the geometric structure and vibration frequencies (see Tables 1-5). Accuracy of the ab initio calculated molecular characteristics depends not only on the quality of description of wave functions, but also on numerical methods. In particular, spectral constants are strongly influenced by the effects of basis sets and electronic correlation. For optimisation of geometry and for construction of matrices of force constants the method using a combination of Newton-Raphson, steepest descent and Marquardt algorithms (similar to the algorithm [19]) has been chosen. The choice of a step ΔS (Euclidian distance between the starting point and the optimised point) is important. Using different nonrigid system calculations as an example [20] the optimum is found to be ΔS = 0.1 a.u. which causes numerical errors in the determination of a harmonic force constant of the order ~ 0.001 mdyne/Å. For the required accuracy of gradient of length $5\cdot10^{-4}$ we can get a value of energy E accurate to the five decimal digits. Vibration frequencies and eigenvectors of vibration problem were computed using the known force constant matrix F and kinetic energy matrix T. The accuracy of the calculated geometric parameters in comparison with the experimental data and calculations with Møller-Plesset perturbation theory of second order (MP2) is ~ 0.01 - 0.02 Å for intramolecular distances r, and ~ 0.12 Å for intermolecular distances R, for intramolecular deformation angles the accuracy is ~ 1.5°, for intermolecular deformation angles accuracy is ~ 1.5° - 10°.

TABLE 1.Calculated properties of H_2O and HCl [14] (distances in Å and angles in degree)

H_2O	method	6-31G**	+VP^s	+VP^s(2d)^s	exp.
r(OH)	SCF	0.943	0.943	0.943	0.9575 [30]
∠(HOH)	SCF	106.0	106.4	106.0	104.5 [30]
E, a.u.	SCF MP2	-76.02361 -0.19544	-76.03576 -0.20956	-76.03885 -0.22362	
μ, D	SCF MP2	2.147 -0.085	2.226 -0.048	1.991 -0.036	1.847 [31]
HCl	method	6-31G**	+VP^s	+VP^s(2d)^s	exp.
r(HCl),Å	SCF MP2	1.266	1.269	1.267 1.274	1.274 [30]
E, a.u.	SCF MP2	-460.06603 -0.14932	-460.06816 -0.15146	-460.07012 -0.17212	
μ, D	SCF MP2	1.469 -0.049	1.511 -0.046	1.231 -0.040	1.093[32]
H_2O...HCl	method	6-31G**	+VP^s	+VP^s(2d)^s	exp.
R(O...Cl)	SCF MP2	3.250	3.303	3.369 3.235	
θ(OHCl)	SCF MP2	2.2	2.9	3.2 3.5	
r(OH)	SCF MP2	0.943	0.944	0.943 0.943	
α(HOH)	SCF MP2	107.1	107.1	106.7 106.7	
r(HCl)	SCF MP2	1.277	1.278	1.275 1.286	

TABLE 2. Calculated geometrical parameters of H_2O...HCl (this paper, 6-31G**, R(Å), ∠(degree)).

r(OH)	R(O...H)	r(HCl)	R(ClO)
0.943	1.975	1.277	3.25
α(HOH)	θ(O...HCl)	β(HO...H)	χ(HO...HH)
107.0	178.8	119.9	115.0

TABLE 3. Vibration frequencies (cm^{-1}) of water and HCl molecules

	H_2O			HCl
	ν_1	ν_2	ν_3	
6-31G**[14]	4149	1770	4267	3178
6-31G** [this paper]	4065	1754	4164	3073
+VP^s [14]	4129	1727	4242	3174
+VP^s(2d)^s [14]	4139	1759	4244	3174
Exp. [33]	3657	1595	3756	3141
exp. $_{harm.fr.}$ [24]	3832	1648	3942	2990
$\nu_{calc.}$, see Eq. (2)	4065	1754	4164	3073
ν_{corr}	3883	1702	3977	2947

a Ref. [25], Ref. [33] gives 3042 cm^{-1}

TABLE 4. Intramolecular frequencies of H_2O...HCl complex (cm^{-1})

	ν(6-31G**)	Δν	Δν*	ν_{corr}	$\Delta\nu_{corr}$
HCl	2991	-82	-105	2870	-78
$(OH)_1$	4065	0	-4	3883	0
$(OH)_2$	4164	0	-3	3977	0
HOH	1760	+6	+1	1707	+5

* - + VP^s(2d)^s [14]

TABLE 5. Intermolecular frequencies, interaction energies and dipole moment of H_2O...HCl complex in comparison with other complexes

	ΔE, 6-31G** kcal/mol	ν_a cm^{-1}	ν_b cm^{-1}	ν_c cm^{-1}	$\nu_{\beta1}$ cm^{-1}	$\nu_{\beta2}$ cm^{-1}	μ, D	μ, D monomer
H_2O...HCl	-5.2	139 100[14]	436 460[14]	289	131	73	4.26	1.09
H_2O...H_2O	-4.63	186	645	345	157	98	2.74	1.85
H_2O...HF	-9.2	190 180[9]	740 696[9]	672 666[9]	220 170[9]	169 145[9]	4.3	1.91

Harmonic force constants calculated at the RHF level for molecular systems are overestimated by ~ 10 - 20% and harmonic vibration frequencies by 5 - 10%. If allowance is made for electronic correlation using the method of configuration interactions (CI) [21] or the Møller-Plesset perturbation theory to second order (MP2) [14,22,23], the vibration frequencies are reduced by ~ 4 %. However, the values of the stretching vibration frequencies remain overestimated (for instance, for H_2O molecules and HF difference of frequencies is ~ 200 cm^{-1}) [23].

The results of this work suggest that the vibration frequencies should be corrected using the following relation

$$\nu_{corr} = a \cdot \nu_{calc} + b \quad (2)$$

where - ν_{calc} are the ab initio calculated frequencies, ν_{corr} – are the corrected values of frequencies, and a and b – are the parameters determined by a least squares fit to the experimental data.

3. Results and discussions

In Table 3 are listed the calculated harmonic frequencies $\nu_{calc.}$, the experimental [24,25] ν_{exp}. and corrected frequencies ν_{corr}. for H_2O and HCl monomers. The parameters a and b determined using the experimental values are 0.944 and 46 cm^{-1}, respectively. The correlation coefficient between experimental and calculated values of harmonic frequencies is 0.9991. The value of average absolute deviation of the predicted values from the experimentally observed harmonic frequencies of H_2O and HCl is 46 cm^{-1}.

The calculated structural parameters of H_2O...HCl heterodimer are presented in Table 2. The calculated frequencies ν_{calc} (6-31G**) corresponding to the geometric parameters, corrected ν_{corr} harmonic intramolecular frequencies, and offset $\Delta\nu$ for monomers of H_2O and HCl molecules are presented in Table 4. In contrast to OH stretch modes of water molecule the frequency ν of HCl stretch involved in hydrogen bonding has a significant low-frequency shift ~ 100 cm^{-1}. Ault and Pimentel [11] have observed a red shift 216 cm^{-1} for H_2O...HCl, but in the N_2 - matrix the measurements are expected to lead to a substantially larger shift than in vacuum (the shift in the HCl stretching frequency in the N_2 matrix is 1.8 times larger than the shift observed in the gas phase for similar complexes).

The next group of frequencies corresponding to the intermolecular vibrations is presented in Table 5. Intermolecular stretch which pulls the two subunits apart is characterized by frequency ν = 139 cm^{-1} (118 cm^{-1} (+$VP^s(2d)^s$) [14]), which agrees reasonably well with the result in Ref. [11] (ν ~ 100 cm^{-1}) and with the result ν ~ 119 cm^{-1} for the similar complex $(CH_3)_2O$...HCl in the gas phase [26]. The frequencies of intermolecular H-bond bends O...H-Cl are 436 cm^{-1} (6-31G**) and 459 cm^{-1} (+ $VP^s(2d)^s$) [14], while 289 cm^{-1} (6-31G**) and 351cm^{-1}(+VPs(2d)s) [14] correspond to in-plane and out-of-plane wagging of proton-donor molecule of the complex. These data agree well with the experimental work of Ault and Pimentel [11]. They observed a broad band at 460 cm^{-1} and assigned it to the H-bond bend.

The wagging of the proton-acceptor H_2O occur at much lower frequencies:31 cm^{-1} and 73 cm^{-1} (6-31G**), 143 cm^{-1} and 94 cm^{-1} (+$VP^s(2d)^s$) [14] and are consistent with the gas-phase spectrum of H_2O...HF where the experimental frequencies are 157 cm^{-1} and 64 cm^{-1} [27].

Summing up the results for H_2O...HCl it is possible to say that for intermolecular complex frequencies with weak hydrogen bond, a sufficiently good agreement is observed already at RHF level (6-31G**) of calculations.

On the basis of the results obtained for H_2O...HCl, we have extended our studies also to structure, stability and vibration IR spectra of 1:2 and 2:1 H_2O and HCl complexes. The results of energy optimization show that there are three stable configurations: one of them corresponds to the complex with C_{2v} symmetry, with a plane of symmetry passing through the molecule H_2O, and a plane formed by HCl molecules and perpendicular to the first one (E = -996.17165 a.u., ΔE = -9.8 kcal/mol (RHF, 6-31G**), and -8.5 kcal/mol (RHF – BSSE (basis set superposition error)), two other configurations correspond to the complexes with the symmetry C_1 with E = -996.17011 a.u., _E = -8.84 kcal/mol (RHF, 6-31G**) and 7.64 kcal/mol (RHF - BSSE), E = -612.13414 a.u. and _E = -13 kcal/mol (RHF, 6-31G**) and -11.1 kcal/mol, respectively (RHF - BSSE). The geometric parameters after optimization (RHF, 6-31G**) are provided in Tables 6 -8.

TABLE 6. Values of nonequivalent geometric parameters of optimum structure of configuration 1, in Å and degree.

r(OH)	R(O...H)	R(HCl)	α(HOH)	θ(Cl-H...O)	β(H...O...H)
0.945	2.08	1.27	106.6	170.2	110.2

TABLE 7. Values of geometric parameters of optimum structure of configuration 2, in Å and degrees.

$r(OH_4)$	$r(OH_6)$	$r(OH_7)$	$r(HCl_1)$	$r(HCl_3)$	$R(HCl_2)$
2.017	0.945	0.944	1.270	1.276	2.696
$\angle Cl_3H_2Cl_1$	$\angle H_6C_1H_2$	$\angle O_5H_6Cl_1$	$\angle H_7O_5H_4$	$R(H_6...Cl_1)$	$\angle H_4Cl_3H_2$
145.5	86.12	126.3	136.87	3.0	82.8
$\angle O_5H_4CL_3$	$\angle H_6O_5H_4$	$\angle H_7O_5H_6$	$\angle H_4Cl_3H_2Cl_1$	$\angle O_5H_4CL_3H_2$	$\angle H_6O_5H_4Cl_3$
164.7	109.4	106.8	-1.5	1.8	-1.24
$\angle H_7O_5H_4Cl_1$	138.04				

TABLE 8. Values of geometric parameters of optimum structure of configuration 3, in Å and degrees.

Bond lengths		Bond angles		Torsion angles	
Cl_1H_2	1.279	$H_5O_3H_4$	106.4	$H_4O_3H_2Cl_1$	-1.88
O_3H_2	2.111	$H_8O_6H_7$	106.4	$H_5O_3H_4H_2$	123.3
H_4O_3	0.957	$O_3H_2Cl_1$	160.94	$O_6H_4O_3H_2$	2.44
H_5O_3	0.943	$H_4O_3H_2$	97.1	$H_7O_6H_4O_3$	-8.52
O_6H_4	2.228	$O_6H_4O_3$	150.86	$H_8O_6H_7Cl_1$	-113.7
H_7O_6	0.944	$H_7O_6H_4$	100.84		
H_8O_6	0.943				
H_7CL_1	2.790				

The complex in configuration 3 is the most stable according to the calculated energy of the interaction ΔE = 11.1 kcal/mol. The complex in configuration 1 has two Cl-O...H bonds with D_e= -3.71 kcal/mol (RHF, 6-31G**) and -3.053 kcal/mol (RHF - BSSE). Structure 2 has one Cl-H...O bond and two weaker bonds H-Cl...H ~ -1.4 kcal/mol (RHF, 6-31G**) and -1.3 kcal/mol (RHF - BSSE). This explains the fact that the complex with an open structure 1 is more stable than the complex with a closed structure 2.

The calculated (6-31G**) and corrected harmonic intramolecular frequencies of the considered complexes are presented in Tables 9 - 11. The frequencies corresponding to the intermolecular O...H-Cl stretch of the optimum structure 1 are 124 cm^{-1} and 138

cm^{-1}. The following frequencies correspond to the intermolecular bends: 306, 269 cm^{-1} (Cl-H...O), 220, 159 cm^{-1} (H...O...H), and ν = 182, 174 cm^{-1}. Frequencies 175 cm^{-1} and 169 cm^{-1} belong to torsion vibrations.

TABLE 9. Intramolecular frequencies of $H_2O...(HCl)_2$ complex (open form, cm^{-1})

	ν (6-31G**)	ν_{corr}	Δν	$\Delta\nu_{corr}$
$(OH)_1$	4046	3865	-19	-18
$(OH)_2$	4135	3949	-29	-28
HOH	1771	1718	+11	+11
$(HCl)_1$	2990	2869	-83	-78
$(HCl)_2$	2995	2873	-78	-74

TABLE 10. Intramolecular frequencies of $H_2O...(HCl)_2$ complex (closed form, cm^{-1})

	ν(6-31G**)	ν_{corr}	Δν	$\Delta\nu_{corr}$
$(OH)_1$	4024	3845	-41	-38
$(OH)_2$	4123	3938	-41	-39
HOH	1750	1698	-4	-4
$(HCl)_1$	2917	2800	-156	-147
$(HCl)_2$	2927	2809	-146	-138

TABLE 11. Intramolecular frequencies of $HCl...(H_2O)_2$ complex (cm^{-1})

	ν(6-31G**)	ν_{corr}	Δν	$\Delta\nu_{corr}$
O_3H_4	3842	3673	-223	-210
O_3H_5	3947	3772	-217	-295
O_6H_7	4014	3835	-51	-48
O_6H_8	4113	3929	-51	-48
$H_8O_6H_7$	1727	1676	-42	-26
$H_5O_3H_4$	1729	1678	-25	-24
Cl_1H_2	2982	2861	-91	-86

The complex with the optimum structure 2 possesses the following set of intermolecular vibration frequencies: 61, 84, 154 cm^{-1} – O_5-H_6...Cl, Cl_1-H_2...Cl_3, and O_5...H_4-Cl_3 stretches; 285, 292, 314, 473, 523, 778, 1264 cm^{-1} – H-bond bends Cl_1-H_2...Cl_3, Cl_1...H_6-O_5, Cl_3-H_4...O_5, H_7-O_5...H_4, H_4-O_5...H_6, H_2...Cl_3-H_4, and H_2-Cl_1...O_6 H-bond bends, respectively.

The following intermolecular frequencies correspond to the complex with the optimum structure 3: 87, 153, and 154 cm^{-1} - Cl_1...H_7-O_6, O_6...H_4-O_3, and O_3...H_2-Cl_1 H-bond stretches; 304, 384, 421, 507, 617, 649, 917 and 1322 cm^{-1} - H_8-O_6...H_4, H_7-O_6...H_4, O_6...H_4-O_3, O_3...H_2-Cl_1, H_4-O_3...H_2, O_6-H_7...Cl_1, H_5-O_3...H_2, and H_2-Cl_1...H_7 H-bond bend, respectively.

The calculations were performed for following systems $H_2O...HF$, $H_2O...(HF)_2$, $(H_2O)_2...HF$, $(H_2O...HF)_2$ and $(H_2O...HF)_3$ (geometric structures of these complexes are presented in [39]). The results of experimental IR-spectroscopy study [9] point out the existence of complexes of the $(H_2O)_n(HF)_m$ ($n,m \geq 2$) type under high pressure, but measurements were only done for the complex $H_2O...HF$ of 1:1 composition. The IR spectrum of HF and H_2O mixture in the Ar matrix at 12 K showed the presence of three different complexes with hydrogen bond [34]: $H_2O...(HF)_2$, $H_2O...HF$, and HF...HOH in decreasing order of stability. The complex HF...HOH identified by the stretching vibration H-F (~ 3915.5 cm^{-1}) is referred to as a "reverse" complex. Though the data on $(H_2O)_2...HF$ complex are not available, on the basis of strong absorption observed in IR-spectrum in the range of dimer vibration $(H_2O)_2$, it was suggested in Ref. [34] that the complex $(H_2O)_2...HF$ exists.

In Ref. [35] structure and stability as well as IR-spectra of the complexes of H_2O and HF with 1:2 and 2:1 composition were studied systematically. Three equilibrium structures from all possible complexes were found. However, the existence of a "reverse" complex was not proven.

For the $(H_2O)_n(HF)_m$ complexes, we have corrected the values of vibration frequencies using the calibration function (2). The determined parameters b and a are 0.86 and 66.9 cm^{-1} respectively and the correlation coefficient between the experimental and calculated frequencies is 0.99994.

In Table 12 are listed the experimental and calculated vibration frequencies used for determining the parameters a and b and the average absolute deviation between $\nu_{corr.}$ and ν_{exp} is 16.7 cm^{-1}. The shift $\Delta\nu = \nu_{monomer} - \nu_{complex}$ of HF vibration frequency calculated using a Hartree-Fock method (HF) is 182 cm^{-1} for the complex $HF...H_2O$, while the experimental value is $\Delta\nu$ = 354 cm^{-1} [9] and $\Delta\nu$ = 364,5 cm^{-1} [38]. Therefore, the frequency shift of $\Delta\nu$ caused by hydrogen bonding predicted by the Hartree-Fock method is underestimated by ~ 200 cm^{-1}.

TABLE 12. Experimental [9] and calculated vibration frequencies ν (cm^{-1}) of $H_2O...HF$ complex.

ν_{calc}	$\nu_{exp.}$	$\nu_{corr.}$	*
4121	3608	3611	ν(HF)
4269	3756	3740	ν_{as}(OH)
4208	3657	3685	ν_s(OH)
1794	1600	1609	δ(HOH)

* - ν_s – bond symmetric vibrations; ν_{as} - bond antisymmetric vibrations; δ – bend vibrations

The calculation of frequencies of $H_2O...HF$ (6-31G**) complex has shown that its intermolecular vibrations fall into the interval of frequencies: 665 cm^{-1}, 633 cm^{-1} - δ(O-H..F), 175 cm^{-1} – $\delta_o(HOH_b)$, 189 cm^{-1} – $\delta_i(HOH_b)$, and ν(O...H-F) – 226 cm^{-1}. Note that in Ref. [9] the observed frequency of bending vibrations $HF...H_2O$ is ~ 700 cm^{-1} and intermolecular bond vibration is ~ 180 cm^{-1} (in Ref. [37] ν(O...H-F) = 198 cm^{-1}).

In Tables 13-15 are listed the calculated (HF/6-31G**) frequencies ν, $\nu_{corr.}$ and ν (MP2 [36]) for complexes $(H_2O)_n(HF)_m$ of 2:1 and 1:2 composition. Except for the frequencies presented in Table 13 there are ten frequencies (HF/6-31G**, cm^{-1}) corresponding to the intermolecular modes of $H_2O...(HF)_2$ complex with open structure - 699**, 699**, 648**, 560**, 258**, 208*; 193*, 141**, 107**and 32 (χ –torsion vibrations), here * - ν(O...H-F); ** - bending vibrations of donor-acceptor type _. Calculations by MP2 [35] provide the following values of intermolecular frequencies (cm^{-1}) and their intensities (km/mol): 782 (390), 753 (0), 712 (253), 653 (296), 314 (85), 231 (0), 212 (1), 184 (0), 144 (1), and 33 (10). Intermolecular frequencies (cm^{-1}) of the $H_2O...(HF)_2$ complex with closed structure are: 1058, 850, 732, 536, 512, 349, 260 - $\delta(\gamma,\beta)$, 222, 216, and 187 – ν(F-H...O, O...H-F, F...H –F). According to the MP2 calculation [35], intermolecular frequencies (cm^{-1}) and their intensities (km/mol) are: 1117 (144), 902 (300), 654 (344), 583 (99), 477 (204), 355 (82), 275 (15), 218 (94), 209 (4), and 111 (17).

TABLE 13. Intramolecular frequencies (ν, cm^{-1}) of $H_2O...(HF)_2$ open structure complex

6-31G**/HF	$\nu_{corr.}$	MP2/6-31++G**, A(km/mol)	vibration
1768	1587	1636 (98)	δ(HOH) (A_1)
4185	3666	3820 (160)	ν(OH) (A_1)
4213	3690	3851 (1066)	ν(OH) (B_1)
4382	3663	3897 (229)	ν(HF) (B_2)
4388	3669	3957 (144)	ν(HF) (A_1)

TABLE 14. Intramolecular frequencies (ν, cm^{-1}) of $H_2O...(HF)_2$ closed structure complex

6-31G**/HF	$\nu_{corr.}$	MP2/6-31++G**, A (km/mol)	vibration
1742	1565	1646 (89)	δ(HOH)
4016	3349	3374 (1111)	ν(H_4F_3)
4161	3652	3807 (168)	ν(OH_6)
4212	3710	3823 (465)	ν(OH_7)
4236	3517	3969 (154)	ν(H_2F_1)

TABLE 15. Intramolecular frequencies (ν, cm^{-1}) of $(H_2O)_2...HF$ complex

6-31G**/HF	$\nu_{corr.}$	MP2/6-31++G**, A (km/mol)	vibration
1721	1546	1640 (154)	δ(HOH)
1738	1562	1653 (21)	δ(HOH)
4037	3367	3440 (1001)	ν(HF)
4138	3625	3703 (345)	ν(H_7O_6)
4148	3634	3805 (117)	ν(H_8O_6)
4159	3643	3954 (129)	ν(H_4O_3)
4236	3710	3974 (126)	ν(H_5O_3)

Intermolecular vibration frequencies (cm^{-1}) of the $(H_2O)_2...HF$ complex according to our calculations (6-31G**/HF) are: 956, 824, 650, 576, 408, 317, 287, and 244 – _, 233, 188, and 177 – ν(O...H-F, O...H-O;F...H-O). The MP2 [35] calculation gives the following intermolecular frequencies (cm^{-1}) and their intensities (km/mol): 127 (11), 196 (164), 215 (11), 266 (5), 293 (111), 326 (69), 388 (72), 500 (189), 658 (249), 903 (274), and 1077 (162).

The calculations show that for the $(H_2O)_2...(HF)_2$ system the optimum configuration is a cyclic structure. The vibration frequencies and structural parameters of this complex are presented in Table 16. Intermolecular vibration frequencies (cm^{-1}) (6-31G**/HF) for this complex are: 97 – χ($H_5F_4H_3O_2$), 164 – χ($O_8H_7F_6H_5$), 397 – χ($H_3O_2H_1O_8$), 468 – χ ($H_9O_8H_{10}F_4$), 930, 715, 711, 665, 643, 461, 451, and 384 – δ and 307, 263, 215, and 180 – ν(O...H-F, H-F...H, O-H...F and O-H...O).

For the $(H_2O)_3...(HF)_3$ system the equilibrium configuration is an open flat oligomer structure. Vibration frequencies of this system are provided in Table 17. Intermolecular vibration frequencies (cm^{-1}) of the $(H_2O)_3...(HF)_3$ complex are (6-31G**/HF): 19, 39, 56, 56, 87, 97, 97, 105, and 108 – χ, 222 – $\delta(\gamma_1)$, 264 – $\delta(\gamma)$, 273 – δ (χ), 362 – δ (σ_1), 405 – δ (σ), 353 – δ (Ω), 362 – δ (Ψ), 615 – δ (ϕ_1), 634 – δ (ϕ_2), 172 – $\nu(F_{14}...H_{13}\text{-}O_{11})$, 210 – $\nu(F_9 ...H_8\text{-}O_6)$, 214 – $\nu(F_4...H_3\text{-}O_2)$, 266 – $\nu(O_{11}...H_{10}\text{-}F_9)$, 360 – $\nu(O_6 ...H_5\text{-}F_4)$. The calculations show that the interaction energy ΔE of the $(H_2O)_3...(HF)_3$ complex (-35.97 kcal/mol) is comparable with ΔE of $(H_2O)_2...(HF)_2$ complex (32.67 kcal/mol) while the binding energy of the $(H_2O)_3...(HF)_3$ complex $D_e = E_3 - (E_2 + E_1) = -7.6$ kcal/mol is far less than $D_e((H_2O)...HF)_2) = E_{cycl.} - E_{open} = -16.2$ kcal/mol.

TABLE 16. Intramolecular frequencies (ν, cm^{-1}) of $(H_2O)_2...(HF)_2$ complex

6-31G**/HF	$\nu_{corr.}$	vibration
1950	1743	δ(HOH)
2074	1851	δ(HOH)
3682	3061	$\nu(H_7F_6)$
3974	3313	$\nu(H_5F_4)$
4173	3657	$\nu(H_1O_2)$
4176	3658	$\nu(H_3O_8)$
4190	3670	$\nu(H_{10}O_8)$
4243	3716	$\nu(H_3O_2)$

TABLE 17. Intramolecular frequencies (ν, cm^{-1}) of $(H_2O)_3...(HF)_3$ complex

6-31G**/HF	$\nu_{corr.}$	vibration
1701	1530	δ(HOH)
1708	1536	δ(HOH)
1743	1567	δ(HOH)
3546	2945	$\nu(H_5F_4)$
3708	3256	$\nu(H_8O_6)$
3871	3396	$\nu(H_3O_2)$
3895	3417	$\nu(H_{12}O_8)$
4041	3542	$\nu(H_7O_6)$
4167	3479	$\nu(F_{14}H_{13})$
4211	3688	$\nu(O_2H_1)$
4330	3619	$\nu(F_3H_8)$
4260	3731	$\nu(O_8H_{11})$

ACKNOWLEDGEMENT

This work was supported in part by the Russian Foundation for Basic Research (Project No. 00-05-64919)

4. References

1. Dyke, T.R. (1984) *Top.Current.Chem.* **120**, 85.
2. Celli, F.G., Janda, K.S. (1986) *Chem.Rev.* **86**, 507.
3. Miller, R.E. (1986) *J.Phys.Chem.* **90**, 3301.
4. Knözinger, E., Schrems, O. (1987) in J.R. During (ed.), *Vibrational Spectra and Structure*, Elsevier, Amsterdam, Vol.16.

5. Barnes, A.J. (1980) in H. Ratajczak and W.J. Orville-Thomas (eds.), *Molecular Interactions*, Vol.1., pp.273 – 279, Wiley, Chichester, Vol.1., pp.273 –279.
6. Barnes, A.J. (1983) *J.Mol.Struct.* **100**, 259.
7. Howard, J., Waddington, T.C. (1980) in R.H. Clark and R.E. Hester (eds.), *Advances in Infrared and Raman Spectroscopy*, Heyden, London,Vol.7, p.86.
8. *Chemical Applications of Thermal Neutron Scattering* (1973), B.T.M. Willes (ed.), Oxford University, London.
9. Thomas, R.K. (1975) Hydrogen bonding in the vapor phase between water and hydrogen fluoride: the infrared spectrum of the 1:1 complex", *R.Soc.Lond.A* **344**, 579 – 593.
10. Legon, A.C., Willoughby, L.C. (1983), *Chem.Phys.Lett.* **95**, 37.
11. Ault, B.S., Pimentel, G.C. (1973), *J.Phys.Chem.* **77**, 37.
12. Ayers, G.P., Pullin, A.D.E. (1976), *Spectrichim.Acta.Part A.* **32**, 1641.
13. Schriver, A., Silvi, B., Maillard, D., Perchard, J.P. (1977), *J.Chem.Phys.* **81**, 2095.
14. Latajka, Z., Scheiner, S., (1987) Structure, energetics, and vibrational spectrum of H_2O-HCl, *J.Chem.Phys.* **87**, 5928 – 5936.
15. Zülike, L. (1972) *Quantum Chemistry*, Vol.1, Mir, Moskwa.
16. Lunichev, V.N. (1979) *Structural features of non-rigid molecules with large nucleus amplitudes*, Candidate Thesis, Moskwa.
17. Lunichev, V.N. (1979), J. Structurnoj Chimii **20**, 20 – 25.
18. Peterson, M.R., Poirer, R. (1990), *Monstergauss*. Department of Chemistry, University of Toronto, Toronto, Canada, Memorial University of Newfoundland, Newfoundland, Canada.
19. M.J.D. Powel, M.J.D. (1970), in , P. Rabinowitz (ed.), *Numerical methods for nonlinear algebraic equations*, P. Rabinowitz, Gordon and Breach, London, p.87.
20. Suchanov, L.P. (1983), *Theoretical Investigation of Physical-Chemical Properties of light metals hydride olygomers*, Candidate Thesis, Moskwa.
21. Yamaguchi, Y., Schaefer III, H.F.A. (1980), *J.Chem.Phys.* **73**, 2310 – 2318.
22. Hout, R.F., Levi, B.A., Hehre, W.J. (1982), *J.Comput.Chem.* **3**, 234 – 250.
23. Frisch, M.J., Del Bene, J.E., J.S.Binkley, J.S., Schaefer III, H.F. (1986), *J.Chem.Phys.* **84**, 2279.
24. Barnes, A.J., Orville-Thomas, W.J. (1980), *J.Mol.Spectrosc.* **84**, 391.
25. Shimanouchi, T.(1972), *Tables of Molecular Vibrational Frequencies*, Vol.1, Natl.Stand.Ref.Data.Scr.Natl.Bur.Stand.No39 (National Bureau of Standarts), Washington.
26. Berti, J.E., Falk, M.V. (1973), *Can.J.Chem.* **51**, 1713.
27. Kisiel, Z., Legon, A.C., Millen, D.J. (1982), *Proc.R.Soc.London.A.* **381**, 419.
28. Coker, D.F., Miller, R.E., Watts, R.O. (1985), *J.Chem.Phys.* **82**, 3554 – 3562.
29. Wuelfert, S., Herren, D., Leutwylen, S. (1987), *J.Chem.Phys.* **86**, 3751 – 3753.
30. Huber, K.P., Herzberg, G. (1979), in *Molecular Spectra and Molecular Structure*, Van Nostrand Reinhold, New York,Vol.14
31. Clough, S.A., Bears, Y., Klein, G.P., Rothman, L.S. (1973), *J.Chem.Phys.* **59**, 2254.
32. Deleeuw, F.H., Dymanus, A. (1973), *J.Mol.Spectrosc.* **48**, 427.
33. Stull, D.R., Prophet, J. (1971), JANAF Thermochemical Tables. – Natl.Stand.Ref.Data.Scr.Natl.Bur.Stand (National Bureau of Standarts), Washington, Vol.37.
34. Andrews, L. and Johnson, G.L. (1983), *J.Chem.Phys.* 3670.
35. Rovira, C., Constants, P., Whangbo, M.H, Novoa, J.J. (1994), *International Journal of Quantum Chemistry* **52**, 177.
36. Sokolov, N.D. (1981), Dynamics of hydrogen bonding, in N.D.Sokolov (ed.), *Hydrogen bonding*, Nauka, Moscow, pp. 63 – 88.
37. Szczesniak, M.M., Scheiner, S., Bouteiller, Y. (1984), *J.Chem.Phys.* **81**, 5024.
38. Hannachi, Y., Silvi, B., Bouteiller, Y. (1991), *J.Chem.Phys.* **94**, 2915.
39. Zvereva, N.A., Nabiev, Sh.Sh., Ponomarev, Yu.N. (1999), Energies of vertical S_0_S_1 transitions of optical active complexes with hydrogen bonds, *Optics of atmosphere and ocean* **12**,843-846.

LIST OF CONTRIBUTORS

Numbers in parentheses indicate the pages on which authors' contributions begin.

Wesley D. Allen, Center for Computational Quantum Chemistry, Department of Chemistry, University of Georgia, Athens, GA 30602, USA (317)

Jean Vander Auwera, Laboratoire de Chimie Physique Moléculaire Cpi 160/09, Université Libre de Bruxelles, 50 Roosevelt ave, 1050 Bruxelles, Belgium (201)

Kurt Beier, DLR, Remote Sensing Technology Institute, Münchner Str. 20, D-82234 Wessling, Germany (259)

Sergey Belov, Institute of Applied Physics of Russian Academy of Sciences, 46, Uljanova st. , Nizhny Novgorod 603600, Russia (73)

Peter Bernath, Department of Chemistry, University of Waterloo, 200 University Avenue West, Waterloo, Ontario, Canada N2L 3G1 (147)

Manfred Birk, DLR, Remote Sensing Technology Institute, Münchner Str. 20, D-82234 Wessling, Germany (219)

Kurt Bottcher, DLR, Remote Sensing Technology Institute, Münchner Str. 20, D-82234 Wessling, Germany (259)

Bruno Carli, Gruppo Stratosfera - Reparto Scienze dell'Atmosfera, Istituto di Ricerca sulle Onde Elettromagnetiche del C.N.R. "Nello Carrara", Via Panciatichi, 64, 50127 Firenze, Italy (171)

Jose Cernicharo, CSIC IEM Department of Molecular Physics, C/Serrano 121, 28006 Madrid, Spain (23)

Svatopluk Civiš, J. Heyrovsky Institute of Physical Chemistry, Academy of Sciences, Dolejškova 3, 182 23 Praha 8, Czech republic (275)

Edward A. Cohen, M/S 183-301, Jet Propulsion Laboratory, California Institute of Technology, Pasadena, CA 91109-8099, USA (59)

Ugo Cortesi, Gruppo Stratosfera - Reparto Scienze dell'Atmosfera, Istituto di Ricerca sulle Onde Elettromagnetiche del C.N.R. "Nello Carrara", Via Panciatichi, 64, 50127 Firenze, Italy (171)

Attila G. Császár, Department of Theoretical Chemistry, Eötvös University, P.O. Box 32, 1518 Budapest 112, Hungary (317)

Jean Demaison, Laboratoire de Physique des Lasers, Atomes et Molécules, Université de Lille 1, 59655 Villeneuve d'Ascq Cedex, France (107)

Herbert Fischer, Forschungszentrum Karlsruhe GmbH, IMK (Institut für Meteorologie und Klimaforschung), Postfach 36 40, D-76021 Karlsruhe, Germany (161)

Jean-Marie Flaud, Laboratoire de Photophysique Moléculaire, CNRS, Université Paris Sud, bât 350, 91405 Orsay Cedex, France (187)

Peter Haschberger, German Remote Sensing Data Center, Münchner Str. 20, D-82234 Wessling, Germany (259)

Dieter Hausamann, DLR, Remote Sensing Technology Institute, Münchner Str. 20, D-82234 Wessling, Germany (219)

Eric Herbst, Department of Physics, The Ohio State University, 174 W. 18th Ave., Columbus, OH 43210-1106,USA (1)

Michel Herman, Laboratoire de Chimie Physique Moléculaire Cpi 160/09, Université Libre de Bruxelles, 50 Roosevelt ave, 1050 Bruxelles, Belgium (201)

Heinz-Wilhelm Hübers, German Aerospace Center, Institute of Space Sensor Technology and Planetary Exploration, Rutherfordstrasse 2, D-12489 Berlin, Germany (43)

Daniel Hurtmans, Laboratoire de Chimie Physique Moléculaire Cpi 160/09, Université Libre de Bruxelles, 50 Roosevelt ave, 1050 Bruxelles, Belgium (201)

Zdenek Jaňour, Institute of Thermomechanics, Academy of Sciences of the Czech Republic, Dolejškova 5, 182 23 Prague 8, Czech Republic (275)

Daniel A. Jelski, Department of Chemistry, Rose-Hulman Institute of Technology, 5500 Wabash Avenue, Terre Haute, Indiana, 47803, USA (301)

Zbigniew Kisiel, Laboratory of Mm- and Submm-Spectroscopy, Institute of Physics, Polish Academy of Sciences, Al.Lotnikow 32/46, 02-668, Warszawa, Poland (91)

Matthew L. Leininger, Sandia National Laboratory, MS 9217, Livermore, CA 94551-0969, USA (317)

Erwin Lindenmeir, DLR, Remote Sensing Technology Institute, MünchnerStr. 20, D-82234 Wessling, Germany (259)

Laszlo Nemes, Research Laboratory for Inorganic Chemistry, Hungarian Academy of Sciences, Budaörsi ut 45, 1112 Budapest, Hungary (301)

Dieter Oertel, DLR Institute of Space Sensor Technology and Planetary Exploration, Rutherfordstrasse 2, D-12489 Berlin, Germany (259)

Luca Palchetti, Gruppo Stratosfera - Reparto Scienze dell'Atmosfera, Istituto di Ricerca sulle Onde Elettromagnetiche del C.N.R. "Nello Carrara", Via Panciatichi, 64, 50127 Firenze, Italy (171)

Agnès Perrin, Laboratoire de Photophysique Moléculaire, CNRS, Université Paris Sud, bât 350, 91405 Orsay Cedex, France (235)

Oleg L. Polyansky, Department of Physics and Astronomy, University College London, Gower St., London WC1E 6BT, United Kingdom (317)

Hans Peter Röser, German Aerospace Center, Institute of Space Sensor Technology and Planetary Exploration, Rutherfordstrasse 2, D-12489 Berlin, Germany (43)

Kamil Sarka, Department of Physical Chemistry, Faculty of Pharmacy, Comenius University, 832 32 Bratislava, Slovakia (107)

Franz Schreier, DLR, Remote Sensing Technology Institute, Münchner Str. 20, D-82234 Wessling, Germany (219, 259)

Michal Střižik, J. Heyrovsky Institute of Physical Chemistry, Academy of Sciences, Dolejškova 3, 182 23 Praha 8, Czech republic (275)

György Tarczay, Department of Theoretical Chemistry, Eötvös University, P.O. Box 32, 1518 Budapest 112, Hungary (317)

Jonathan Tennyson, Department of Physics and Astronomy, University College London, Gower St., London WC1E 6BT, United Kingdom (317)

Mikhail Tretyakov, Institute of Applied Physics of Russian Academy of Sciences, 46, Uljanova st., Nizhny Novgorod 603600, Russia (73)

Georg Wagner, DLR, Remote Sensing Technology Institute, Münchner Str. 20, D-82234 Wessling, Germany (219, 259)

Li-Hong Xu, University of New Brunswick, P.O. Box 5050, Saint John, NB, Canada (131)

Zdenek Zelinger, J. Heyrovsky Institute of Physical Chemistry, Academy of Sciences, Dolejškova 3, 182 23 Praha 8, Czech republic (275)

Natalia Zvereva, SFTI at Tomsk State University, Novo-Sobornaja sq.1, 634050, Tomsk, Russia (341)

Index of Subjects